Excel

Get the Results You Want!

SmartStudy 10

Mathematics

Lyn Baker

PASCAL PRESS

Reprinted 2017, 2019, 2020, 2022, 2024

ISBN 978 1 74125 476 1

Pascal Press
PO Box 250 Glebe NSW 2037
(02) 9198 1748
www.pascalpress.com.au

Publisher: Vivienne Joannou
Project editor: Rosemary Peers
Answers checked by Peter Little
Typeset by Precision Typesetting (Barbara Nilsson)
Cover design by DiZign Pty Ltd
Printed by Vivar Printing/Green Giant Press

Students

All care has been taken in compiling this book, but please check with your teacher about the exact requirements of the course as these can change from year to year.

TABLE OF CONTENTS

TAKE THESE REVISION STEPS TO SUCCESS!

Step 1 Study Notes

- This section contains valuable **suggestions**, **tips** and **essential points** about the topic.
- When reading this section, **highlight** the points that are new to you or that you consistently forget. You could also **rewrite** these points so that you are making a summary of the Study Notes. Keep this summary in a prominent place at home—you could stick it on a mirror, wall or door.
- Work carefully through the **checklist** at the end of the section to ensure that you have mastered each of the listed skills.
- If there is something that you are **not sure about**, take time out to read through your class notes, textbooks or another relevant ***Excel*** book. (Check our website at www.pascalpress.com.au for more titles.)

Step 2 Skills Check

- Check that you know the **basic skills** you need to successfully complete the topic.
- Once you have completed the check, **mark your work** quickly by looking at the answers at the bottom of the page. This is **instant feedback** for you. The **worked solutions** are located at the back of the book if you want to check the working of any answer.
- If you have **scored less than 50%** in this section then you should revise your basic skills.

Step 3 Intermediate Test

- This test has very similar questions to the ones you will get in your **class test** or **exam**.
- Look carefully for the **Hints** which are provided for the trickiest questions—these appear at the bottom of the page.
- **Marks** are allocated for each question. These are similar to the marks in your tests.
- Time yourself—**check** how much **time** you have got to complete the test. Also look at the **total marks** of the test to calculate approximately how much time you should spend on each question. For example, if there are twenty marks in total and twenty minutes have been allocated for completion of the test, then spend about one minute on each mark. If you cannot complete all the questions within the suggested time, you may need to revise the topic.
- Fill in the **Your Feedback** panel once you have marked your work in order to calculate your percentage mark. Then complete the **Test & Exam Results** on page 176 to keep a running total of all your test marks.

Step 4 Advanced Test

- This test features **extension questions** such as problem-solving questions.
- This test is not like a class test, as all the questions in it are challenging. Mastery of questions in this test, however, will ensure you gain top marks in your class tests and exams, and will also prepare you for the mathematics you will learn next year.
- **Marks** are allocated for each question. These are similar to the marks in your tests.
- Time yourself—**check** how much **time** you have to complete the test. Also look at the **total marks** of the test to calculate approximately how much time you should spend on each question. For example, if there are forty marks in total and forty minutes have been allocated for completion of the test, then spend about one minute on each mark.

- Fill in the **Your Feedback** panel once you have marked your work in order to calculate your percentage mark. Then complete the **Test & Exam Results** on page 176 to keep a running total of all your test marks.

Step 5 Check Your Solutions

- **Worked solutions** to all questions are found at the back of the book. Work through the solutions to any questions that you got wrong.
- **Longer questions** are usually worth two or more marks and will involve some working. You should set out all your working, because in Maths you may get some marks for your working even if your answer is wrong.
- The **ticks** that appear in the worked solutions indicate those parts of the working which receive marks. Therefore, even if your answer is wrong, you may be entitled to some marks for what you have written. Compare the worked solution to your own working to find out whether you are entitled to any marks for the question. But remember there may be more than one way to find the answer.
- If you still **cannot understand** how the correct answer was obtained, revise that part of the topic and, if necessary, refer to your class notes or ask your teacher for help. It is important to learn from your mistakes.

Step 6 Test & Exam Results

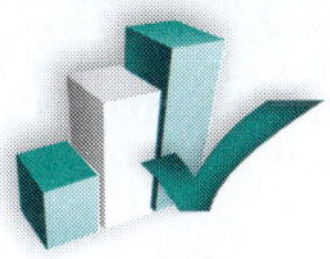

- Go to the **Test & Exam Results** section on page 176 to record your test score as a percentage. When you have completed all topics you will be able to determine your areas of weakness and your areas of strength.
- It is important that you know which **areas need further work**—to 'know what you don't know'. The more you prepare for the topic tests, the more successful you will be and the more you will remember when you sit your end-of-term/semester/year test or exam.

Step 7 Tips for the Sample Exam Papers

- These **useful tips** appear on page 91. Read them before you start one of the Sample Exam Papers (see Step 8 below).

Step 8 Sample Exam Papers

- Three **Sample Exam Papers** are provided at the end of the book. These are of three levels of difficulty: Average, Above Average and Difficult. The Above Average paper will be very similar to your final examination.
- Before attempting the Sample Exam Papers, make sure that you have completed all of the **Tests** and have worked through the solutions to all questions that you answered incorrectly.
- Set aside the **time allowed** for the paper and complete it under **exam conditions**—no sneaking a look at your notes or textbooks! That way you will be better prepared for your final exam.
- **Worked solutions** to the Sample Exam Papers are found at the back of the book. Work through the solutions to any questions that you got wrong. Carefully note the **ticks** in the worked solutions and remember to give yourself marks for correct working. Write down your total marks for each section in the **Your Feedback** boxes at the end of each part of the paper, then add them up to get a total percentage for each test.

HOW TO USE THIS BOOK TO STUDY FOR A CLASS TEST, HALF-YEARLY OR END-OF-YEAR EXAM

Depending on your teacher or school, you will be given a variety of tests and exams each year. There may be a single-topic test, a test that covers a number of topics, a semester test or exam, or even a half-yearly or yearly exam.

Step 1

Find out which topics will be covered in the class test.

- To do this, look at your class workbook/textbook, laptop/tablet or online study program, and ask your teacher.
- For example, your class test may be on Geometric Reasoning.

Step 2

Match the topics that your test is on to the topics in this book.

- For example, the three units on pages 57 to 68 of this book cover Geometric Reasoning.

Step 3

Use this book to study the topics being tested.

- For example, the three units on pages 57 to 68 contain the topics you will study for your class test on Geometric Reasoning!

Note:

- When you are using this book to study for a **half-yearly** test, follow the same steps as above—the only difference being that you will have more topics to revise of course!
- When you are using this book to study for an **end-of-year** test, you will more than likely need to study the whole book!

COMPOUND INTEREST

Financial Mathematics

STUDY NOTES

1 **Interest** is money paid as the cost of borrowing money. (You must pay interest if you take out a loan but interest will be paid to you if you invest money.)

2 p.a. means **per annum** or **per year**. Generally interest rates are expressed as percentages per annum but they could be given as a rate over any given time period. Other common rates are per quarter (¼ of a year) or per month.

3 **Simple interest** (or flat rate of interest) is always calculated on the original amount that is borrowed or invested so that the same amount of interest is charged each period. Simple interest is given by $I = PRN$ where P is the **principal** or original amount, R is the **rate of interest** ($\frac{r}{100}$ where the interest rate is $r\%$ per period) and N is the **number of periods**.

4 **Compound interest** is charged on the amount owing (or invested) at the particular time and is then added to that amount, so that the sum on which interest is charged will be a different amount each time the interest is calculated.

5 The formula $A = P(1 + R)^n$ gives the **amount** to which the principal has **accumulated** after n periods. P is the principal or original sum of money. R $(= \frac{r}{100})$ is the rate of interest per compound period, expressed as a decimal. For example, if \$2000 is invested at 8% p.a. for 3 years then at the end of those three years its value will be given by $\$2000(1 + 0.08)^3$.

6 The **amount of compound interest earned** is found by subtracting the principal from the accumulated amount. $I = A - P$

7 Care must be taken to match the **number of periods** with the **rate per compound period**. For example, if an investment is earning interest at 2% per quarter for 4 years, then n must be the number of quarters not the number of years. In this case $n = 16$. (4 quarters per year over 4 years = 4×4 quarters.) If the interest rate is given as a percentage per annum but the compounding period is not yearly then the interest rate must also be changed. For example, a rate of 6% p.a. is 0.5% per month. (Note: although 0.5% is a decimal it is still a percentage and must be divided by 100 to give R. $R = 0.5 \div 100 = 0.005$).

8 When money is borrowed, compound interest is sometimes referred to as **reducible interest** because as the loan is repaid the amount on which interest is paid reduces and hence the interest reduces.

9 **Tables and graphs** can be used to show accumulated amounts or amount of interest earned.

10 **Depreciation** is when the value of an asset decreases over time. The value of an item that is being depreciated can be found using the formula $A = P(1 - R)^n$ where P is the original value of the item, R is the rate of depreciation per period and n is the number of periods. (Sometimes the formula is given as $A = P(1 + R)^n$ and R given as a negative rate.) For example, if an item originally valued at \$2000 depreciates at 15% p.a. then its value after 4 years is given by $\$2000(1 - 0.15)^4$.

Checklist

Can you:

1 *Understand the difference between simple interest and compound interest?* ☐
2 *Solve problems involving simple interest?* ☐
3 *Use the compound interest formula and solve problems involving compound interest?* ☐
4 *Understand the need to match interest rates and compounding periods?* ☐
5 *Understand what is meant by depreciation and solve problems involving depreciation?* ☐

COMPOUND INTEREST
Financial Mathematics

SKILLS CHECK

1 Find:

a 10% of \$800   b 7% of \$6000   c 3% of \$200 000

2 How many:

a months in 7 years?
b quarters in 5 years?

3 A rate of 6% pa is what rate:

a per month?
b per quarter?

4 An amount of money earned \$360 p.a. simple interest. How much interest was earned in 4 years?

5 An amount of money earned \$1860 simple interest when invested for 6 years. How much interest was earned each year?

6 Write as a decimal:

a 8%   b 12%   c $2\frac{1}{2}\%$ d 0.5%

7 Find the simple interest on:

a \$5000 at 7% p.a. for 3 years
b \$240 000 at 8% p.a. for 5 years
c \$12 500 at 0.8% per month for 2 years

8 \$10 000 was invested. It earned interest at the rate of 9% p.a. At the end of each year the interest earned is added to the investment.

a How much interest was earned in the first year?
b How much was in the account at the end of the first year (immediately after the interest has been paid)?
c How much interest was earned in the second year?
d How much was in the account at the end of the second year?
e How much was in the account at the end of the third year?

9 Give $A = P(1 + R)^n$, find A to the nearest dollar when:

a $P = \$5000$, $R = 0.06$ and $n = 4$
b $P = \$8750$, $R = 0.1$ and $n = 7$
c $P = \$125\,000$, $R = 0.0075$ and $n = 144$

10 Use the formula $A = P(1 + R)^n$ to find the amount, to the nearest dollar, to which \$25 000 will accumulate if invested at 4% p.a. interest compounded annually for 10 years.

11 \$8500 is invested at 5% p.a. interest compounded annually for four years. Find, to the nearest cent:

a the amount to which it accumulates
b the amount of compound interest earned

12 Use the formula $A = P(1 - R)^n$ to find the value of an item after three years if it was originally worth \$7000 and if it depreciates at 10% p.a.

PAGE 106

Answers **1 a** \$80 **b** \$420 **c** \$6000 **2 a** 84 **b** 20 **3 a** 0.5% **b** 1.5% **4** \$1440 **5** \$310 **6 a** 0.08 **b** 0.12 **c** 0.025 **d** 0.005
7 a \$1050 **b** \$96 000 **c** \$2400 **8 a** \$900 **b** \$10 900 **c** \$981 **d** \$11 881 **e** \$12 950.29 **9 a** \$6312 **b** \$17 051 **c** \$366 605
10 \$37 006 **11 a** \$10 331.80 **b** \$1831.80 **12** \$5103

COMPOUND INTEREST

Financial Mathematics

30 MINUTES

INTERMEDIATE TEST

Part A Multiple Choice

1 $7000 is placed in an account earning 10% p.a. compound interest. How much is in the account at the end of three years?

A $9100 **B** $9317 **C** $9170 **D** $9549 (1 mark)

2 Which investment will earn the most interest after 4 years?

A $12 000 at 8% p.a. simple interest
B $8000 at 12% p.a. simple interest
C $9000 at 9.5% p.a. compound interest
D $9500 at 9% p.a. compound interest (1 mark)

3 $1000 is invested for 2 years. Which will produce the most interest?

A 8.5% p.a. simple interest
B 2% per quarter, compounded quarterly
C 0.6% per month, compounded monthly
D 7.5% p.a. compounded yearly (1 mark)

4 $50 000 is invested for 25 years at 1% per month interest, compounded monthly. The total interest earned, to the nearest dollar, is: *Hint 1*

A $64 122 **B** $67 483 **C** $491 735 **D** $939 423 (1 mark)

5 Which amount of money is closest to the amount that would need to be invested at 2.25% per quarter compound interest to accumulate to $20 000 after 3 years? *Hint 2*

A $15 315 **B** $14 870 **C** $18 570 **D** $10 240 (1 mark)

Part B Short Answer

6 **a** Find the amount of interest earned when $5000 is invested for 4 years at 6% p.a. interest, compounded annually. (2 marks)

b How much more interest would be earned if $5000 is invested for 4 years at 6% p.a. interest, compounded monthly? (3 marks)

7 **a** Lily borrows $90 000. 8% p.a. simple interest is charged on the loan. She repays the loan with three equal yearly payments. How much is each payment? *Hint 3* (3 marks)

b Mike also borrows $90 000. He is charged 8% reducible interest over three years. How much does Mike owe on the loan at the end of the first year after the interest has been added and the first repayment of $35 000 made? (2 marks)

c How much does Mike owe at the end of the second year if another $35 000 is repaid? *Hint 4* (2 marks)

d How much must Mike repay in the third year to repay the loan? *Hint 5* (2 marks)

e How much more in total does Lily repay than Mike? (2 marks)

8 A car is valued at $32 000 when it is new. It depreciates at 10% p.a.

a What will it be worth after 4 years? (1 mark)

b After how many years will its value be approximately $10 000? *Hint 6* (2 marks)

9 Neil takes out a loan for $30 000. Simple interest of 9% p.a. is charged on the loan. He pays the loan off over 5 years with equal monthly instalments.

a How much is each monthly instalment? (3 marks)

b If Neil invested the money at 0.7% per month interest, compounded monthly, would he have made a profit or loss at the end of five years? Justify your answer. *Hint 7* (3 marks)

Hint 1: The interest is compounded monthly while the time has been given in years.
Hint 2: Although the interest rate is given in decimal form it is still a percentage.
Hint 3: Add the interest to the amount borrowed to get the total amount to be repaid.
Hint 4: The principal will be the answer to part **b**.
Hint 5: What is the total of the (new) principal plus interest?
Hint 6: You will need to take some guesses until you find the required number of years.
Hint 7: Don't just make an assumption. You need to do some calculations.

Your Feedback

$\frac{\square}{30} \times 100\% = \square\%$

PAGES 106–107

PAGE 176

COMPOUND INTEREST

Financial Mathematics

35 MINUTES

ADVANCED TEST

1 Find the amount to which \$50 000 will accumulate if it is invested:

- **a** for 7 years at 9% p.a. compounded monthly (2 marks)
- **b** for 12 years at 6% p.a. compounded quarterly (2 marks)
- **c** for 25 years at 7.5% p.a. compounded half-yearly (2 marks)
- **d** for 418 days at 0.05% per day compounded daily (1 mark)

2 Find the amount of interest earned if:

- **a** \$20 000 is invested for 7 years at 8% p.a. interest, compounded quarterly (2 marks)
- **b** \$350 000 is invested for 30 years at 7% p.a. interest, compounded monthly (2 marks)

3 An amount of \$20 000 earns \$6620 in interest, compounded annually, in 3 years.

- **a** What rate of interest is paid? (2 marks)
- **b** How much interest will be earned in the next 3 years? (1 mark)

4 An amount of money is invested at 6% p.a. compounded monthly. How long, in years and months, will it take to double in value? (2 marks)

5 \$35 000 is invested at 9% p.a. interest compounded quarterly for 8 years.

- **a** How much interest is earned? (2 marks)
- **b** What annual rate of simple interest will produce the same result? (2 marks)

6 What amount of money needs to be invested at 4% p.a. interest compounded monthly to accumulate to \$8000 in 6 years? (2 marks)

7 Thalia borrows \$20 000 and repays it with equal yearly instalments. Interest, at 6% p.a. is charged on the amount owing at the end of each year, just before the instalment is paid.

- **a** How much interest will Thalia pay in the first year? (1 mark)
- **b** Thalia owes \$13 717.80 at the beginning of the second year after the first lot of interest has been paid and the first instalment repaid. How much was the instalment? (1 mark)
- **c** How much will Thalia owe at the beginning of the third year? (2 marks)
- **d** Show that the loan will be repaid after 3 years. (2 marks)
- **e** Briefly explain why the simple interest rate is not 6% p.a.? (1 mark)
- **f** Find the annual rate of simple interest. (1 mark)

8 A machine originally valued at \$12 800 depreciates by the same amount every year. After 16 years its value is \$512.

- **a** By how much does the machine depreciate each year? (1 mark)
- **b** Explain why the machine cannot continue to depreciate by this amount. (1 mark)
- **c** What percentage of the original value is the annual amount of depreciation? (1 mark)
- **d** If, instead, the machine depreciated by that percentage each year,
  - **i** what would its value be after 16 years? (1 mark)
  - **ii** Will the machine ever be worthless under this method? (1 mark)

Your Feedback

$\frac{\square}{35} \times 100\% = \square\%$

PAGES 107–108

PAGE 176

PRODUCTS AND QUOTIENTS

Algebraic Techniques

STUDY NOTES

1 **Pronumerals** are symbols, usually letters, used in place of numbers in an expression.
For example, $3x$ means $3 \times x$ x^2 means $x \times x$ $3x^2$ means $3 \times x^2$ or $3 \times x \times x$

2 When **multiplying expressions involving pronumerals** we multiply any numbers and multiply the pronumerals separately. For example, $2x \times 3y = 6xy$. (Note: xy means $x \times y$. The multiplication sign is left out.)

3 x^4 is written using **index notation**. x is the **base**. 4 is the **index** or **power**. The plural of index is **indices**. The index tells us how many times the base is multiplied by itself. For example, $x^4 = x \times x \times x \times x$ (there are 4 factors of x), $p^2q^3 = p \times p \times q \times q \times q$ (there are 2 factors of p and 3 factors of q).

4 A symbol without an index actually has a **power of one**. $x = x^1$
We don't usually write a power of one. For example, $a^1 = a$ $e = e^1$ $2y^1 = 2y$

5 When **multiplying**, if the base is the same, **add the indices**. $x^m \times x^n = x^{m+n}$
For example, $x^2 \times x^3 = x^{2+3} = x^5$ $a^3 \times a^8 = a^{3+8} = a^{11}$.

6 When **multiplying**, if the **base is not the same**, we cannot add the indices.
For example, $x^2 \times y^3 = x^2y^3$

7 If a **term** involves a **number and a pronumeral**, or more than one pronumeral, we **multiply each separately**. For example, $2x^5 \times 3x^2 = 6x^7$ $4p^3q^4 \times 5pq^2 = 20p^4q^6$

8 When **dividing**, if the **base is the same**, **subtract the indices**. $x^m \div x^n = x^{m-n}$
For example, $a^8 \div a^2 = a^{8-2} = a^6$ $e^9 \div e^2 = e^{9-2} = e^7$ $6^5 \div 6^3 = 6^{5-3} = 6^2$.

9 If a term involves a **number and a pronumeral**, or more than one pronumeral, we **divide each separately**.
For example, $8a^8 \div 2a^2 = 4a^6$ $12y^4z^9 \div 4y^2z^8 = 3y^2z$

10 **Fractions involving pronumerals** can be **simplified** by dividing both numerator and denominator by any **common factors**. For example, $\frac{8ab}{2b} = 4a$, $\frac{10n^5}{2n^8} = \frac{5}{n^3}$

11 When a **power is raised to a power**, we multiply the indices. $(x^m)^n = x^{mn}$.
For example, $(a^2)^4 = a^{2\times4} = a^8$, $(m^5)^3 = m^{5\times3} = m^{15}$. Only those terms inside the bracket are raised to the power but **every part of a term** in the bracket must be raised to the power.
For example, $(7x^3)^2 = 49x^6$ but $7(x^3)^2 = 7x^6$

12 Any number raised to the **zero power** equals **one**. For example, $x^0 = 1$, $5^0 = 1$, $37^0 = 1$. The answers to questions involving the **zero index** are not always one. For example, $3x^0 = 3 \times x^0 = 3 \times 1 = 3$.

13 Raising a number or term to the **power of negative one** is the same as taking the **reciprocal**:
$a^{-1} = \frac{1}{a}$. For example, $7^{-1} = \frac{1}{7}$, $(3m)^{-1} = \frac{1}{3m}$, $3m^{-1} = \frac{3}{m}$.

14 Raising a term to the **power** of **negative two** is the same as taking the **square** of the reciprocal and raising a term to the power of **negative three** is the same as taking the **cube** of the reciprocal. For example,
$5^{-2} = \frac{1}{5^2} = \frac{1}{25}$, $(2a)^{-3} = \frac{1}{(2a)^3} = \frac{1}{8a^3}$ but $2a^{-3} = \frac{2}{a^3}$.

Checklist

Can you:

1 *Understand that separate numbers and pronumerals in a product are multiplied separately?*
2 *Understand the Index Laws?*
3 *Multiply terms together and divide one term by another?*
4 *Simplify an algebraic fraction?*
5 *Raise a power to a power and understand and use the zero index and negative powers?*

PRODUCTS AND QUOTIENTS

Algebraic Techniques

SKILLS CHECK

1 Simplify:

a $2 \times 3x$ **b** $x + x$ **c** $x \times x$ **d** $a \times b$

e $2x \times y$ **f** $3a \times 2b$ **g** $5x \times 4x$ **h** $4 \times 2x^2$

i $q \times 7p$ **j** $2xy \times 2z$ **k** $2a \times 3b \times 4c$ **l** $3k \times 2k \times 4t$

2 Simplify:

a $8a \div 2$ **b** $8a \div a$ **c** $8a \div 2a$ **d** $20xy \div 4$

e $20xy \div x$ **f** $20xy \div 4y$ **g** $20xy \div xy$ **h** $20xy \div 4xy$

3 Simplify:

a $\frac{4p}{2}$ **b** $\frac{3ab}{3}$ **c** $\frac{3ab}{a}$ **d** $\frac{3ab}{3b}$

e $\frac{4x}{8y}$ **f** $\frac{2t^2}{5t}$ **g** $\frac{6pqr}{9pq}$ **h** $\frac{6x}{12x}$

4 If $x = 4$, find the value of:

a $x + 3$ **b** $3x$ **c** $4x - 1$ **d** $24 \div x$

e $8 - 2x$ **f** x^2 **g** $2x^2$ **h** $\frac{3x}{2}$

5 Simplify:

a $x^4 \times x^5$ **b** $t^6 \times t$ **c** $5p \times p^3$ **d** $2a^2 \times 3a^3$

e $-8m^3 \times 2m^7$ **f** $ab^2 \times 3b$ **g** $3x^2 \times 3x^2$ **h** $x^3y^2 \times x^4y^6$

i $6x^2y^3 \times 2x^3y^5$ **j** $4p^3q^7 \times 4pq^2$ **k** $3p^2qr^3 \times 5pqr^4$ **l** $4a^2 \times 3a^7 \times 2a^3$

6 Simplify:

a $x^{12} \div x^2$ **b** $12m^{12} \div 2m^2$ **c** $16x^6 \div 4x$ **d** $9m^4n^3 \div 3m^2n^2$

e $\frac{7n^6}{14n^3}$ **f** $\frac{x^4y^5}{x^3y^7}$ **g** $\frac{9a^2b^7}{6a^2b^3}$ **h** $\frac{8a^3b^6c^2}{4abc^2}$

7 Simplify:

a $(x^4)^2$ **b** $(3t^6)^3$ **c** $5(p^4)^9$ **d** $(2m^2n^4)^4$

PAGES 108–109

Answers **1 a** $6x$ **b** $2x$ **c** x^2 **d** ab **e** $2xy$ **f** $6ab$ **g** $20x^2$ **h** $8x^2$ **i** $7pq$ **j** $4xyz$ **k** $24abc$ **l** $24k^2t$ **2 a** $4a$ **b** 8 **c** 4 **d** $5xy$ **e** $20y$ **f** $5x$ **g** 20 **h** 5 **3 a** $2p$ **b** ab **c** $3b$ **d** a **e** $\frac{x}{2y}$ **f** $\frac{2t}{5}$ **g** $\frac{2r}{3}$ **h** $\frac{1}{2}$ **4 a** 7 **b** 12 **c** 15 **d** 6 **e** 0 **f** 16 **g** 32 **h** 6 **5 a** x^9 **b** t^7 **c** $5p^4$ **d** $6a^5$ **e** $-16m^{10}$ **f** $3ab^3$ **g** $9x^4$ **h** x^7y^8 **i** $12x^5y^8$ **j** $16p^4q^9$ **k** $15p^3q^2r^7$ **l** $24a^{12}$ **6 a** x^{10} **b** $6m^{10}$ **c** $4x^5$ **d** $3m^2n$ **e** $\frac{n^3}{2}$ **f** $\frac{x}{y^2}$ **g** $\frac{3b^4}{2}$ **h** $2a^2b^5$ **7 a** x^8 **b** $27t^{18}$ **c** $5p^{36}$ **d** $16m^8n^{16}$

PRODUCTS AND QUOTIENTS

Algebraic Techniques

25 MINUTES

INTERMEDIATE TEST

Part A Multiple Choice

1 $2a^2 \times 6a^6 = ?$

A $8a^8$ **B** $8a^{12}$ **C** $12a^8$ **D** $12a^{12}$ (1 mark)

2 $5n^8 \div 5n^4 = ?$

A n^2 **B** n^4 **C** $5n^2$ **D** $5n^4$ (1 mark)

3 $(2x^4)^3 = ?$

A $2x^7$ **B** $2x^{12}$ **C** $8x^7$ **D** $8x^{12}$ (1 mark)

4 $a^2b^4c^3 \times a^2bc^5 = ?$

A $a^4b^4c^{15}$ **B** $a^4b^4c^8$ **C** $a^4b^5c^8$ **D** $a^4b^5c^{15}$ (1 mark)

5 $4a^7 + 2a^3 \times a^4 = ?$ *Hint 1*

A $4a^7 + 2a^{12}$ **B** $6a^7$ **C** $6a^{14}$ **D** $6a^{19}$ (1 mark)

6 $\dfrac{15e^6}{25e^7} = ?$

A $\dfrac{3}{5e}$ **B** $\dfrac{3e}{5}$ **C** $\dfrac{e}{2}$ **D** $\dfrac{1}{2e}$ (1 mark)

7 $(x^3y^2)^4 = ?$

A x^7y^6 **B** $x^{12}y^8$ **C** $x^{81}y^{16}$ **D** $(xy)^{24}$ (1 mark)

8 $9p^3 + 3p^3 \times 5p = ?$

A $24p^3$ **B** $60p^4$ **C** $60p^7$ **D** $9p^3 + 15p^4$ (1 mark)

9 $5x^2y^2(3x^3 - 4y^4) = ?$

A $15x^5y^2 - 20x^2y^6$ **B** $15x^6y^2 - 20x^2y^6$ **C** $15x^5y^2 - 20x^2y^8$ **D** $15x^6y^2 - 20x^2y^8$ (1 mark)

10 $\dfrac{6a^7b^4c^9}{(abc)^5} = ?$ *Hint 2*

A $6a^6b^3c^4$ **B** $6a^2bc^4$ **C** $\dfrac{6a^2c^4}{b}$ **D** $6(abc)^4$ (1 mark)

Part B Short Answer

11 Simplify $3x^2 - 2x \times 4x$ (1 mark)

12 Simplify $8x^4 \div 2x^6$ giving the answer in fractional form. *Hint 3* (2 marks)

13 Simplify $2a^8 \times 6a^4 \div 4a^5$ (2 marks)

14 Simplify $(4k^3)^2 \div (24k^7 \div 3k)$ (3 marks)

15 Expand and simplify $4x^2(3x^5 + 7) - 3x(4x - 9)$ (2 marks)

16 Simplify $(8x^6 \div 4x^2)^5$ *Hint 4* (2 marks)

17 Simplify $\dfrac{15x^{15} \div 3x^3}{10x^{10} \div 2x^2}$ *Hint 5* (3 marks)

Hint 1: Take care with the order of operations.
Hint 2: Work on the denominator first.
Hint 3: What does it mean if a power is negative?
Hint 4: Work out the part in brackets first.
Hint 5: Simplify the numerator and simplify the denominator, then simplify the fraction.

Your Feedback

$\dfrac{\square}{25} \times 100\% = \square\%$

PAGES 109–110

PAGE 176

PRODUCTS AND QUOTIENTS

Algebraic Techniques

40 MINUTES

ADVANCED TEST

1 Simplify, leaving answers in index form:

a $9x^{-2} \times 5x^4$ (1 mark)

b $3a^7 \times 4a^{-4}$ (1 mark)

c $7m^{-5} \times 2m^{-6}$ (1 mark)

d $3d^5 \times 8d^{-11}$ (1 mark)

e $24k^3 \div 6k^8$ (1 mark)

f $40p^5 \div 5p^{-3}$ (1 mark)

g $36m^{-3} \div 9m^7$ (1 mark)

h $70u^{-8} \div 10u^{-7}$ (1 mark)

i $18y^{-4} \div 3y^{-9}$ (1 mark)

2 Simplify:

a $(5p^4)^3$

b $2(x^2y^3)^4$

c $(11a^0)^2$

d $(3x^{-2})^2$

e $5(a^5)^{-2}$

f $\left(\frac{1}{2}e^{-3}\right)^{-4}$

(1 mark each)

3 Simplify:

a $\dfrac{12x^3y^2 \times 4x^5y}{16x^7y^3 \div 2xy^2}$ (2 marks)

b $\dfrac{32a^5b^3c \div 4abc^2}{6a^2bc^4 \times 4abc^2}$ (2 marks)

c $(9x^3y^2)^2 \times 5(xy^3)^3$ (2 marks)

d $15p^6q^4 \div 3pq \div p^8q^9 \times 2p^3q^3$ (2 marks)

4 Expand and simplify:

a $4a^2(a^2 + 3a + 1) - 5a^2$ (2 marks)

b $7x^2y^2(x^2 - y) + 2xy^3(3x - 2y)$ (2 marks)

c $3p(5q - 2r) - 4q(2p + 3r) - r(8p - 7q)$ (2 marks)

d $(2k)^2(2k - 3m) - k(8k^2 - 12km)$ (2 marks)

5 Write without negative indices:

a $3a^{-1}$

b $(4x)^{-1}$

c $\left(\frac{x}{2}\right)^{-1}$

d $5\left(\frac{1}{x}\right)^{-1}$

e $\frac{1}{2}m^{-1}$

(1 mark each)

6 a Simplify $x(x^2 + ax + a^2) - ax(x + a)$ (1 mark)

b Evaluate $x(x^2 + ax + a^2) - ax(x + a)$ when $x = 5$ (1 mark)

c Briefly explain why it is not necessary to know the value of a in order to evaluate the expression in part **b**. (1 mark)

d Michael said 'You can always check whether the simplification is correct by substituting into both parts. If the answer is the same in both, it must be correct.' Is Michael correct? Justify your answer. (1 mark)

Your Feedback

$\dfrac{\square}{40} \times 100\% = \square\%$

PAGE 110

PAGE 176

PRODUCTS AND QUOTIENTS

Algebraic Techniques

25 MINUTES

INTERMEDIATE TEST

Part A Multiple Choice

1 $2a^2 \times 6a^6 = ?$

A $8a^8$ **B** $8a^{12}$ **C** $12a^8$ **D** $12a^{12}$ (1 mark)

2 $5n^8 \div 5n^4 = ?$

A n^2 **B** n^4 **C** $5n^2$ **D** $5n^4$ (1 mark)

3 $(2x^4)^3 = ?$

A $2x^7$ **B** $2x^{12}$ **C** $8x^7$ **D** $8x^{12}$ (1 mark)

4 $a^2b^4c^3 \times a^2bc^5 = ?$

A $a^4b^4c^{15}$ **B** $a^4b^4c^8$ **C** $a^4b^5c^8$ **D** $a^4b^5c^{15}$ (1 mark)

5 $4a^7 + 2a^3 \times a^4 = ?$ *Hint 1*

A $4a^7 + 2a^{12}$ **B** $6a^7$ **C** $6a^{14}$ **D** $6a^{19}$ (1 mark)

6 $\dfrac{15e^6}{25e^7} = ?$

A $\dfrac{3}{5e}$ **B** $\dfrac{3e}{5}$ **C** $\dfrac{e}{2}$ **D** $\dfrac{1}{2e}$ (1 mark)

7 $(x^3y^2)^4 = ?$

A x^7y^6 **B** $x^{12}y^8$ **C** $x^{81}y^{16}$ **D** $(xy)^{24}$ (1 mark)

8 $9p^3 + 3p^3 \times 5p = ?$

A $24p^3$ **B** $60p^4$ **C** $60p^7$ **D** $9p^3 + 15p^4$ (1 mark)

9 $5x^2y^2(3x^3 - 4y^4) = ?$

A $15x^5y^2 - 20x^2y^6$ **B** $15x^6y^2 - 20x^2y^6$ **C** $15x^5y^2 - 20x^2y^8$ **D** $15x^6y^2 - 20x^2y^8$ (1 mark)

10 $\dfrac{6a^7b^4c^9}{(abc)^5} = ?$ *Hint 2*

A $6a^6b^3c^4$ **B** $6a^2bc^4$ **C** $\dfrac{6a^2c^4}{b}$ **D** $6(abc)^4$ (1 mark)

Part B Short Answer

11 Simplify $3x^2 - 2x \times 4x$ (1 mark)

12 Simplify $8x^4 \div 2x^6$ giving the answer in fractional form. *Hint 3* (2 marks)

13 Simplify $2a^8 \times 6a^4 \div 4a^5$ (2 marks)

14 Simplify $(4k^3)^2 \div (24k^7 \div 3k)$ (3 marks)

15 Expand and simplify $4x^2(3x^5 + 7) - 3x(4x - 9)$ (2 marks)

16 Simplify $(8x^6 \div 4x^2)^5$ *Hint 4* (2 marks)

17 Simplify $\dfrac{15x^{15} \div 3x^3}{10x^{10} \div 2x^2}$ *Hint 5* (3 marks)

Hint 1: Take care with the order of operations.
Hint 2: Work on the denominator first.
Hint 3: What does it mean if a power is negative?
Hint 4: Work out the part in brackets first.
Hint 5: Simplify the numerator and simplify the denominator, then simplify the fraction.

Your Feedback

$\dfrac{\square}{25} \times 100\% = \square\%$

PAGES 109–110
PAGE 176

PRODUCTS AND QUOTIENTS

Algebraic Techniques

40 MINUTES

ADVANCED TEST

1 Simplify, leaving answers in index form:

a $9x^{-2} \times 5x^4$ (1 mark)
b $3a^7 \times 4a^{-4}$ (1 mark)
c $7m^{-5} \times 2m^{-6}$ (1 mark)
d $3d^5 \times 8d^{-11}$ (1 mark)
e $24k^3 \div 6k^8$ (1 mark)
f $40p^5 \div 5p^{-3}$ (1 mark)
g $36m^{-3} \div 9m^7$ (1 mark)
h $70u^{-8} \div 10u^{-7}$ (1 mark)
i $18y^{-4} \div 3y^{-9}$ (1 mark)

2 Simplify:

a $(5p^4)^3$
b $2(x^2y^3)^4$
c $(11a^0)^2$
d $(3x^{-2})^2$
e $5(a^5)^{-2}$
f $\left(\frac{1}{2}e^{-3}\right)^{-4}$ (1 mark each)

3 Simplify:

a $\dfrac{12x^3y^2 \times 4x^5y}{16x^7y^3 \div 2xy^2}$ (2 marks)

b $\dfrac{32a^5b^3c \div 4abc^2}{6a^2bc^4 \times 4abc^2}$ (2 marks)

c $(9x^3y^2)^2 \times 5(xy^3)^3$ (2 marks)
d $15p^6q^4 \div 3pq \div p^8q^9 \times 2p^3q^3$ (2 marks)

4 Expand and simplify:

a $4a^2(a^2 + 3a + 1) - 5a^2$ (2 marks)
b $7x^2y^2(x^2 - y) + 2xy^3(3x - 2y)$ (2 marks)
c $3p(5q - 2r) - 4q(2p + 3r) - r(8p - 7q)$ (2 marks)
d $(2k)^2(2k - 3m) - k(8k^2 - 12km)$ (2 marks)

5 Write without negative indices:

a $3a^{-1}$
b $(4x)^{-1}$
c $\left(\frac{x}{2}\right)^{-1}$
d $5\left(\frac{1}{x}\right)^{-1}$
e $\frac{1}{2}m^{-1}$ (1 mark each)

6 a Simplify $x(x^2 + ax + a^2) - ax(x + a)$ (1 mark)
b Evaluate $x(x^2 + ax + a^2) - ax(x + a)$ when $x = 5$ (1 mark)
c Briefly explain why it is not necessary to know the value of a in order to evaluate the expression in part **b**. (1 mark)
d Michael said 'You can always check whether the simplification is correct by substituting into both parts. If the answer is the same in both, it must be correct.' Is Michael correct? Justify your answer. (1 mark)

Your Feedback

$\dfrac{\square}{40} \times 100\% = \square\%$

PAGE 110

PAGE 176

FACTORISING

Algebraic Techniques

STUDY NOTES

1 A **factor** is a number, or term, which will divide evenly into another number or term. For example, the factors of 20 are 1, 2, 4, 5, 10 and 20.

2 A **prime number** is a number that has exactly two factors, one and itself. So 2, 3, 5, 7 and 11 are all prime. 9 is not prime because $9 = 3 \times 3$. (A number that is not prime is composite.) 1 is not a prime number as it has only one factor.

3 In **expanded form**, an algebraic term is written as a product of its factors. For example, in expanded form $5x^2y^3 = 5 \times x \times x \times y \times y \times y$.

4 When **expanding**, every term inside the brackets must be multiplied by the term outside.
$a(b + c) = a \times b + a \times c = ab + ac$

5 **Factorising** an algebraic expression involving the sum or difference of two or more terms means rewriting it as the **product of its factors**.

6 **Factorising** is the opposite operation to expanding. When factorising, ask yourself what number (or term), divides into every part. This answer is the **common factor**. The common factor appears outside the brackets. The expression inside the bracket is the answer when the original expression is divided by the common factor. For example, to factorise $2x + 8$ ask yourself what divides into both $2x$ and 8. The answer is 2. Then ask yourself how many times does 2 divide into $2x$ (answer x) and how many times does 2 divide into 8 (answer 4). So $2x + 8 = 2(x + 4)$. Similarly $4x - 12 = 4(x - 3)$ and $x^2 - 11x = x(x - 11)$.

7 You can always **check a factorisation by expanding** the answer.

8 Take extra care if the **common factor divides into a term exactly once**. $4ab + 2a = 2a(2b + 1)$.
Don't forget the one.

9 **An expression** should always be **fully factorised**. $2x^2 + 6x$ has a common factor of 2 and a common factor of x. To factorise it fully, both the 2 and x must be taken outside the brackets.
$2x^2 + 6x = 2x(x + 3)$

10 A **common factor** is a number or term that divides evenly into each term of an algebraic expression. When finding a common factor, always look for the largest common factor.
For example, $2x(4xy + 2y^2)$ is a factorisation of $8x^2y + 4xy^2$, but the part in the brackets can still be factorised. When **factorised fully** $8x^2y + 4xy^2 = 4xy(2x + y)$.

11 When **simplifying fractions involving algebraic terms**, we can only **divide** (or cancel) if there is a **common factor**.
For example, $\dfrac{2x + 4y}{2} = \dfrac{2(x + 2y)}{2} = \dfrac{x + 2y}{1} = x + 2y$. Take extra care if the numerator **cancels out** completely. The numerator will be one after all cancelling has been done.
For example, $\dfrac{3}{3a - 9b} = \dfrac{3}{3(a - 3b)} = \dfrac{1}{a - 3b}$.

Checklist

Can you:

1 *Understand what is meant by a factor and a common factor?* ☐
2 *Find factors of numbers and terms?* ☐
3 *Expand and factorise algebraic expressions?* ☐
4 *Understand the need to fully factorise expressions and identify whether an expression is fully factorised?* ☐
5 *Simplify algebraic expressions involving fractions by first factorising?* ☐

FACTORISING

Algebraic Techniques

SKILLS CHECK

1 List all the factors of 24.

2 Write 90 as a product of its prime factors.

3 Write $3a^2b^3c$ in expanded form.

4 What is the highest common factor of:

a 12 and 18? **b** 36 and 48? **c** 15 and 30?

5 What is the largest common factor of:

a $4x^2$ and $6x$? **b** $3ab^2c$ and $9a^2bc^3$? **c** $8x^2y^3z^5$ and $12x^3y^2z^3$?

6 Expand:

a $5(x + 2)$ **b** $6(2x - 3)$ **c** $4(3x + y)$

7 Factorise:

a $7x + 14$ **b** $10x - 15$ **c** $6x + 3y$

d $2p + 2q$ **e** $5t - 5$ **f** $12m - 15n$

8 Expand:

a $x(x + 4)$ **b** $a(a - 2)$ **c** $p(p + q)$

9 Factorise:

a $x^2 + 5x$ **b** $a^2 - 3a$ **c** $t^2 + tu$

d $2x^2 + 3x$ **e** $5p^2 - 2pq$ **f** $x^3 + x^2 + x$

10 Expand:

a $-2(3x + 5)$ **b** $-k(2k - 3m)$ **c** $-(2t - 7)$

11 Factorise, by taking out a negative factor:

a $-8x - 6$ **b** $-4k + 12$ **c** $-3x^2 + xy$

12 Factorise, by taking out a factor of –1:

a $-5a - 4$ **b** $-9t + 8$ **c** $-2a - 3b + 4c$

13 Fill in the blanks:

a $2x + 8 = 2(x + \_\_\_)$ **b** $6y - 12z = \_\_\_(y - 2z)$

c $a^2b + abc = \_\_\_(a + c)$ **d** $6x^2y + 9xy^2 = 3xy(\_\_\_\_\_\_)$

e $3p^2 + 3p = 3p(p + \_\_\_)$ **f** $12k + 8m - 20n = 4(\_\_\_\_\_\_)$

PAGES 110–111

Answers **1** 1, 2, 3, 4, 6, 8, 12, 24 **2** $2 \times 3 \times 3 \times 5$ **3** $3 \times a \times a \times b \times b \times b \times c$ **4 a** 6 **b** 12 **c** 15 **5 a** $2x$ **b** $3abc$ **c** $4x^2y^2z^3$ **6 a** $5x + 10$ **b** $12x - 18$ **c** $12x + 4y$ **7 a** $7(x + 2)$ **b** $5(2x - 3)$ **c** $3(2x + y)$ **d** $2(p + q)$ **e** $5(t - 1)$ **f** $3(4m - 5n)$ **8 a** $x^2 + 4x$ **b** $a^2 - 2a$ **c** $p^2 + pq$ **9 a** $x(x + 5)$ **b** $a(a - 3)$ **c** $t(t + u)$ **d** $x(2x + 3)$ **e** $p(5p - 2q)$ **f** $x(x^2 + x + 1)$ **10 a** $-6x - 10$ **b** $-2k^2 + 3km$ **c** $-2t + 7$ **11 a** $-2(4x + 3)$ **b** $-4(k - 3)$ **c** $-x(3x - y)$ **12 a** $-(5a + 4)$ **b** $-(9t - 8)$ **c** $-(2a + 3b - 4c)$ **13 a** 4 **b** 6 **c** ab **d** $2x + 3y$ **e** 1 **f** $3k + 2m - 5n$

FACTORISING

Algebraic Techniques

30 MINUTES

INTERMEDIATE TEST

Part A Multiple Choice

1 Which of the following is a factor of $8 - 2x$? (1 mark)

A x B 4 C $4 - x$ D $4 - 2x$

2 When fully factorised, $6x^2 - 18xy = ?$ (1 mark)

A $6(x^2 - 3xy)$ B $3x(2x - 6y)$ C $2x(3x - 9y)$ D $6x(x - 3y)$

3 Which is not a correct factorisation of $9x^2y^3 - 6xy^4$? *Hint 1* (1 mark)

A $3(3x^2y^3 - 2xy^4)$ B $3xy(3xy^2 - 2y^3)$ C $xy^2(9xy - 6y^2)$ D $3xy^3(3x - 2)$

4 $2x^2$ and $8x$ have a common factor of: (1 mark)

A 2 only B x only C neither 2 nor x D both 2 and x

5 Which of these is fully factorised? (1 mark)

A $4x^2 + 8x = 4(x^2 + 2x)$

B $15p - 18q = 3(5p - 6q)$

C $6a - 24 = 2(3a - 12)$

D $16t^2 - 8t = 4t(4t - 2)$

6 Which is not correct? (1 mark)

A $3a^2b - 2b = b(3a^2 - 2)$

B $4p^3q^2 + 2p^2q = 2p^2q(2pq + 1)$

C $16t^2u - 12tu^2 = 4tu(4t - 3u^2)$

D $8x^3y - 4x^2y^2 + 2xy^3 = 2xy(4x^2 - 2xy + y^2)$

7 Which is correct? (1 mark)

A $8b - 3a = -(3a - 8b)$

B $-x^2 + 5 = -(x^2 + 5)$

C $-3x - 2y = -(3x - 2y)$

D $-4k + 7 = -(7 - 4k)$

8 $\frac{5x + 10}{5} = ?$ *Hint 2* (1 mark)

A $3x$ B $x + 2$ C $x + 10$ D $5x + 2$

9 $(x + 2)(x + 3) + (x + 2)(x - 3) = ?$ *Hint 3* (1 mark)

A $2x(x + 2)$

B $(x + 2)(x + 3)(x - 3)$

C $-3x(x + 2)$

D none of these

Part B Short Answer

10 Factorise $6a^6 + 3a$ (2 marks)

11 Why is the following incorrect? $15x^2y^3 - 10xy^4 = 5xy(3x - 2y)$ *Hint 4* (2 marks)

12 Factorise $3a + 9ab - 12a^2$ fully. (2 marks)

13 Fully factorise each of the following: (2 marks each)

a $8x^2 - 12x$ b $3x^3 + 15x^8$ c $ab^2 + a^2b$

d $9xy^2 - 6xyz$ e $12x^2y^4z^3 - 21xy^3z^2$ f $10ab^2 + 20a^2b - 5ab - 25a^2b^2$

14 Simplify $\frac{8x^3 + 24x^2}{2x^2 + 6x}$ *Hint 5* (3 marks)

Hint 1: Some might not be fully factorised but that doesn't mean they are not correct.
Hint 2: Factorise the numerator first.
Hint 3: $(x + 2)$ is a common factor.
Hint 4: If unsure, expand the second part.
Hint 5: Factorise the numerator and factorise the denominator then simplify the fraction.

Your Feedback

$\frac{\square}{30} \times 100\% = \square\%$

PAGES 111–112

PAGE 176

FACTORISING

Algebraic Techniques

40 MINUTES

ADVANCED TEST

1 Factorise:

a $3p^2 - p$

b $x^4 + xy^2$

c $9a^2b - 5b^2c$

d $xe^x + e^x$

e $2000 + 2000(1.08) + 2000(1.08)^2$

f $5\sin\theta - 10\cos\theta$

(1 mark each)

2 Jarod was asked to factorise an expression. His solution is shown below.

$9x^3y^4z^2 + 12x^2y^3z^2 = 3xyz^2(3x^2y^3 + 4xy^2)$

Is Jarod correct? Justify your answer. (1 mark)

3 Factorise fully:

a $x^3 - x^2y^2$

b $8a^2b^3c - 12ab^4c^3$

c $24x^9 - 36x^{12}$

d $12x^3y^2 - 18xy^3$

e $-4a^2 - 18ab$

f $2\pi r^2 + 2\pi rh$

g $\frac{1}{2}a^2 - \frac{ab}{2}$

h $16a^2b^2 - 8ab - 2b$

i $15pq^2 - 3pq + 5p^2q$

j $15a^4b^9cd^3 + 30ab^5cd^7$

k $35x^7y^4z^3 - 49x^5y^5z^6 + 28x^4y^6z^4$

(1 mark each)

4 Briefly explain why the following factorisations are not complete and give the full factorisation.

a $40a^2b - 24ab^2 = 4ab(10a - 6b)$ (1 mark)

b $9m^2n^3 + 12mn^4 = 3mn(3mn^2 + 4n^3)$ (1 mark)

5 Briefly explain why the following factorisations are incorrect and give the correct factorisation.

a $4x^2yz^3 + 12xyz^3 = 4xy^3z^3(x + 3)$ (1 mark)

b $-25p^2q^2r^3 - 15pqr^2 = -5pqr^2(5pqr - 3)$ (1 mark)

6 Factorise:

a $x(x + 4) + 7(x + 4)$

b $2x(3x - 5) - 3(3x - 5)$

c $(a + b)^2 + 6(a + b)$

d $t(t - 1) + t - 1$

(1 mark each)

7 Factorise by first factorising those parts of the expression in brackets:

a $(x^3 - 7x^2) + (5x - 35)$ (1 mark)

b $(pq - 2pr) - (3q^2 - 6qr)$ (1 mark)

8 Factorise, then simplify, where possible:

a $\frac{4x^2 - 8x}{2y}$

b $\frac{7x - 14y}{7x + 21y}$

c $\frac{3ab - 9b^2}{6a^2b}$

d $\frac{10mn - 5n}{2m - 1}$

e $\frac{3t - 4}{6t^2 - 8t}$

f $\frac{12 - 2q}{3q - 18}$

(2 marks each)

Your Feedback

$\frac{\square}{40} \times 100\% = \square\%$

PAGES 112–113

PAGE 176

ALGEBRAIC FRACTIONS

Algebraic Techniques

STUDY NOTES

1 Fractions can be **simplified** by cancelling or dividing both numerator and denominator by the same number or term. A fraction is in **simplest form** when the numerator and denominator have **no common factors**.

2 **Equivalent fractions** are fractions that **have the same value**. $\frac{1}{2}$ and $\frac{2}{4}$ are equivalent fractions because $\frac{1}{2} = \frac{2}{4}$. $\frac{x}{3} = \frac{5x}{15}$ so $\frac{x}{3}$ and $\frac{5x}{15}$ are equivalent fractions. Equivalent fractions are formed by multiplying both the numerator and denominator by the same number or term or by dividing both numerator and denominator by the same number or term. Adding the same number to, or subtracting the same number from, both numerator and denominator will NOT produce an equivalent fraction.

3 If fractions have a **common denominator** they can be **added** (or **subtracted**), by adding (or subtracting) the **numerators**.
If there is **no common denominator**, the fractions must be **changed to equivalent fractions** with the same **denominator**. Always look for the lowest common denominator.
For example, to add $\frac{x}{3}$ and $\frac{3x}{7}$, the common denominator is 21. So change both fractions to equivalent fractions with denominator 21. $\frac{x}{3} + \frac{3x}{7} = \frac{7x}{21} + \frac{9x}{21} = \frac{16x}{21}$. To subtract $\frac{5a}{8}$ from $\frac{11a}{12}$, the lowest common denominator is 24. $\frac{11a}{12} - \frac{5a}{8} = \frac{22a}{24} - \frac{15a}{24} = \frac{7a}{24}$

4 To **multiply fractions** multiply the **numerators** and multiply the **denominators**.

$\frac{x}{2} \times \frac{y}{3} = \frac{x \times y}{2 \times 3} = \frac{xy}{6}$ $\qquad$ $\frac{a}{b} \times \frac{c}{d} = \frac{ac}{bd}$

5 If there are **common factors** in the numerator and denominator these can be **cancelled** before multiplying. Only cancel a term in the numerator with one in the denominator.

$\frac{x}{3} \times \frac{6a}{xy} = \frac{2a}{y}$ [dividing the numerator and denominator by both 3 and x]

6 **Two fractions** are **reciprocals** if they multiply to 1. The reciprocal of any fraction is another fraction whose numerator was the denominator of the first fraction and whose denominator was the numerator of the first fraction. The reciprocal of $\frac{x}{3}$ is $\frac{3}{x}$. The reciprocal of $\frac{2a}{b}$ is $\frac{b}{2a}$.

7 To **divide by a fraction**, multiply by the reciprocal. $\frac{a}{6} \div \frac{3}{5a} = \frac{a}{6} \times \frac{5a}{3} = \frac{5a^2}{18}$

8 An **improper fraction** has its numerator greater than its denominator. With algebraic fractions, answers are usually left in improper form rather than as mixed numbers. We say $\frac{3x}{2}$ not $1\frac{1}{2}x$.

Checklist

Can you:

1 *Simplify algebraic fractions?* ☐
2 *Understand what is meant by equivalent fractions, improper fractions and reciprocals?* ☐
3 *Add, subtract, multiply and divide fractions involving algebraic expressions?* ☐
4 *Cancel correctly?* ☐

ALGEBRAIC FRACTIONS

Algebraic Techniques

SKILLS CHECK

1 Find:

a $\frac{1}{3}+\frac{1}{3}$ b $\frac{a}{3}+\frac{a}{3}$ c $\frac{a}{3}+\frac{b}{3}$

d $\frac{2}{7}+\frac{3}{7}$ e $\frac{2x}{7}+\frac{3x}{7}$ f $\frac{2x}{7}+\frac{3y}{7}$

2 Find:

a $\frac{4}{5}-\frac{1}{5}$ b $\frac{4t}{5}-\frac{t}{5}$ c $\frac{4t}{5}-\frac{u}{5}$

d $\frac{8}{9}-\frac{7}{9}$ e $\frac{8x}{9}-\frac{7x}{9}$ f $\frac{8}{9a}-\frac{7}{9a}$

3 Simplify:

a $\frac{4}{6}$ b $\frac{4x}{6}$ c $\frac{4a}{6b}$ d $\frac{12e}{20}$

e $\frac{3u}{5u}$ f $\frac{8m}{2}$ g $\frac{15}{20a}$ h $\frac{7n^2}{3n}$

i $\frac{8e}{6e}$ j $\frac{3}{6k}$ k $\frac{4a}{8a}$ l $\frac{9e}{3e}$

4 Complete:

a $\frac{2}{3}=\frac{}{6}$ b $\frac{2x}{3}=\frac{}{6}$ c $\frac{2ab}{3}=\frac{}{6}$

d $\frac{5a}{8}=\frac{}{24}$ e $\frac{x}{5}=\frac{}{15}$ f $\frac{3}{a}=\frac{}{2a}$

5 Find:

a $\frac{2}{3}\times\frac{4}{5}$ b $\frac{2a}{3}\times\frac{4b}{5}$ c $\frac{2x}{3}\times\frac{4x}{5}$

d $\frac{a}{4}\times\frac{b}{5}$ e $\frac{x}{2}\times\frac{x}{3}$ f $\frac{3a}{4}\times\frac{2bc}{9}$

6 Find:

a $\frac{3}{5}\div\frac{2}{3}$ b $\frac{3a}{5}\div\frac{2}{3}$ c $\frac{3a}{5}\div\frac{2a}{3}$

PAGE 113

Answers 1 a $\frac{2}{3}$ b $\frac{2a}{3}$ c $\frac{a+b}{3}$ d $\frac{5}{7}$ e $\frac{5x}{7}$ f $\frac{2x+3y}{7}$ 2 a $\frac{3}{5}$ b $\frac{3t}{5}$ c $\frac{4t-u}{5}$ d $\frac{1}{9}$ e $\frac{x}{9}$ f $\frac{1}{9a}$ 3 a $\frac{2}{3}$ b $\frac{2x}{3}$ c $\frac{2a}{3b}$ d $\frac{3e}{5}$ e $\frac{3}{5}$ f $4m$ g $\frac{3}{4a}$ h $\frac{7n}{3}$ i $\frac{4}{3}$ or $1\frac{1}{3}$ j $\frac{1}{2k}$ k $\frac{1}{2}$ l 3 4 a 4 b $4x$ c $4ab$ d $15a$ e $3x$ f 6 5 a $\frac{8}{15}$ b $\frac{8ab}{15}$ c $\frac{8x^2}{15}$ d $\frac{ab}{20}$ e $\frac{x^2}{6}$ f $\frac{abc}{6}$ 6 a $\frac{9}{10}$ b $\frac{9a}{10}$ c $\frac{9}{10}$

ALGEBRAIC FRACTIONS
Algebraic Techniques

20 MINUTES
INTERMEDIATE TEST

Part A Multiple Choice

1 $\frac{x}{3}+\frac{x}{4}=?$

A $\frac{2x}{7}$ B $\frac{x}{6}$

C $\frac{7x}{12}$ D none of these (1 mark)

2 In simplest form, $\frac{8ab}{12a}=?$

A $\frac{4b}{6}$ B $\frac{2ab}{3a}$

C $\frac{4ab}{6a}$ D $\frac{2b}{3}$ (1 mark)

3 $\frac{7a}{8}-\frac{3a}{5}=?$

A $\frac{4a}{3}$ B $\frac{a}{10}$

C $\frac{11a}{40}$ D $\frac{41a}{40}$ (1 mark)

4 Which fraction is equivalent to $\frac{3x}{5y}$?

A $\frac{12x}{20y}$ B $\frac{27x}{45y}$

C $\frac{36x}{60y}$ D all of these (1 mark)

5 $\frac{2x}{3}\times\frac{3}{4}=?$

A $\frac{x}{2}$ B $\frac{5x}{12}$

C $\frac{2x+3}{12}$ D $6x$ (1 mark)

6 $\frac{2a}{3}\div\frac{1}{9}=?$

A $\frac{2a}{27}$ B $\frac{27}{2a}$

C $\frac{1}{6a}$ D $6a$ (1 mark)

Part B Short Answer

7 Simplify $\frac{m}{6}+\frac{n}{5}$ (2 marks)

8 Simplify $\frac{3a}{5}-\frac{a}{2}$ (2 marks)

9 Find $\frac{2x}{3}+\frac{5x}{6}$ *Hint 1* (2 marks)

10 Find $\frac{3a}{4b}\div\frac{2a}{5c}$ *Hint 2* (2 marks)

11 Simplify $\frac{5pq}{2}\times\frac{4}{10p}$ *Hint 3* (2 marks)

12 Simplify $\frac{7e}{12}-\frac{5e}{9}$ *Hint 4* (2 marks)

13 Find the sum of $\frac{7a}{20}$ and $\frac{2a}{15}$ *Hint 5* (2 marks)

Hint 1: Simplify, but leave in improper form.
Hint 2: Multiply by the reciprocal. The *a*s will cancel.
Hint 3: Cancel before multiplying.
Hint 4: Look for the lowest common denominator.
Hint 5: You need to add the fractions.

Your Feedback

$\frac{\square}{20}\times 100\% = \square\,\%$

PAGE 114
PAGE 176

ALGEBRAIC FRACTIONS

Algebraic Techniques

40 MINUTES

ADVANCED TEST

1 Simplify:

a $\dfrac{12x^6}{3x^3}$

b $\dfrac{2ab}{6a^2b^3}$

c $\dfrac{8p^3q^4}{4p^5q^2}$

d $\dfrac{16ab^5c^6}{24a^5bc^7}$

(1 mark each)

2 Find the sums:

a $\dfrac{2x}{5} + \dfrac{3x}{4}$

b $\dfrac{3a}{5} + \dfrac{2b}{7}$

c $\dfrac{4x^2}{3} + \dfrac{x}{6}$

d $\dfrac{5}{2m} + \dfrac{2}{3m}$

e $\dfrac{3}{a} + \dfrac{7}{b}$

f $\dfrac{5a}{2b} + \dfrac{3a}{b}$

(1 mark each)

3 Find these differences:

a $\dfrac{8a}{5} - \dfrac{3a}{2}$

b $\dfrac{7n}{2} - \dfrac{n}{3}$

c $\dfrac{5a^2b^3}{9} - \dfrac{2ab^4}{3}$

d $\dfrac{4}{a} - \dfrac{1}{4a}$

e $\dfrac{6pq}{r} - \dfrac{5pr}{q}$

f $\dfrac{4a}{3b} - \dfrac{2a}{5b}$

(1 mark each)

4 Find these products:

a $\dfrac{3mn}{4} \times \dfrac{2}{5m}$

b $\dfrac{6x^2}{5} \times \dfrac{4x}{9}$

c $\dfrac{8a^2b^2}{9} \times \dfrac{1}{12ab^2}$

d $\dfrac{2a}{5b} \times \dfrac{3b^2}{8}$

e $\dfrac{4x^2y^3}{7} \times \dfrac{3x}{10y^2}$

f $\dfrac{5p^2q}{3r} \times \dfrac{9qr}{10p^3}$

(1 mark each)

5 Find:

a $\dfrac{3}{a} \div \dfrac{6}{a}$

b $\dfrac{x^2}{5} \div \dfrac{x}{4}$

c $\dfrac{2x}{3y} \div \dfrac{x}{y}$

d $\dfrac{8a^2}{5b^2} \div \dfrac{2a}{3b^2}$

e $\dfrac{6a^2b^3}{7c^4} \div \dfrac{3a^2bc}{7}$

f $\dfrac{p^2q}{pq^3} \div \dfrac{4a^2p^3}{8a^5}$

(1 mark each)

6 Find:

a $\dfrac{a^2b^3}{5} \times \dfrac{6}{ab} \times \dfrac{2a}{3b}$

b $\dfrac{4p^2}{5q^2} \times \dfrac{6q^3}{p} - \dfrac{pq}{8}$

c $\dfrac{x}{3} + \dfrac{5x^2}{2} \times \dfrac{8}{x}$

d $\dfrac{a}{5} \div \dfrac{b}{5} + \dfrac{b}{a}$

e $\dfrac{9a^2}{3b} \div \dfrac{2a^2}{6b} + \dfrac{3x^3}{14y^2} \div \dfrac{9x^3}{7y^2}$

f $\dfrac{5x - 4y}{2} - \dfrac{3x + 2y}{5}$

(2 marks each)

Your Feedback

$\dfrac{\square}{40} \times 100\% = \square\ \%$

PAGES 114–115

PAGE 176

BINOMIAL PRODUCTS

Algebraic Techniques

STUDY NOTES

1 Expressions are **simplified** by **collecting like terms**. (We only add or subtract those terms whose pronumeral part is exactly the same.) For example, $x^2 + 5x - 7x - 35 = x^2 - 2x - 35$.

2 When **expanding**, every term inside the bracket must be multiplied by the number, or term, that is outside the brackets. For example, $4(x + 2y) = 4 \times x + 4 \times 2y = 4x + 8y$. Special care must be taken if the term outside the brackets is negative.

$-3(2a - 7b) = -6a + 21b$

$-x^2(3x + 2y) = -3x^3 - 2x^2y$

3 A **binomial** is an algebraic expression involving two terms added together or subtracted. Examples of binomials involving the sum of two terms are $3x + 7$, $x^2 + 8x$ and $9ab^2 + 7abc$. Some binomials involving the difference of two terms are $5x - 1$, $m^2 - 6m$ and $3p^2q - 4qr$.

4 When finding a **binomial product**, we multiply one binomial by another. We can use the area of rectangles to demonstrate the process.

	x	6
x	x^2	$6x$
3	$3x$	18

$(x + 6)(x + 3)$
$= x^2 + 3x + 6x + 18$
$= x^2 + 9x + 18$

	x	a
y	xy	ay
b	bx	ab

$(x + a)(y + b)$
$= xy + bx + ay + ab$

Note: The diagrams are only to illustrate the result. You are not expected to draw them in order to find a binomial product. $\mathbf{(a + b)(c + d) = ac + ad + bc + bd}$

5 Another way of considering a binomial product is that **each term** in the second brackets must be **multiplied by each term** in the first brackets. For example, $(x + 5)(x + 3) = x(x + 3) + 5(x + 3)$
$= x^2 + 3x + 5x + 15$
$= x^2 + 8x + 15$

6 The **square of a binomial** is simply a special binomial product. If a binomial is multiplied by itself, the two middle terms will be the same. You can always find the square of a binomial by multiplying it by itself.

$\mathbf{(a + b)^2 = a^2 + 2ab + b^2}$ $\mathbf{(a - b)^2 = a^2 - 2ab + b^2}$

7 If in a binomial product, one binomial is the sum of two terms and the other binomial is the difference of the same two terms, then we have another special binomial product. In the **sum by difference** the two middle terms will be the same but with opposite signs (one will be positive and one will be negative), so they will always cancel out. $\mathbf{(a + b)(a - b) = a^2 - b^2}$

8 When **expanding and simplifying** algebraic expressions involving binomial products, the correct order of operations must be followed and care must be taken with any negative signs.
For example, $(2x + 3)(x - 5) - (x - 2)(x - 3) = 2x^2 - 10x + 3x - 15 - (x^2 - 3x - 2x + 6)$
$= 2x^2 - 7x - 15 - x^2 + 5x - 6$
$= x^2 - 2x - 21$

Checklist

Can you:

1. *Simplify algebraic expressions?* ☐
2. *Understand what is meant by a binomial and by a binomial product?* ☐
3. *Find binomial products?* ☐
4. *Find special products such as the square of a binomial and the sum by difference?* ☐
5. *Expand and simplify more complicated algebraic expressions?* ☐

BINOMIAL PRODUCTS
Algebraic Techniques

SKILLS CHECK

1 Simplify:

a $-5a + 3b + 2a - 6b + 3$ b $x^2 + 7x - 3x + 4$ c $2a^2 - 6a + 8a - 5$

2 Expand:

a $2(3x - 5)$ b $x(x + 6)$ c $2p(3p - 7q)$

3 Expand and simplify:

a $5(3x + 2) + 4(x - 2)$ b $2(4a + 3) + 3(5a + 1)$
c $6(4p - q) + 5(2p + 3q)$ d $3(2e - 7) - 2(3e - 2)$
e $-4(5a + 2b) - 3(6a + 5)$ f $-2(8x - y) - (3x + y)$

4 Expand and simplify:

a $x(x + 3) + 4(x + 3)$ b $a(a + 5) - 2(a + 5)$
c $m(m - 4) + 7(m - 4)$ d $n(n - 2) - 5(n - 2)$

5 Find the area of each rectangle:

a A_1 b A_2
c A_3 d A_4

	x	7
x	A_1	A_3
5	A_2	A_4

6 Using the diagram in question 5, $(x + 7)(x + 5) = ?$

7 Complete:

a $(x + 2)(x + 5) = x^2 + 5x + 2x + \ldots\ldots$ b $(a + 4)(a - 3) = a^2 - 3a + \ldots\ldots - 12$
c $(t - 7)(t + 1) = t^2 + \ldots\ldots - 7t - 7$ d $(e - 8)(e - 5) = e^2 - 5e - 8e + \ldots\ldots$
e $(2p + 3)(p + 4) = \ldots\ldots + 8p + 3p + 12$ f $(3k - 1)(2k + 7) = \ldots\ldots + 21k - 2k - \ldots\ldots$

8 Find these binomial products:

a $(x + 4)(x + 2)$ b $(a + 3)(a - 2)$ c $(2a + b)(3c - 4d)$
d $(p - 6)(p + 4)$ e $(3m - 8)(m - 5)$ f $(5x - 3)(2x + 5)$
g $(x - 1)(x + 1)$ h $(x + 7)(x + 7)$ i $(3y + 2)(3y - 2)$

9 Find these products of a sum by difference:

a $(x - 6)(x + 6)$ b $(a + 4)(a - 4)$ c $(2n + 9)(2n - 9)$ d $(5x - 1)(5x + 1)$

10 Find these squares of binomials:

a $(x + 9)^2$ b $(y - 6)^2$ c $(4x - 11)^2$ d $(3p + 4q)^2$

PAGES 115–116

Answers **1 a** $-3a - 3b + 3$ **b** $x^2 + 4x + 4$ **c** $2a^2 + 2a - 5$ **2 a** $6x - 10$ **b** $x^2 + 6x$ **c** $6p^2 - 14pq$ **3 a** $19x + 2$ **b** $23a + 9$ **c** $34p + 9q$ **d** -17 **e** $-38a - 8b - 15$ **f** $-19x + y$ **4 a** $x^2 + 7x + 12$ **b** $a^2 + 3a - 10$ **c** $m^2 + 3m - 28$ **d** $n^2 - 7n + 10$ **5 a** x^2 **b** $5x$ **c** $7x$ **d** 35 **6** $x^2 + 12x + 35$ **7 a** 10 **b** $4a$ **c** t **d** 40 **e** $2p^2$ **f** $6k^2, 7$ **8 a** $x^2 + 6x + 8$ **b** $a^2 + a - 6$ **c** $6ac - 8ad + 3bc - 4bd$ **d** $p^2 - 2p - 24$ **e** $3m^2 - 23m + 40$ **f** $10x^2 + 19x - 15$ **g** $x^2 - 1$ **h** $x^2 + 14x + 49$ **i** $9y^2 - 4$ **9 a** $x^2 - 36$ **b** $a^2 - 16$ **c** $4n^2 - 81$ **d** $25x^2 - 1$ **10 a** $x^2 + 18x + 81$ **b** $y^2 - 12y + 36$ **c** $16x^2 - 88x + 121$ **d** $9p^2 + 24pq + 16q^2$

BINOMIAL PRODUCTS

Algebraic Techniques

30 MINUTES

INTERMEDIATE TEST

Part A Multiple Choice

1 $(x-5)(y+2) = ?$

A $xy - 5x + 2y - 10$ B $xy - 5x + 2y + 10$
C $xy + 2x - 5y - 10$ D $xy + 2x - 5y + 10$

(1 mark)

2 $(x+4)(x-5) = ?$

A $x^2 - x - 20$ B $x^2 + x - 20$
C $x^2 - 9x - 20$ D $x^2 + 9x - 20$

(1 mark)

3 $(n-4)(n-2) = ?$

A $n^2 - 2n - 8$ B $n^2 - 2n + 8$
C $n^2 - 6n - 8$ D $n^2 - 6n + 8$

(1 mark)

4 $(3a-5)(2a-1) = ?$

A $6a^2 - 13a + 5$ B $6a^2 - 13a - 5$
C $6a^2 - 7a + 5$ D $6a^2 + 7a - 5$

(1 mark)

5 $(2m-5n)(2m+5n) = ?$

A $4m^2 + 25n^2$ B $4m^2 - 25n^2$
C $4m^2 - 20mn + 25n^2$ D $4m^2 - 20mn - 25n^2$

(1 mark)

6 $(4x-3)^2 = ?$ *Hint 1*

A $16x^2 - 12x - 9$ B $16x^2 - 12x + 9$
C $16x^2 - 24x - 9$ D $16x^2 - 24x + 9$

(1 mark)

Part B Short Answer

7 Find $(3u+2)(b-3)$ (1 mark)

8 Find $(4x^2-3)(2x+1)$ (1 mark)

9 Find $(6m-5p)(2m-3p)$ (2 marks)

10 a Find $(x-7)(x+3)$ (1 mark)

b Find $(x+7)(x-3)$ (1 mark)

c Expand and simplify
$(x-7)(x+3) - (x+7)(x-3)$
Hint 2 (1 mark)

11 Expand and simplify:
$(3x+2)(x+5) + (2x+5)(x+3)$
Hint 3 (2 marks)

12 Expand and simplify:
$(4x+5)(2x-3) - (3x-1)(x+7)$
Hint 4 (3 marks)

13 Consider the expression
$(2n-3)(n+5) + (3n-1)(n-1)$

a Substitute $n = 2$ into the expression (without expanding) and evaluate. (1 mark)

b Expand and simplify the expression. (2 marks)

c Substitute $n = 2$ into the simplified expression. (1 mark)

d The answer to part **c** should be the same as the answer to part **a**. Is there a reason for this, or is it a coincidence? Briefly comment. (1 mark)

14 a Write down the value of:

i 7^2 and 6×8
ii 2^2 and 1×3
iii 11^2 and 10×12
iv 43^2 and 42×44 (1 mark)

b Briefly describe any pattern found in part **a**. (1 mark)

c Write down three consecutive numbers beginning with x. *Hint 5* (1 mark)

d Find an expression for the product of the first of these consecutive numbers with the third number. (1 mark)

e Find an expression for the square of the second number. (1 mark)

f Explain any significance of your answer to the pattern described in part **b**. (2 marks)

Hint 1: You need to find $(4x-3)(4x-3)$.
Hint 2: Use the results from parts **a** and **b**.
Hint 3: Find each binomial product separately and then add them together.
Hint 4: Be careful to subtract all of the second binomial product.
Hint 5: Consecutive numbers are following numbers like 6, 7 and 8. If the first number is x, the next number is one more than x.

Your Feedback

$\frac{\square}{30} \times 100\% = \square\%$

PAGES 116–117
PAGE 176

BINOMIAL PRODUCTS

Algebraic Techniques

40 MINUTES

ADVANCED TEST

1 Expand:

- a $(x+7)(x+8)$ (1 mark)
- b $(x-y)(x+11)$ (1 mark)
- c $(x+4)(x-12)$ (1 mark)
- d $(a-12)(b-20)$ (1 mark)
- e $(2e+3f)(5e-f)$ (1 mark)
- f $(9x-5y)(5x-9y)$ (1 mark)
- g $(x^2-1)(x-4)$ (1 mark)
- h $(3pq+5r)(9p-2qr)$ (1 mark)
- i $2(7m-2n)(m+n)$ (1 mark)
- j $(5-a^3)(a+1)$ (1 mark)
- k $3(2m^2-5n)(4n^2-3m)$ (1 mark)
- l $(4a^2b^3c+5abc)(2ab^2c^2+3b^2c)$ (1 mark)

2 Expand:

- a $x(x+4)(x-3)$ (2 marks)
- b $3a^2(2a-5)(a-5)$ (2 marks)

3 Expand and simplify:

- a $(x+6)(x-3)+(x+5)(x-2)$ (2 marks)
- b $(a-3)(a-4)-(a-5)(a-6)$ (2 marks)
- c $(3m+2n)(m-2n)+(4m-n)(5n-m)$ (2 marks)
- d $(5p-2qr)(3p+2qr)-4qr(1-qr)$ (2 marks)

4 Expand and simplify:

- a $(x+3)(x+y+5)$ (2 marks)
- b $(2x-1)(x^2+x-2)$ (2 marks)

5

- a Expand and simplify $2(m+8)(m-5)$ (1 mark)
- b Expand and simplify $(m+8)(m-7)+(m+8)(m-3)$ (2 marks)
- c The answers to parts **a** and **b** should be the same. Briefly explain why that is the case. (1 mark)

6 Expand, giving answers in simplest form:

- a $x^2(x-3)(x-2)-7x(x+5)(x-1)+2(x+3)(x+4)$ (3 marks)
- b $bc^4(4a^2b^2c-3)\times ab^2(5a^2c^2+2b)$ (2 marks)
- c $(x+1)(x+2)(x+3)$ (3 marks)

Your Feedback

$\dfrac{\square}{40}\times 100\% = \square\%$

PAGES 117–118

PAGE 176

FACTORISATION OF QUADRATIC EXPRESSIONS

Algebraic Techniques

STUDY NOTES

1 **Factorising** an algebraic expression means rewriting it as the **product of its factors**.

2 A **common factor** is a number or term that divides evenly into each term of an algebraic expression. When finding a common factor, always look for the largest common factor.
For example, $6(x + x^2)$ is a factorisation of $6x + 6x^2$, but the part in the brackets can still be factorised.
When **factorised fully** $6x + 6x^2 = 6x(1 + x)$.

3 A **quadratic expression** is an algebraic expression where the highest power of x is 2. So, for example, $x^2 + 7x + 3$ is a quadratic expression.

4 The **binomial product** $(x + a)(x + b) = x^2 + ax + bx + ab = x^2 + (a + b)x + ab$ so, as factorising is the opposite of expanding, any expression of the form $x^2 + (a + b)x + ab$ can be factorised to give $(x + a)(x + b)$.

5 To **factorise** a **quadratic expression** of the **type $x^2 + px + q$**, where the coefficient (the number in front) of x^2 is 1, look for two numbers which add to the coefficient of the middle term (p) and which multiply to the end term (q). For example, to factorise $x^2 + 9x + 20$ look for two numbers that add to 9 and multiply to 20. The numbers are 4 and 5 so $x^2 + 9x + 20 = (x + 4)(x + 5)$ [or $(x + 5)(x + 4)$, the factors may be written in either order.] To factorise $x^2 - 5x - 14$ look for two numbers that add to –5 and multiply to –14. The numbers are –7 and 2 so $x^2 - 5x - 14 = (x - 7)(x + 2)$.

6 Some **quadratic expressions cannot be factorised**. For example, no two numbers multiply to 7 and add to 5 so $x^2 + 5x + 7$ cannot be factorised.

7 The two **signs** in the **quadratic expression** can be used when finding the factors. If the expression is of the form:

- **$x^2 + px + q$** then the two numbers multiply to give a positive number so they must both have the same sign. They add to a positive number so must both be positive.
- **$x^2 - px + q$** then again the two numbers multiply to give a positive number and have the same sign. They add to a negative number so both must be negative.
- **$x^2 + px - q$ or $x^2 - px - q$** then the two numbers multiply to a negative number and must have different signs (one will be positive and one will be negative).

8 If an algebraic expression consists of two terms both of which are perfect squares and if one is subtracted from the other then we have the **difference of two squares**. When factorising the difference of two squares the result is the sum by difference:

$$a^2 - b^2 = (a + b)(a - b)$$

For example, $16 - p^2 = 4^2 - p^2$ so, when factorised, $16 - p^2 = (4 + p)(4 - p)$. Note: the difference of two squares is just a special case of a quadratic expression. For example, $x^2 - 9$ is the same thing as $x^2 + 0x - 9$. It can still be factorised by looking for two numbers that multiply to –9 and add to zero.
So $x^2 - 9 = (x + 3)(x - 3)$

9 **Perfect squares** are another special type of quadratic expression.

$$a^2 + 2ab + b^2 = (a + b)^2 \qquad a^2 - 2ab + b^2 = (a - b)^2$$

The factors can still be found in the usual way. For example, $x^2 + 14x + 49 = (x + 7)(x + 7) = (x + 7)^2$, $m^2 - 10m + 25 = (m - 5)(m - 5) = (m - 5)^2$.

Checklist

Can you:

1 *Understand what is meant by factorising?* ☐
2 *Find common factors?* ☐
3 *Find pairs of numbers that add to one number and multiply to another?* ☐
4 *Factorise quadratic expressions?* ☐
5 *Recognise perfect squares and the difference of two squares?* ☐

FACTORISATION OF QUADRATIC EXPRESSIONS

Algebraic Techniques

SKILLS CHECK

1 Fully factorise, by taking out a common factor:

- **a** $9x - 3y$
- **b** $x^2 - 6x$
- **c** $10xy - 5y$
- **d** $12x^2 - 8x$
- **e** $x^2y + xy^2$
- **f** $8ab^2 - 4abc$
- **g** $15x^7 - 3x^2$
- **h** $16p^3q^2 - 2p^4q + 4pq^2$

2 Find two numbers (both positive), that:

- **a** add to 7 and multiply to 10
- **b** add to 5 and multiply to 6
- **c** add to 12 and multiply to 35
- **d** add to 11 and multiply to 24

3 Find two numbers (one negative and one positive), that:

- **a** add to 2 and multiply to –15
- **b** add to 3 and multiply to –28
- **c** add to 4 and multiply to –5
- **d** add to 0 and multiply to –16

4 Find two numbers (one negative and one positive), that:

- **a** add to –6 and multiply to –55
- **b** add to –7 and multiply to –30
- **c** add to –3 and multiply to –18
- **d** add to –7 and multiply to –60

5 Find two numbers (both negative), that:

- **a** add to –9 and multiply to 20
- **b** add to –5 and multiply to 6
- **c** add to –15 and multiply to 36
- **d** add to –12 and multiply to 36

6 Factorise:

- **a** $x^2 + 10x + 21$
- **b** $x^2 + 7x + 12$
- **c** $x^2 + 12x + 32$
- **d** $a^2 + 2a - 24$
- **e** $p^2 + 3p - 10$
- **f** $x^2 + 7x - 18$
- **g** $x^2 - 4x - 12$
- **h** $a^2 - 6a - 27$
- **i** $m^2 - 5m - 14$
- **j** $n^2 - 8n + 7$
- **k** $x^2 - 13x + 36$
- **l** $b^2 - 10b + 25$

7 Factorise these quadratic expressions:

- **a** $x^2 + 8x + 15$
- **b** $a^2 - 14a + 48$
- **c** $p^2 + 2p - 35$
- **d** $m^2 - 3m - 28$
- **e** $n^2 + n - 6$
- **f** $k^2 - 5k - 6$

8 Factorise:

- **a** $x^2 + 18x + 81$
- **b** $y^2 + 12y + 36$
- **c** $b^2 - 14b + 49$

9 Factorise:

- **a** $x^2 - 49$
- **b** $a^2 - 1$
- **c** $36 - t^2$
- **d** $x^2 - y^2$

PAGE 118

Answers **1 a** $3(3x - y)$ **b** $x(x - 6)$ **c** $5y(2x - 1)$ **d** $4x(3x - 2)$ **e** $xy(x + y)$ **f** $4ab(2b - c)$ **g** $3x^2(5x^5 - 1)$ **h** $2pq(8p^2q - p^3 + 2q)$ **2 a** 2 and 5 **b** 2 and 3 **c** 5 and 7 **d** 3 and 8 **3 a** –3 and 5 **b** –4 and 7 **c** –1 and 5 **d** –4 and 4 **4 a** –11 and 5 **b** –10 and 3 **c** –6 and 3 **d** –12 and 5 **5 a** –4 and –5 **b** –2 and –3 **c** –3 and –12 **d** –6 and –6 **6 a** $(x + 3)(x + 7)$ **b** $(x + 3)(x + 4)$ **c** $(x + 4)(x + 8)$ **d** $(a - 4)(a + 6)$ **e** $(p - 2)(p + 5)$ **f** $(x - 2)(x + 9)$ **g** $(x - 6)(x + 2)$ **h** $(a - 9)(a + 3)$ **i** $(m - 7)(m + 2)$ **j** $(n - 7)(n - 1)$ **k** $(x - 4)(x - 9)$ **l** $(b - 5)(b - 5)$ **7 a** $(x + 5)(x + 3)$ **b** $(a - 8)(a - 6)$ **c** $(p + 7)(p - 5)$ **d** $(m - 7)(m + 4)$ **e** $(n + 3)(n - 2)$ **f** $(k - 6)(k + 1)$ **8 a** $(x + 9)^2$ **b** $(y + 6)^2$ **c** $(b - 7)^2$ **9 a** $(x + 7)(x - 7)$ **b** $(a + 1)(a - 1)$ **c** $(6 + t)(6 - t)$ **d** $(x + y)(x - y)$

FACTORISATION OF QUADRATIC EXPRESSIONS

Algebraic Techniques

25 MINUTES

INTERMEDIATE TEST

Part A Multiple Choice

1 When factorised $a^2 - 2a - 63 = ?$

A $(a + 9)(a + 7)$ **B** $(a + 9)(a - 7)$ **C** $(a - 9)(a + 7)$ **D** $(a - 9)(a - 7)$ (1 mark)

2 Which is the correct factorisation of $x^2 - 5x + 6$?

A $(x - 3)(x - 2)$ **B** $(x - 6)(x + 1)$ **C** $(x - 5)(x - 1)$ **D** none of these (1 mark)

3 $x^2 + 17x - 60 = ?$

A $(x - 5)(x - 12)$ **B** $(x - 5)(x + 12)$ **C** $(x - 3)(x - 20)$ **D** $(x - 3)(x + 20)$ (1 mark)

4 $(x + 5)(x - 4)$ is the correct factorisation of which expression?

A $x^2 - x - 20$ **B** $x^2 + x - 20$ **C** $x^2 - 9x - 20$ **D** $x^2 - 9x + 20$ (1 mark)

5 $x^2 - 18x + 81 = ?$

A $(x + 9)^2$ **B** $(x - 9)^2$ **C** $(x + 9)(x - 9)$ **D** none of these (1 mark)

6 Which of the following has a factor of $x + 3$?

A $x^2 - 4x + 3$ **B** $x^2 + 4x - 12$ **C** $x^2 + 9x + 18$ **D** $x^2 - 3x + 2$ (1 mark)

7 $m^2 - 1 = ?$ *Hint 1*

A $(m + 1)^2$ **B** $(m - 1)^2$ **C** $(m + 1)(m - 1)$ **D** $m(m - 1)$ (1 mark)

8 Which of these does not have a factor of $x - 2$?

A $x^2 - 13x + 22$ **B** $x^2 + 2x - 8$ **C** $x^2 + 18x - 40$ **D** $x^2 - 6x - 16$ (1 mark)

9 Which of the following is a factor of $x^2 + 10x - 24$?

A $x - 2$ **B** $x - 4$ **C** $x - 6$ **D** $x - 12$ (1 mark)

10 $\dfrac{g^2 + 7g + 6}{g + 6} = ?$ *Hint 2*

A $7g$ **B** $8g$ **C** $g + 7$ **D** $g + 1$ (1 mark)

Part B Short Answer

11 Factorise $x^2 + x - 90$ *Hint 3* (1 mark)

12 Factorise $21 - 10x + x^2$ *Hint 4* (1 mark)

13 Four friends were asked to factorise $x^2 + x - 6$. Joe wrote $(x + 2)(x - 3)$. Megan put $(x + 3)(x - 2)$. Ali wrote $(x - 2)(x - 3)$ and Xavier's answer was $(x - 2)(x + 3)$.

a Who had the correct answer? (2 marks)

b Briefly explain why the other answers were incorrect. What binomial would factorise to give each of those answers? (2 marks)

14 Factorise each of the following:

a $x^2 + 2x - 48$ **b** $a^2 + 15a + 56$ **c** $h^2 - 5h - 24$

d $k^2 - 6k + 72$ **e** $x^2 - 17x + 72$ **f** $q^2 + q - 42$

g $x^2 + 8x + 16$ **h** $p^2 - 64$ **i** $x^2 - 2x + 1$ (1 mark each)

Hint 1: $m^2 - 1 = m^2 + 0m - 1$
Hint 2: Factorise the numerator.
Hint 3: What two numbers multiply to –90 but add to 1? (So the difference in magnitude is 1.)
Hint 4: If having difficulty, rearrange the equation into the usual form.

Your Feedback

$\dfrac{\square}{25} \times 100\% = \square\,\%$

PAGES 118–119

PAGE 176

FACTORISATION OF QUADRATIC EXPRESSIONS

Algebraic Techniques

40 MINUTES

ADVANCED TEST

1 Factorise:

a $x^2 + 9x + 20$

b $a^2 - 12a + 32$

c $x^2 - 2x - 24$

d $m^2 + 5m - 6$

e $t^2 + 17t - 60$

f $h^2 - 32h - 144$ (1 mark each)

2 Factorise fully:

a $3n^2 - 9n - 120$ (1 mark)

b $5p^2 - 55p + 140$ (1 mark)

3 Simplify:

a $\dfrac{x^2 + 8x - 9}{x - 1}$ b $\dfrac{a^2 + 10a + 24}{a^2 + 4a - 12}$ c $\dfrac{n - 2}{n^2 - 11n + 18}$ (2 marks each)

4 Factorise by grouping in pairs:

a $x^2 + xy + 5x + 5y$ (2 marks)

b $a^2 - 7a + ab - 7b$ (2 marks)

c $pq + 3p - 8q - 24$ (2 marks)

d $5x^3 - 5xy - x^2y + y^2$ (2 marks)

5 Factorise:

a $2m^2 + 13m + 15$ (2 marks)

b $3a^2 - ab - 2b^2$ (2 marks)

c $21p^2 + 29p - 10$ (2 marks)

6 Factorise and simplify (leaving answers in factorised form where necessary):

a $\dfrac{m^2 + 8m + 15}{m + 5} \times \dfrac{m - 3}{m^2 - 5m + 6}$ (2 marks)

b $\dfrac{x^2 + 17x + 60}{x^2 - 16x + 60} \times \dfrac{x^2 + 2x - 48}{x^2 + 20x + 96}$ (2 marks)

c $\dfrac{n^2 + 3n - 4}{n^2 - 10n + 21} \div \dfrac{n^2 - 5n - 36}{n^2 - 6n - 7}$ (2 marks)

7 Simplify, leaving answers in factorised form:

a $\dfrac{1}{x^2 + 6x + 8} + \dfrac{1}{x^2 + 7x + 12}$ (2 marks)

b $\dfrac{3}{x^2 + 5x + 4} - \dfrac{2}{x^2 + 4x + 3}$ (2 marks)

c $\dfrac{x^2 + 7x + 10}{x^2 + 4x - 5} + \dfrac{x^2 - 9x + 18}{x^2 - 4x + 3}$ (2 marks)

Your Feedback

$\dfrac{\square}{40} \times 100\% = \square\,\%$

PAGES 119–120

PAGE 176

SPECIAL BINOMIAL PRODUCTS AND FACTORISATIONS

Algebraic Techniques

STUDY NOTES

1 In a **binomial product**, each term in the second binomial is multiplied by each term in the first binomial.

2 Finding the square of a binomial is just a **special binomial product**.

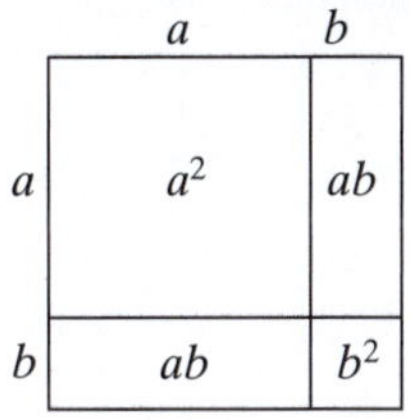

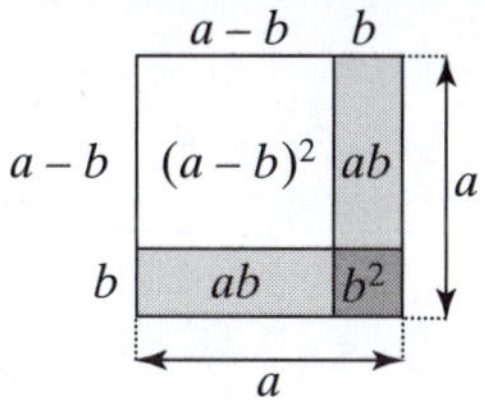

$(a + b)^2 = a^2 + 2ab + b^2$ and $(a - b)^2 = a^2 - 2ab + b^2$
(The two ab rectangles must be subtracted from a^2 but this subtracts the square in the bottom right corner twice so b^2 must be added back on.)

3 To **square a binomial** 'square the first (term), square the last (term) and twice the product of the two'.
For example, $(3x + 4)^2 = (3x)^2 + 2 \times 3x \times 4 + 4^2$
$= 9x^2 + 24x + 16$
$(5k - 1)^2 = (5k)^2 - 2 \times 5k \times 1 + 1^2$
$= 25k^2 - 10k + 1$

4 Another special binomial product is the **sum by difference**. The product of a sum by difference will always be the difference of two squares.

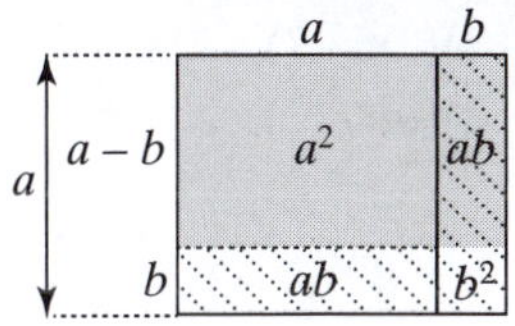

[From a^2 we must subtract the bottom rectangle (ab) and add the side rectangle (ab) but without the bottom square (b^2) so $(a + b)(a - b) = a^2 - ab + ab - b^2$
$= a^2 - b^2$]

$$\mathbf{(a + b)(a - b) = a^2 - b^2}$$

For example, $(x - 4)(x + 4) = x^2 - 4^2$
$= x^2 - 16$
$(2x + 7y)(2x - 7y) = (2x)^2 - (7y)^2$
$= 4x^2 - 49y^2$.

5 Because **factorising is the opposite operation of expanding**, every special binomial product has an equivalent special factorisation.

6 The squares of binomials are called **perfect squares**.

$$\mathbf{a^2 + 2ab + b^2 = (a + b)^2 \qquad a^2 - 2ab + b^2 = (a - b)^2}$$

For example, $x^2 + 14x + 49 = (x + 7)^2$, $x^2 - 30x + 225 = (x - 15)^2$

7 If an algebraic expression consists of two terms both of which are squares and if one is subtracted from the other then we have the **difference of two squares**. When factorising the difference of two squares the result is the sum by difference:

$$\mathbf{a^2 - b^2 = (a + b)(a - b)}$$

For example, $36 - x^2 = 6^2 - x^2$ so when factorised $36 - x^2 = (6 + x)(6 - x)$.

8 Some expressions involving four terms can be factorised by **grouping in pairs**. Find a common factor for the first 2 terms, and a different common factor for the last 2 terms. Then you may find that you have a new common factor for these two factorised expressions.
For example, $x^2 + ax + bx + ab = x(x + a) + b(x + a)$
$= (x + a)(x + b)$

Checklist

Can you:

1 *Find special products such as the square of a binomial and the sum by difference?* ☐
2 *Recognise perfect squares?* ☐
3 *Factorise quadratic expressions involving perfect squares and differences between two squares?* ☐
4 *Expand and simplify more complicated algebraic expressions?* ☐

SPECIAL BINOMIAL PRODUCTS AND FACTORISATIONS

Algebraic Techniques

SKILLS CHECK

1 Expand and simplify:

a $(x+3)(x+8)$ b $(a-5)(a+9)$ c $(2a+3)(5b+4)$
d $(6x-1)(x-8)$ e $(5n+3)(2n-3)$ f $(x+y)(a+b)$

2 Fill in the blanks to complete these squares:

a $(x+3)^2 = x^2 + 6x + \_\_\_\_$ b $(x-6)^2 = x^2 - \_\_\_\_ + 36$
c $(3x+2)^2 = 9x^2 + \_\_\_\_ + 4$ d $(2x-5)^2 = \_\_\_\_ - 20x + 25$

3 Find these squares of binomials:

a $(x+4)^2$ b $(x-9)^2$ c $(3a-10)^2$
d $(2p+3q)^2$ e $(7-x)^2$ f $(a+b)^2$

4 Fill in the blanks:

a $(p+3)(p-3) = p^2 - \_\_\_\_$ b $(2x-1)(2x+1) = \_\_\_\_ - 1$
c $(3x+2y)(3x-2y) = 9x^2 - \_\_\_\_$ d $(x^2+2)(x^2-2) = \_\_\_\_ - 4$

5 Find these products of a sum by difference:

a $(x-9)(x+9)$ b $(3x-1)(3x+1)$ c $(2n+7)(2n-7)$
d $(8-x)(8+x)$ e $(7x+2y)(7x-2y)$ f $(ab+2)(ab-2)$

6 Factorise:

a $x^2+13x+42$ b $x^2+3x-40$ c $x^2-10x+16$ d $x^2-3x-54$

7 Complete:

a $x^2+8x+16 = (x + \_\_\_\_)^2$ b $a^2-10a+25 = (a - \_\_\_\_)^2$
c $n^2+2n+1 = (n + \_\_\_\_)^2$ d $x^2-16x+64 = (x - \_\_\_\_)^2$

8 Express as a perfect square:

a $x^2+20x+100$ b y^2+4y+4 c $49+14n+n^2$
d a^2-2a+1 e $9-6m+m^2$ f $x^2-2xy+y^2$

9 Complete:

a $x^2-49 = (x+7)(x - \_\_\_\_)$ b $a^2-9 = (a+3)(\_\_\_\_\_\_)$
c $25-m^2 = (\_\_\_\_\_\_)(5-m)$ d $x^2-y^2 = (x + \_\_\_\_)(x - \_\_\_\_)$

10 Factorise these differences of two squares:

a x^2-81 b $121-y^2$ c m^2-900 d p^2-q^2

11 Find a common factor for the first pair of terms and a common factor for the second pair of terms and hence factorise the whole expression:

a $x^2+xy+2x+2y$ b $ab+3a-5b-15$ c $a^2-ab-ac+bc$
d $2mn-2m+3n-3$ e $xz-yz-x+y$ f $6kn-3k-2n+1$

PAGES 120–121

Answers **1 a** $x^2+11x+24$ **b** $a^2+4a-45$ **c** $10ab+8a+15b+12$ **d** $6x^2-49x+8$ **e** $10n^2-9n-9$ **f** $ax+bx+ay+by$ **2 a** 9 **b** $12x$ **c** $12x$ **d** $4x^2$ **3 a** $x^2+8x+16$ **b** $x^2-18x+81$ **c** $9a^2-60a+100$ **d** $4p^2+12pq+9q^2$ **e** $49-14x+x^2$ **f** $a^2+2ab+b^2$ **4 a** 9 **b** $4x^2$ **c** $4y^2$ **d** x^4 **5 a** x^2-81 **b** $9x^2-1$ **c** $4n^2-49$ **d** $64-x^2$ **e** $49x^2-4y^2$ **f** a^2b^2-4 **6 a** $(x+6)(x+7)$ **b** $(x+8)(x-5)$ **c** $(x-2)(x-8)$ **d** $(x+6)(x-9)$ **7 a** 4 **b** 5 **c** 1 **d** 8 **8 a** $(x+10)^2$ **b** $(y+2)^2$ **c** $(7+n)^2$ **d** $(a-1)^2$ **e** $(3-m)^2$ **f** $(x-y)^2$ **9 a** 7 **b** $a-3$ **c** $5+m$ **d** y, y **10 a** $(x+9)(x-9)$ **b** $(11+y)(11-y)$ **c** $(m+30)(m-30)$ **d** $(p+q)(p-q)$ **11 a** $(x+y)(x+2)$ **b** $(b+3)(a-5)$ **c** $(a-b)(a-c)$ **d** $(n-1)(2m+3)$ **e** $(x-y)(z-1)$ **f** $(2n-1)(3k-1)$

SPECIAL BINOMIAL PRODUCTS AND FACTORISATIONS

Algebraic Techniques

25 MINUTES

INTERMEDIATE TEST

Part A Multiple Choice

1 $(2x-5)^2 = ?$

A $4x^2-10x+25$ B $4x^2+10x-25$ C $4x^2-20x+25$ D $4x^2+20x-25$ (1 mark)

2 $(3x-y)(3x+y) = ?$

A $3x^2-y^2$ B $3x^2+y^2$ C $9x^2-y^2$ D $9x^2+y^2$ (1 mark)

3 $xy-4x-6y+24 = ?$

A $(x-6)(y-4)$ B $(x-6)(y+4)$ C $(x+6)(y-4)$ D $(x+6)(y+4)$ (1 mark)

4 Which is a perfect square?

A $x^2-12x-36$ B $x^2+34x+289$ C $x^2-26x+159$ D $x^2+28x-196$ (1 mark)

5 $6k(k+2)-(k+2) = ?$

A $5k(k+2)$ B $(k+2)(5k-1)$ C $6k(k+2)$ D $(6k-1)(k+2)$ (1 mark)

6 When simplified $\dfrac{x^2-64}{8x-64} = ?$ *Hint 1*

A $\dfrac{x}{8}$ B $\dfrac{(x+1)(x-1)}{8x-1}$ C $\dfrac{x+8}{8}$ D x (1 mark)

7 Which factorisation is correct?

A $x^2-49=(x-7)^2$ B $x^2-10x-24=(x-4)(x-6)$

C $x^2-24x+144=(x+12)^2$ D $x^2-625=(x+25)(x-25)$ (1 mark)

8 Which is <u>not</u> a perfect square?

A $x^2+40x+400$ B $x^2-38x+361$ C $x^2+42x+441$ D $x^2-44x+464$ (1 mark)

9 $(3x+5)^2-(3x-5)^2 = ?$ *Hint 2*

A $18x^2+60x+50$ B $18x^2-60x+50$ C $18x^2-50$ D $60x$ (1 mark)

Part B Short Answer

10 Expand and simplify $(3x+8)^2+(x-8)(x+8)$ *Hint 3* (2 marks)

11 Factorise $a^2+ab-a-b$ *Hint 4* (2 marks)

12 Find $a^2b[(2a+b)^2-(2a-b)^2]$ *Hint 5* (2 marks)

13 Expand $(2a-1)^2$ (1 mark)

14 a Factorise fully: $7p^2-63$ (2 marks)

b Factorise fully: $6x^2-48x+6$ (2 marks)

15 a Find the value of $2x(x+3)-(x+2)(x-2)-(x+3)^2$ when:

i $x=2$ ii $x=4$ iii $x=-1$ (3 marks)

b By expanding and simplifying $2x(x+3)-(x+2)(x-2)-(x+3)^2$ show that it has the same value for every value of x. (2 marks)

Hint 1: Factorise both the numerator and denominator first.
Hint 2: Use the difference of two squares.
Hint 3: Find each binomial product separately and then add them together.
Hint 4: Be careful with the negative values. Remember that *a* divides into *a* once.
Hint 5: Be careful to subtract all of the second binomial product.

Your Feedback

$\dfrac{\square}{25} \times 100\% = \square\,\%$

PAGE 121
PAGE 176

SPECIAL BINOMIAL PRODUCTS AND FACTORISATIONS

Algebraic Techniques

40 MINUTES

ADVANCED TEST

1 Expand:

a $(4p + 5q)^2$

b $(12 - p)(12 + p)$

c $(5x - 3y)(5x + 3y)$

d $2(x - 4)^2$

e $3x(x + 3)(x - 3)$

f $(4x^2y^3 - 7z)(4x^2y^3 + 7z)$

(1 mark each)

2 Factorise:

a $81 - 4d^2$

b $36 - 12c + c^2$

c $4n^2 - 12n + 9$

d $x^4 - 25y^2$

e $x^4 + 4x^2 + 4$

f $p^2 - p + \frac{1}{4}$

(1 mark each)

3 Factorise fully:

a $2x^2 - 18$

b $3x^2 + 30x + 75$

c $x^4 - 1$

d $p - p^3$

e $2x^3 - 12x^2 + 18x$

f $16a^4 - 72a^2 + 81$

(1 mark each)

4 Expand and simplify:

a $(5x - 2)(7 - x) + (3x - 2)^2$

b $(x + y)(x - y) - (x - y)^2$

c $9(m + 5)(m - 2) - (3m - 10)(3m + 10)$

d $(5a - 2b)^2 - (5a + 2b)^2$

(2 marks each)

5 Factorise and simplify (leaving answers in factorised form if necessary):

a $\dfrac{9p^2 + 9}{9p^2 + 18p + 9}$

b $\dfrac{n^2 - 100}{n^2 - 12n + 20}$

c $\dfrac{x^3 - 4x}{x^3 - 4x^2 + 4x}$

d $\dfrac{x^2 - 2x + 1}{1 - x^2}$

(2 marks each)

6 Simplify:

a $\dfrac{1}{x^2 + 20x + 100} - \dfrac{1}{x^2 - 100}$

b $\dfrac{1}{a^3 - 1} \div \dfrac{1}{a^6 - 1}$

(2 marks each)

7 a Given that $1637^2 = 2\,679\,769$, explain why it is not difficult to find the answer to the product 1636×1638 without a calculator. (1 mark)

b Find 897×903 without a calculator. (1 mark)

Your Feedback

$\dfrac{\square}{40} \times 100\% = \square\,\%$

PAGES 121–122

PAGE 176

SUBSTITUTION, EQUATIONS AND FORMULAE

Algebraic Techniques

STUDY NOTES

1 In an algebraic expression **pronumerals** are used as variables. A pronumeral is a symbol, often a letter, used in place of a number.

2 When **substituting** into an algebraic expression we replace the pronumerals with the numbers they represent. For example, if $m = 3$, $m + 6 = 3 + 6 = 9$, $2x$ means $2 \times x$, so when $x = 4$, $2x = 2 \times 4 = 8$

3 A **linear equation** is an equation involving a single power of an unknown (x, say). $3x + 7 = 19$, $8(2x - 5) - 4(x + 8) = 5(2x - 1)$ and $\frac{x+3}{4} = \frac{x-1}{5}$ are all examples of linear equations. Equations involving higher powers of an unknown (e.g. x^2, x^7) are not linear.

4 Linear equations are **solved** by performing the same operation on **both sides** of the equation. An equation may be solved by:
- **adding** the same number **to both sides** of the equation,
- **subtracting** the same number **from both sides** of the equation,
- **multiplying both sides** of the equation by the same amount, or
- **dividing each side** of the equation by the same amount.

5 The **solution to an equation** is an algebraic expression, not a number. For example, the solution to the equation $x + 2 = 5$ is not 3, it is $x = 3$.

6 Always **simplify the expressions** on either side of the equals sign before attempting to solve.

7 Any equation involving fractions can be simplified by multiplying both sides of the equation by the **lowest common denominator.** Care must be taken to multiply every part of the equation by this common denominator. For example, if asked to solve $\frac{x}{4} + 5 = 2x$, every part must be multiplied by 4 (giving $x + 20 = 8x$).

8 The solution to an equation can always be **checked by substitution** into the original equation. If a value substituted into an equation makes the equation correct, we say the value **satisfies** the equation.

9 A **formula** is simply an algebraic expression connecting different values.

10 The **subject of a formula** is the value on the left hand side, before the equals sign. For example, the subject of the formula $A = 2\pi r^2 + 2\pi rh$ is A.

11 When substituting into a formula always **write the formula down first**, even if it is given in the question. Then **show the substitution** and finally **work out the answer**.

12 **Equations arising from substitution** into a formula are solved like any other equation after substituting the given values.

13 Many **problems** involve finding the answer to an equation. Think carefully about what is required and form an equation to solve.

Checklist

Can you:

1 *Substitute values in algebraic expressions?* ☐
2 *Solve simple linear equations?* ☐
3 *Know how to check the solution to an equation?* ☐
4 *Understand what is meant by a formula?* ☐
5 *Substitute into a formula and solve equations arising from substitution into a formula?* ☐

SUBSTITUTION, EQUATIONS AND FORMULAE

Algebraic Techniques

SKILLS CHECK

1 If $x = 4$, find the value of:

a $x + 3$ b $x - 6$ c $9 - x$ d $5x$

e $2x + 1$ f $3x - 2$ g x^2 h $2x^2$

2 If $a = 2$, $b = 5$ and $c = 6$, find the value of:

a $a + b$ b $c - b$ c $b + c - a$ d $3a$

e bc f $ab + c$ g $3ac$ h $a^2 + b^2$

3 Solve:

a $x + 13 = 25$ b $y - 9 = 37$ c $5a = 325$ d $\frac{h}{6} = 12$

e $2x + 1 = 17$ f $3x - 4 = 5$ g $\frac{k}{2} + 3 = 7$ h $\frac{d-4}{3} = -9$

4 The given value of x satisfies the equation. True or false?

a $5x + 8 = 29 - 2x$ $[x = 3]$ b $\frac{x}{3} + \frac{x}{4} = 7$ $[x = 12]$

c $4(2x + 1) - 3(x + 5) = 28$ $[x = 8]$ d $\frac{7x+1}{5} = \frac{2x-9}{3}$ $[x = -4]$

5 Solve:

a $9p = 15 - p$

b $7k + 5 = 2k + 25$

6 If $V = Ah$, find V when $A = 120$ and $h = 12.5$

7 If $P = 2l + 2b$, find P when $l = 8$ and $b = 5$.

8 If $v = u + at$, find v when $u = 28$, $a = -4$ and $t = 5$.

9 If $c^2 = a^2 + b^2$ find c $(c > 0)$, when $a = 24$ and $b = 45$.

10 If $A = \pi r^2$ find A, to one decimal place, when $r = 12$.

11 Given $m = \frac{y_2 - y_1}{x_2 - x_1}$, find m when $x_1 = -2$, $x_2 = 1$, $y_1 = 8$ and $y_2 = -4$.

PAGES 122–123

Answers **1** a 7 b –2 c 5 d 20 e 9 f 10 g 16 h 32 **2** a 7 b 1 c 9 d 6 e 30 f 16 g 36 h 29 **3** a $x = 12$ b $y = 46$ c $a = 65$ d $h = 72$ e $x = 8$ f $x = 3$ g $k = 8$ h $d = -23$ **4** a true b true c false d false **5** a $p = 1.5$ b $k = 4$ **6** 1500 **7** 26 **8** 8 **9** 51 **10** 452.4 **11** –4

SUBSTITUTION, EQUATIONS AND FORMULAE

Algebraic Techniques

25 MINUTES

INTERMEDIATE TEST

Part A Multiple Choice

1 If $a = 5$ and $b = -3$ then $3a - b = ?$

A 12 **B** 18 **C** 32 **D** 38 (1 mark)

2 The solution to the equation $2(3x + 4) = 2x - 4$ is:

A $x = -2$ **B** $x = -3$ **C** $x = 2$ **D** $x = 3$ (1 mark)

3 If $x = -2$, then $3x^2 = ?$

A -36 **B** -12 **C** 12 **D** 36 (1 mark)

4 The formula $A = (n - 2) \times 180°$ gives the sum of the angles of a polygon when n is the number of sides. What will be the sum of the angles of a 15-sided polygon?

A 345° **B** 1140° **C** 1980° **D** 2340° (1 mark)

5 Given that $y = mx + b$, and that $m = 2$ and $b = -6$, for what value of x will $y = 8$?

A $x = 0$ **B** $x = 1$ **C** $x = 4$ **D** $x = 7$ (1 mark)

6 Which is a correct rearrangement of the formula $v = u + at$?

A $u = at - v$ **B** $a = \dfrac{v - u}{t}$ **C** $t = \dfrac{u - v}{a}$ **D** $t = \dfrac{a}{v - u}$ (1 mark)

Part B Short Answer

7 Solve $6a + 9 - a = 8 - 3a + 1$ *Hint 1* (3 marks)

8 Solve $4(x - 5) + 3(x - 2) = 16$ *Hint 2* (3 marks)

9 Solve $\dfrac{2t + 4}{7} - 3 = 5$ (3 marks)

10 Solve $\dfrac{2m - 5}{3} = \dfrac{3m + 2}{5}$ *Hint 3* (3 marks)

11 **a** The given solution to the equation has at least one mistake. Between which lines do mistakes occur? Briefly justify your answer. *Hint 4* (4 marks)

$3x + 5 - 5x = 9$ *Line 1*

$8x + 5 = 9$ *Line 2*

$8x = 4$ *Line 3*

$x = 2$ *Line 4*

b What is the correct solution to the equation $3x + 5 - 5x = 9$? (1 mark)

12 Young's Rule is a formula for changing an adult dose of medicine to a child's dose. Young's Rule is $D = \dfrac{yA}{y + 12}$ where D is the child's dose, y is the age of the child in years and A is the adult dose. Find the dose of medicine required for an 8-year-old child when the adult dose is 20 mL. (2 marks)

Hint 1: Collect like terms before attempting to solve.
Hint 2: Be careful to get the operation signs correct when expanding.
Hint 3: Multiply by the common denominator to get rid of the fractions.
Hint 4: There may be more than one mistake. Work carefully through each line of the solution.

Your Feedback

$\dfrac{\square}{25} \times 100\% = \square\%$

PAGES 123–124

PAGE 176

SUBSTITUTION, EQUATIONS AND FORMULAE

Algebraic Techniques

40 MINUTES

ADVANCED TEST

1 Solve:

a $5(x-3)+2(x-7)=6$

b $4(3k-2)=3(1-2k)$

c $3(4p-5)=2p-(7-p)$

d $11(7-2g)-5(3g-8)=8g$

e $\frac{3a}{4}+7=2a+1$

f $\frac{9n-3}{7}=4n+5$

g $\frac{5x-3}{4}=\frac{4x-2}{3}$

h $\frac{m}{5}+\frac{m}{3}=16$

i $\frac{3u+2}{5}-\frac{4u-1}{2}=\frac{2}{3}$

(2 marks each)

2 Solve:

a $x^2=0.25$

b $9a^2=1$

c $t^2-4=0$

d $\frac{y^2}{2}-5=3$

(1 mark each)

3 a Solve $5x^2-2=7$ $(x>0)$, giving the answer correct to three decimal places. (2 marks)

b What difference does it make having the restriction that x is greater than zero? (1 mark)

4 If $p=-2$ and $q=-3$, evaluate:

a $5p-3pq$

b $4p^2$

c $\frac{p+q}{p-q}$

d pq^2-p+q

(1 mark each)

5 Given that $s=ut+\frac{1}{2}at^2$ find:

a s when $u=32$, $a=9.8$ and $t=7$ (1 mark)

b a when $s=1599$, $u=58$ and $t=13$ (2 marks)

6 Given that $V=\frac{4}{3}\pi r^3$ find, to two decimal places:

a V when $r=9$ (1 mark)

b r when $V=222\,665$ (2 marks)

7 The formula $F=\frac{9C}{5}+32$ can be used to change temperatures in degrees Celsius to Fahrenheit.

a At what temperature should an oven be set in °F if a recipe states 180 °C? (1 mark)

b Tina read about a place where the temperature was 68 °F. She thought that was very hot. Was she correct? Justify your answer. (1 mark)

8 a Given that $d=\sqrt{(x_2-x_1)^2+(y_2-y_1)^2}$ and that $d=10$ when $x_1=-3$, $y_1=2$ and $x_2=5$, find two possible values for y_2. (2 marks)

b Given that the above formula finds the distance between the points (x_1, y_1) and (x_2, y_2) explain the significance of the above result. (1 mark)

Your Feedback

$\frac{\square}{40}\times 100\% = \square\%$

PAGES 124–125

PAGE 176

FURTHER EQUATIONS AND INEQUALITIES

Algebraic Techniques

STUDY NOTES

1 Many **problems** involve **finding the answer to an equation**. Think carefully about what is required and form an equation to solve. Let x (or some other variable) represent the unknown quantity and express the problem algebraically. Make sure that you answer the question that has been asked.

2 When **solving equations to answer problems** we should always consider whether any answers are **physically possible**. For example, if finding the hypotenuse of a right-angled triangle, the length of a side of a triangle cannot be negative so we only choose the positive answer. Writing 'but $a > 0$' illustrates that you understand and have considered the negative value.

3 An **inequality** is an expression that involves an inequality sign (such as >, ≤, ≠ for example) instead of an equals sign.

4 There are different **inequality signs**, the meaning of which should be understood. A line (/) through the symbol means 'not', e.g. ≠ means 'is not equal to' and ≯ means 'is not greater than'.

Sign	Meaning	Sign	Meaning
>	is greater than	≥	is greater than or equal to
<	is less than	≤	is less than or equal to

5 **Inequalities** are solved in a similar way to equations. Perform the same operation on both sides of the inequality.

6 An **inequality remains unchanged** if:
- the same number is added to both sides of the inequality
- the same number is subtracted from both sides of the inequality
- both sides of the inequality are multiplied by a <u>positive</u> number
- both sides of the inequality are divided by a <u>positive</u> number.

7 If both sides of an inequality are multiplied (or divided) by a negative number the inequality does not remain unchanged. (For example, $3 > 2$ but if both numbers are multiplied by –1, we get –3 and –2 and $-3 < -2$). **The inequality sign must be reversed when multiplying or dividing by a negative number.**

8 The **solution to an inequality** is an algebraic expression. For example, $x > 3$ or $y \leq 9$. The solution is usually a range of values all of which will satisfy the inequality.

9 **Check by substituting** a value from the solution range into the inequality to see if it works. For example, if the solution to an inequality is $x > 3$, substitute $x = 4$ (or $x = 5$ or $x = 10$ or any value greater than three), to see whether it satisfies the inequality.

10 The **solution to an inequality** can be graphed on a **number line**. The boundary point is shown with a circle. The circle is empty if the boundary point is not actually part of the solution (if the inequality sign is > or <). The circle is solid if the boundary point is part of the solution (if the inequality sign is ≥ or ≤). A line is then drawn to show the direction of the set of values that satisfy the inequality. For example, to graph $x > 1$, place an empty circle at 1, and draw a line showing those values larger than one.

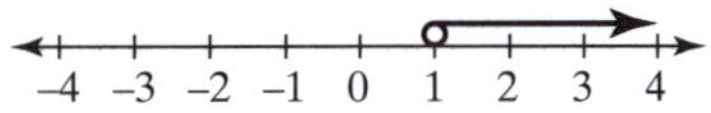

Checklist

Can you:

1 *Solve problems by forming equations and solving?* ☐
2 *Understand the need to consider whether answers are physically possible?* ☐
3 *Understand the different inequality signs?* ☐
4 *Solve linear inequalities?* ☐
5 *Graph inequalities on a number line?* ☐

FURTHER EQUATIONS AND INEQUALITIES

Algebraic Techniques

SKILLS CHECK

1 Determine whether each statement is true or false:

a $3 > 2$ b $-3 > -2$ c $6 < 15$ d $-6 < -15$

e $4 \geq -1$ f $-4 \geq 1$ g $-2 \leq 2$ h $2 \leq -2$

2 Determine whether the following statements are true or false:

a $-3 \geq 2$ b $5 < 2 \times 3$ c $12 + 8 \neq 35 - 5 \times 3$

d $19 > 17 - 3 \times -2$ e $6^2 \leq 2^6$ f $8 + 5 \not> 5 + 8$

3 The given value of the pronumeral satisfies the inequality. True or false?

a $5a - 4 < 2a \quad [a = -2]$ b $\frac{9e - 1}{11} \geq 7 \quad [e = 5]$

c $3p + 1 \leq 3 - 3p \quad [p = -1]$ d $\frac{x}{3} + 8 \geq -2 \quad [x = -30]$

4 Solve:

a $-4p = 8$ b $4p < -8$ c $-4p > 8$ d $3x + 17 = 74$

e $2n - 1 > 15$ f $5x \leq x + 8$ g $\frac{a}{4} + 7 \geq -2$ h $\frac{t - 3}{2} > -1$

5 Solve:

a $x + 4 \leq -2$ b $a - 5 > 5$ c $4e < -2$ d $\frac{n}{3} \geq 6$

6 Solve:

a $2x + 5 \geq 11$ b $3a - 1 < -2$ c $\frac{m + 2}{3} \leq 4$

7 On separate number lines, graph:

a $x \geq 1$ b $x < 2$ c $x \leq -2$

8 $x = -1$ satisfies the equation $5(x + 1) - 2(3 - 2x) = 10$. True or false?

9 $x = -1$ satisfies the inequality $7x - 3 > x - 8$. True or false?

10 a If $A = \frac{1}{2}bh$, find h when $A = 36$ and $b = 8$.

b If the area of a triangle is 84 cm² and the perpendicular height is 7 cm, find the length of the base of the triangle.

PAGES 125–126

Answers **1 a** true **b** false **c** true **d** false **e** true **f** false **g** true **h** false **2 a** false **b** true **c** false **d** false **e** true **f** true **3 a** true **b** false **c** true **d** true **4 a** $p = -2$ **b** $p < -2$ **c** $p < -2$ **d** $x = 19$ **e** $n > 8$ **f** $x \leq 2$ **g** $a \geq -36$ **h** $t > 1$ **5 a** $x \leq -6$ **b** $a > 10$ **c** $e < -\frac{1}{2}$ **d** $n \geq 18$ **6 a** $x \geq 3$ **b** $a < -\frac{1}{3}$ **c** $m \leq 10$ **7** see worked solutions **8** false **9** false **10 a** 9 **b** 24 cm

FURTHER EQUATIONS AND INEQUALITIES

Algebraic Techniques

25 MINUTES

INTERMEDIATE TEST

Part A Multiple Choice

1 Which value of x satisfies the inequality $11 < 2x - 15$?

A $x = -10$ **B** $x = 0$ **C** $x = 9$ **D** none of these (1 mark)

2 Which value of m does not satisfy the inequality $\frac{7m}{3} - 8 \leq 9 - \frac{m}{2}$? *Hint 1*

A $m = -2$ **B** $m = 0$ **C** $m = 6$ **D** $m = 9$ (1 mark)

3 If $-2x > -6$, which is correct?

A $x > 3$ **B** $x < 3$ **C** $x > -3$ **D** $x < -3$ (1 mark)

4 The correct solution to the equation $3p - 17 = 5p + 19$ is: *Hint 2*

A $p = -18$ **B** $p = 18$ **C** $p = -1$ **D** $p = 1$ (1 mark)

5 'When 15 is subtracted from three times a certain number the result is 8 more than twice the number.' Let the number be x. Which statement is correct?

A $3x - 15 = 2x + 8$ **B** $15 - 3x = 2x + 8$ **C** $3x - 15 = 2(x + 8)$ **D** $15 - 3x = 2(x + 8)$ (1 mark)

6 The number, 8 more than $2x - 5$ is: *Hint 3*

A $2x + 3$ **B** $2x + 13$ **C** $10x - 5$ **D** $10x + 13$ (1 mark)

Part B Short Answer

7 Rikki bought seven bottles of soft drink and a packet of biscuits from the local store. The total of her purchases was \$15.98. If the packet of biscuits cost \$2.75:

a form an equation which could be solved to find the cost of each bottle of soft drink *Hint 4* (1 mark)

b how much would five bottles of the same soft drink cost? *Hint 5* (2 marks)

8 Solve $5(2x + 3) \leq 20$ *Hint 6* (1 mark)

9 Solve $9n + 4 > 2n - 17$ (2 marks)

10 Solve $3 - 2a > 7$, graphing the solution on a number line. *Hint 7* (3 marks)

11 Solve $3m + 17 \leq 8m$ (2 marks)

12 Solve $2 - \frac{u}{7} \geq 3$ (2 marks)

13 'When 15 is subtracted from a certain number and the result multiplied by 5, the answer is seven more than three times the original number.' Form an equation and solve to find the original number. *Hint 8* (3 marks)

14 A particular art competition has special requirements about the size of the entries. Each entry must be presented on a rectangular background, the length of which is to be 20 cm longer than its breadth. The maximum total perimeter allowed is 150 cm. Find the dimensions of the largest possible entry. *Hint 9*

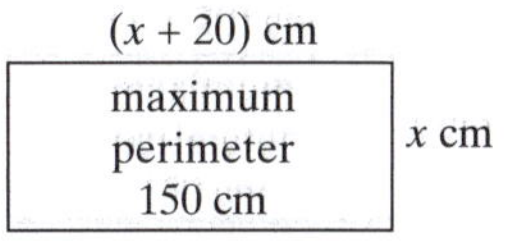

(3 marks)

Hint 1: Substitute each answer to see whether it satisfies the inequality.
Hint 2: Either solve the equation or substitute each value to see which one satisfies the equation.
Hint 3: Add 8 to $2x - 5$.
Hint 4: Let x be the price of a bottle of soft drink.
Hint 5: First find the cost of one bottle.
Hint 6: Solve an inequality in the same way as an equation.
Hint 7: If you divide by a negative number, the inequality sign will be reversed.
Hint 8: Let x be the number.
Hint 9: Form an inequality and solve.

Your Feedback

$\frac{\square}{25} \times 100\% = \square\%$

PAGES 126–127

PAGE 176

FURTHER EQUATIONS AND INEQUALITIES

Algebraic Techniques

40 MINUTES

ADVANCED TEST

1 Graph on a number line:

a $x \geq 3$ b $x < -1$ or $x > 4$ c $-2 \leq x < 5$ (1 mark each)

2 Solve, and graph the solution on a number line:

a $5x + 1 \geq 9 - 3x$ b $3(p + 5) - 1 < 11$ c $9 - 2n > 5$ (2 marks each)

3 Solve:

a $3(2m - 5) \leq 2(m + 7)$ b $\frac{5k}{3} - 2 < 4 - k$

c $\frac{6 - a}{2} > 4$ d $7(4q - 1) - 3(2q + 5) \geq 0$

e $12 - \frac{3d}{4} \leq 18$ f $5(2y - 3) - 4(3y - 7) > 2(y + 5)$ (2 marks each)

4 Jo produced the following when trying to solve the equation $x(x + 5) = x(x + 3)$.

$$\frac{x(x+5)}{x} = \frac{x(x+3)}{x}$$
$$x + 5 = x + 3$$
$$5 = 3$$

Briefly explain what Jo did wrong and give the correct solution to the equation. (2 marks)

5 Change the subject of each formula to b:

a $y = mx + b$ (1 mark)

b $A = \frac{1}{2}h(a + b)$ (2 marks)

c $c = \sqrt{a^2 + b^2}$ (2 marks)

6 a Make r the subject of the formula $V = \pi r^2 h$. (1 mark)

b What restrictions apply to the variables in the formula before and after it is rearranged? (2 marks)

c Find the value of r to one decimal place ($r > 0$), when $V = 179$ and $h = 8.7$ (1 mark)

7 A right-angled triangle has area 27.3 m^2 and base length 19.5 m.

a What is the perpendicular height of the triangle? (1 mark)

b Use Pythagoras' theorem to find the length of the hypotenuse. (1 mark)

c Why do we not consider the negative value when finding the hypotenuse? (1 mark)

8 a Find the value of $\sqrt{3 - 2x}$ when $x = -23$? (1 mark)

b What restrictions must be placed on x in $\sqrt{3 - 2x}$? (1 mark)

c Solve $\sqrt{3 - 2x} = 5$ (1 mark)

d Considering the previous parts of this question, write down the solution to:

i $\sqrt{3 - 2x} > 5$ (1 mark)

ii $\sqrt{3 - 2x} < 5$ (1 mark)

Your Feedback

$\frac{\square}{40} \times 100\% = \square\%$

PAGES 127–128

PAGE 176

QUADRATIC EQUATIONS

Algebraic Techniques

STUDY NOTES

1 A **quadratic equation** is one where the highest power of the unknown is 2. It is an equation involving the square of the unknown (x^2). Some examples of quadratic equations are $x^2 = 9$, $5a^2 = 100$ and $2n^2 - 8n + 13 = 0$.

2 Whenever a **negative number** is **squared**, the **answer is positive**. So a simple quadratic equation will have two solutions—the positive square root and the negative square root, e.g. if $x^2 = 16$, $x = 4$ or $x = -4$.

3 **A quadratic equation** may have **2, 1 or no solutions**. It will have no solutions if a perfect square is equal to a negative number ($x^2 = -4$ for example). It will have one solution if a perfect square is equal to zero. For example, if $(x - 3)^2 = 0$ then $x = 3$ is the only solution. A quadratic equation will have two solutions if a perfect square is equal to a positive number. For example, if $n^2 = 9$ then $n = 3$ and $n = -3$ are both solutions.

4 The **symbol** ± means 'plus or minus' or 'positive or negative'. If we write $x = \pm 4$ we mean $x = 4$ or $x = -4$.

5 **Simple quadratic equations** are solved by taking the **square root** of both sides of the equation, remembering to consider both the **positive and negative** square roots. If $2x^2 = 8$ then $x^2 = 4$ and $x = \pm 2$.

6 **Quadratic equations** may be **solved by factorisation**. This method relies on the fact that if two numbers multiply to zero, one (or both) of the numbers must be zero. Using this method, put the expression equal to zero, factorise (if possible), let each of the factorised parts equal zero and solve these simple equations. For example, consider the equation $x^2 - 6x + 8 = 0$. Now, $x^2 - 6x + 8 = (x - 2)(x - 4)$ so we know that $(x - 2)(x - 4) = 0$ and this means that either $x - 2 = 0$ or $x - 4 = 0$. Solving these two simple equations gives $x = 2$ or $x = 4$.

7 **Not every quadratic expression can be factorised**. This doesn't necessarily mean that the equation has no solution.

8 The **factorisation method** only works if the **equation is equal to zero**. If $(x - 2)(x - 4) = 8$ say, it doesn't mean that one part must be 8. You must first expand, put the expression equal to zero and then factorise this new expression.

$$\begin{aligned}(x-2)(x-4) &= 8\\ x^2 - 6x + 8 &= 8\\ x^2 - 6x &= 0\\ x(x-6) &= 0\\ x = 0 \quad &\text{or} \quad x = 6\end{aligned}$$

9 Like any equations, the **solutions to quadratic equations** can be **checked by substitution**. Take care to substitute the value of the unknown every time it appears in the equation.

10 When **answering problems involving the solution of a quadratic**, we should always consider whether both positive and negative answers are **physically possible**. For example, if finding the length of rectangle with a particular area, the length cannot be negative so we only choose the positive answer. Writing 'but $a > 0$' illustrates that you understand and have considered the negative value.

Checklist

Can you:

1 *Understand what is meant by a quadratic equation?* ☐

2 *Understand that a quadratic equation may have 0, 1 or 2 solutions?* ☐

3 *Understand that if two numbers multiply to zero, at least one of the numbers must be zero?* ☐

4 *Solve quadratic equations by factorising?* ☐

5 *Check the solutions to quadratic equations?* ☐

6 *Understand that not every quadratic equation can be solved by factorisation?* ☐

QUADRATIC EQUATIONS

Algebraic Techniques

SKILLS CHECK

1 Find the value of x^2 if:

a $x = 3$ b $x = -3$ c $x = 7$ d $x = -7$
e $x = -5$ f $x = 5$ g $x = 1$ h $x = -1$

2 The square of 6 is 36 ($6^2 = 36$).

a The square of what other number is also 36?
b Write down two numbers whose square is 100.

3 Solve:

a $x^2 = 9$ b $x^2 = 49$ c $x^2 = 25$ d $x^2 = 1$

4 Find the value of $2p^2$ if:

a $p = 2$ b $p = -2$ c $p = 6$ d $p = -6$

5 Solve:

a $2p^2 = 8$ b $2p^2 = 72$

6 Solve:

a $x^2 = 16$ b $3e^2 = 243$ c $4k^2 = 1$

7 Solve, giving the answers correct to two decimal places:

a $x^2 = 60$ b $x^2 = 12.5$ c $7x^2 = 45$

8 Solve:

a $5x = 0$ b $3a = 0$ c $12t = 0$ d $-37m = 0$

9 Solve:

a $3(x - 7) = 0$ b $4(2x + 5) = 0$ c $6(3x - 2) = 0$

10 Solve:

a $x(x + 4) = 0$ b $(x - 7)(x + 1) = 0$ c $(2a + 5)(3a - 2) = 0$

11 Factorise:

a $x^2 - 9x$ b $x^2 + 8x + 15$ c $t^2 + 3t - 10$
d $m^2 - 7m + 12$ e $n^2 - 4n - 60$ f $a^2 - 2a + 1$

12 The given value is a solution of the quadratic equation. True or false?

a $x^2 + 7x + 12 = 0$ $(x = 3)$
b $y^2 - 4y - 21 = 0$ $(y = -3)$

13 Solve:

a $n^2 - 3n = 0$ b $x^2 + 6x + 8 = 0$ c $a^2 - 4a - 45 = 0$

PAGES 128–129

Answers **1 a** 9 **b** 9 **c** 49 **d** 49 **e** 25 **f** 25 **g** 1 **h** 1 **2 a** −6 **b** ±10 **3 a** $x = \pm 3$ **b** $x = \pm 7$ **c** $x = \pm 5$ **d** $x = \pm 1$ **4 a** 8 **b** 8 **c** 72 **d** 72 **5 a** $p = \pm 2$ **b** $p = \pm 6$ **6 a** $x = \pm 4$ **b** $e = \pm 9$ **c** $k = \pm \frac{1}{2}$ **7 a** $x = \pm 7.75$ **b** $x = \pm 3.54$ **c** $x = \pm 2.54$ **8 a** $x = 0$ **b** $a = 0$ **c** $t = 0$ **d** $m = 0$ **9 a** $x = 7$ **b** $x = -2.5$ **c** $x = \frac{2}{3}$ **10 a** $x = 0$ or $x = -4$ **b** $x = 7$ or $x = -1$ **c** $a = -2\frac{1}{2}$ or $a = \frac{2}{3}$ **11 a** $x(x - 9)$ **b** $(x + 5)(x + 3)$ **c** $(t + 5)(t - 2)$ **d** $(m - 3)(m - 4)$ **e** $(n - 10)(n + 6)$ **f** $(a - 1)^2$ **12 a** false **b** true **13 a** $n = 0$ or $n = 3$ **b** $x = -2$ or $x = -4$ **c** $a = 9$ or $a = -5$

QUADRATIC EQUATIONS

Algebraic Techniques

30 MINUTES

INTERMEDIATE TEST

Part A Multiple Choice

1 A solution to the equation $4a^2 = 16$ is: I $a = 2$ II $a = -2$

A I only B II only C neither I nor II D both I and II (1 mark)

2 Which is a solution to the equation $p^2 + 8p - 9 = 0$?

A $p = -1$ B $p = 8$ C $p = -8$ D $p = -9$ (1 mark)

3 The solutions to the equation $x^2 - 11x + 28 = 0$ are:

A $x = 4$ and $x = 7$ B $x = -4$ and $x = 7$ C $x = 4$ and $x = -7$ D $x = -4$ and $x = -7$ (1 mark)

4 A quadratic equation has solutions $x = 3$ and $x = -2$. The equation is:

A $x^2 + 5x - 6 = 0$ B $x^2 - 5x - 6 = 0$ C $x^2 + x - 6 = 0$ D $x^2 - x - 6 = 0$ (1 mark)

5 The solution to the equation $(x - 5)(x - 2) = 28$ is: *Hint 1*

A $x = 2$ or $x = 5$ B $x = 7$ or $x = 4$ C $x = -2$ or $x = 9$ D $x = -5$ or $x = 3$ (1 mark)

Part B Short Answer

6 Solve $100x^2 = 1$ *Hint 2* (2 marks)

7 Solve $4x^2 - 100 = 0$ *Hint 3* (2 marks)

8 Solve by first factorising:

a $x^2 + 11x = 0$ *Hint 4* (2 marks)

b $x^2 + 2x - 80 = 0$ (2 marks)

c $x^2 - 18x + 81 = 0$ (2 marks)

d $x^2 + 14x + 48 = 0$ (2 marks)

9 Solve $(2x + 3)^2 = 25$ *Hint 5* (2 marks)

10 a A class was asked to solve the equation $(x + 3)(x + 5) = 35$. One student said that because the product is 35 one of the factors must be 35 so the answer was $x = 32$ or $x = 30$. Briefly explain what is wrong with this student's logic. (2 marks)

b Solve $(x + 3)(x + 5) = 35$ *Hint 6* (3 marks)

11 A rectangle is 12 metres longer than it is wide. The area of the rectangle is 45 m^2.

a Let the width of the rectangle be x m. Explain why the width of the rectangle can be found by solving the quadratic equation $x^2 + 12x - 45 = 0$. (3 marks)

b Find the dimensions of the rectangle. *Hint 7* (3 marks)

Hint 1: Check by substitution.
Hint 2: There are two answers.
Hint 3: Add 100 to both sides and then divide both sides by 4.
Hint 4: Take out a common factor.
Hint 5: Take the square root of both sides of the equation. There will be two answers.
Hint 6: Expand the binomial first and then put the equation equal to zero.
Hint 7: Make sure your answers are physically possible.

Your Feedback

$\frac{\square}{30} \times 100\% = \square\,\%$

PAGES 129–130

PAGE 176

QUADRATIC EQUATIONS

Algebraic Techniques

40 MINUTES

ADVANCED TEST

1 Solve:

a $a^2 = \frac{1}{16}$ (1 mark)

b $4t^2 = 25$ (1 mark)

c $12d^2 - 27 = 0$ (1 mark)

2 Explain why there is no real solution to the equation $x^2 + 1 = 0$. (1 mark)

3 Solve:

a $x^2 + 9x + 14 = 0$ (1 mark)

b $a^2 - 2a - 48 = 0$ (1 mark)

c $54 - 3x - x^2 = 0$ (1 mark)

4 Solve:

a $x^2 = 5x$ (2 marks)

b $n^2 = 12n + 64$ (2 marks)

c $s^2 + 5s = 24$ (2 marks)

5 Solve, giving each answer correct to two decimal places:

a $x^2 = 7$ (2 marks)

b $3y^2 = 11$ (2 marks)

c $5m^2 - 13 = 0$ (2 marks)

6 Solve:

a $2x^2 - 28x + 66 = 0$ (2 marks)

b $3m^2 + 30m - 72 = 0$ (2 marks)

c $5y^2 - 5y - 10 = 0$ (2 marks)

7 Solve:

a $y^3 + 7y^2 = 0$ (2 marks)

b $a^3 - 9a = 0$ (2 marks)

c $x^4 - 5x^2 + 4 = 0$ (2 marks)

8 Solve:

a $(x - 5)(x - 2) = -2$ (2 marks)

b $(x + 6)(x - 7) = x + 21$ (2 marks)

c $2x^2 = (x - 5)(x - 12)$ (2 marks)

9 Solve $(x + 4)(x + 6) + (x + 8)(x - 3) = (x - 12)(x + 2)$ (3 marks)

Your Feedback

$\frac{\square}{40} \times 100\% = \square\%$

PAGE 130

PAGE 176

SIMULTANEOUS EQUATIONS

Algebraic Techniques

STUDY NOTES

1 **Simultaneous equations** are equations that are solved together.

2 The **simultaneous solution** of two equations in two unknowns (variables), is the values of those variables which **satisfy both equations** at the **same time**. For example, the simultaneous solution of $y = x + 1$ and $y = 2x$ is $x = 1$ and $y = 2$.

3 Two equations are independent if they are different equations. ($x + y = 4$ and $3x + 3y = 12$ are not independent because the second equation is just three times the first equation.) Whenever the equations are independent, the **solution to two simultaneous linear equations** is **unique**. There are an infinite number of solutions that satisfy each equation separately but only one solution that satisfies both equations at the same time.

4 The **solution of two linear simultaneous equations** can be found by **graphing** each line on a **number plane**, and finding the **point of intersection** of the two graphs. (The disadvantages of this method include the time needed to draw the graphs and the difficulty to find the solution with any accuracy unless the answer involves exact integral values.)

5 The **solution to two simultaneous equations** can be found by **substituting** one equation into the other. This method is particularly appropriate if one of the unknowns is the subject of one of the equations. For example, to solve $3x + 2y - 11 = 0$ and $y = 4x$ simultaneously, substitute $y = 4x$ into $3x + 2y - 11 = 0$, giving $3x + 2(4x) - 11 = 0$. This new equation can easily be solved to give $x = 1$. Substituting this value for x into either of the original equations gives $y = 4$.

6 **Simultaneous equations** can also be **solved by** an **elimination method**. Under this method the equations are added together, or one equation is subtracted from the other, so that one of the unknowns is eliminated. For example, if $3x + y = 8$ and $5x + y = 10$ then subtracting the first equation from the second gives $2x = 2$, so $x = 1$ and, upon substituting $x = 1$ into either equation, $y = 5$.

7 If the **coefficients of an unknown are equal**, and of the same sign (both positive or both negative) then **subtracting** one equation from the other will eliminate that unknown. It doesn't matter whether the first equation is subtracted from the second or vice versa. If the coefficients of an unknown are equal but of different signs, then **adding** the two equations together will eliminate the unknown.

8 A **variable** will only be eliminated if the **coefficients are the same**. It may be necessary to multiply either or both equations so that two coefficients are equal. For example, to solve $2x + 3y - 5 = 0$ and $x + 6y - 8 = 0$ simultaneously, multiply the first equation by 2 so that 6 is the coefficient of y in both equations. (Alternatively, the first equation could be multiplied by 5, and the second equation by 2, so that the coefficients of x are equal.) Care must be taken to **multiply every term** in the equation.

9 Whatever method is used to solve simultaneous equations, the **solution can always be checked** by substitution. If the solution satisfies both equations, it must be correct.

10 It is also possible to solve **three simultaneous equations** in three unknowns by first using two of the equations to eliminate one variable and then another two equations (one used before, one not) to eliminate the same variable and then solving these two new equations simultaneously.

Checklist

Can you:

1 *Understand what is meant by simultaneous equations?* ☐
2 *Solve simultaneous equations graphically?* ☐
3 *Solve simultaneous equations by substituting one equation into another?* ☐
4 *Solve simultaneous equations by the elimination method?* ☐

SIMULTANEOUS EQUATIONS

Algebraic Techniques

SKILLS CHECK

1 Solve:

a $3a + 11 = -1$ b $4x + 3(6x - 5) - 2 = 16$

2 If $q = -2$, find p when:

a $p + q = 5$ b $3p - q = 17$ c $4p + 3q - 10 = 0$

3 Find the values of x and y at the point where the two lines intersect.

a

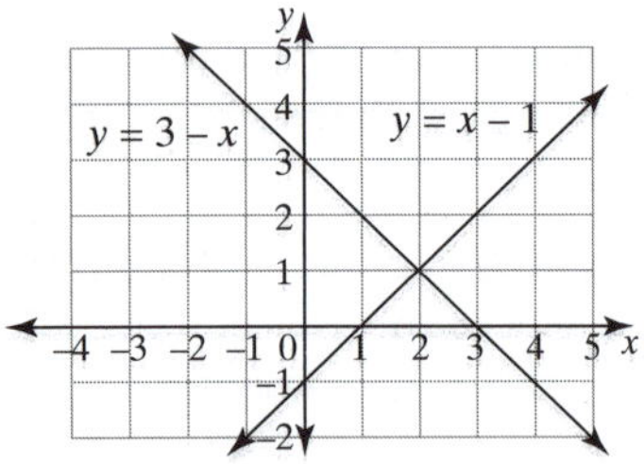

b

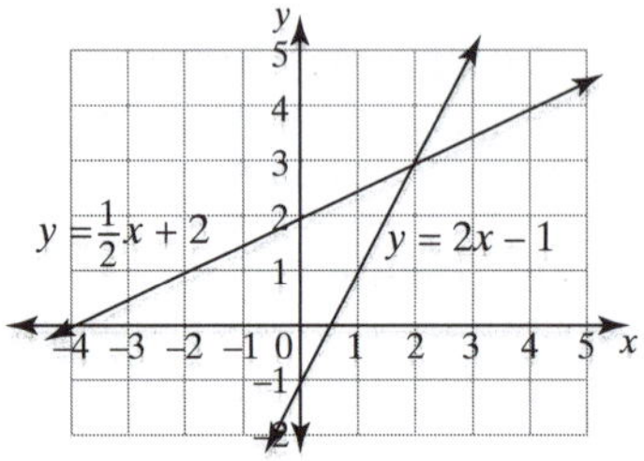

c

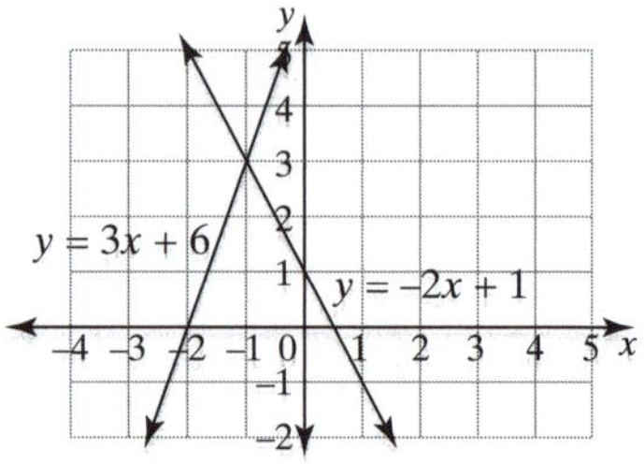

4 Tables of values have been drawn up for the equations $y = 2x - 5$ and $y = x + 1$. By considering the patterns, find the values of x and y which satisfy both equations.

$y = 2x - 5$

x	1	2	3	4
y	–3	–1	1	3

$y = x + 1$

x	1	2	3	4
y	2	3	4	5

5 a Draw up a table of values, for $x = 0, 1, 2, 3, 4, 5$ and 6, and complete for the rule $y = 3x - 8$.
b For what values in the table does $x = y$?
c For what values in the table does $2x + y = 7$?

6 Make y the subject of each equation:

a $7x - y = 10$ b $5x + y - 2 = 0$ c $6x + 2y - 1 = 0$

7 Write down the new equation formed when equation [i] is added to equation [ii]. (There is no need to solve.)

a $2a + b = 5$ [i]
$3a - b = 5$ [ii]

b $8x + 5y - 2 = 0$ [i]
$3x - 5y + 13 = 0$ [ii]

c $7p - 3q = 15$ [i]
$2p + 3q = 18$ [ii]

8 Subtract equation [ii] from equation [i]:

a $6x + 7y = 9$ [i]
$2x + 7y = 13$ [ii]

b $11a - 3b = 8$ [i]
$5a - 3b = 2$ [ii]

c $5a - 3b - 1 = 0$ [i]
$5a + 6b + 10 = 0$ [ii]

9 Determine whether adding both equations or subtracting one equation from the other will eliminate a variable.

a $7a - 4b = 9$ [i]
$3a + 4b = 31$ [ii]

b $8x + 2y = 19$ [i]
$8x - 2y = 15$ [ii]

c $4p - 3q + 7 = 0$ [i]
$p - 3q - 15 = 0$ [ii]

10 A box holds only red and blue pens. Let x be the number of blue pens and y the number of red pens. Write algebraic statements to represent the given information.

a There are 15 pens in the box.
b There are five more blue pens than red pens.
c There are twice as many blue pens as red pens.

PAGES 130–131

Answers **1 a** $a = -4$ **b** $x = 1.5$ **2 a** $p = 7$ **b** $p = 5$ **c** $p = 4$ **3 a** $x = 2, y = 1$ **b** $x = 2, y = 3$ **c** $x = -1, y = 3$ **4** $x = 6, y = 7$
5 a $-8, -5, -2, 1, 4, 7, 10$ **b** $x = 4, y = 4$ **c** $x = 3, y = 1$ **6 a** $y = 7x - 10$ **b** $y = -5x + 2$ **c** $y = -3x + \frac{1}{2}$ **7 a** $5a = 10$ **b** $11x + 11 = 0$
c $9p = 33$ **8 a** $4x = -4$ **b** $6a = 6$ **c** $-9b - 11 = 0$ **9 a** add **b** add or subtract **c** subtract **10 a** $x + y = 15$ **b** $x = y + 5$ **c** $x = 2y$

SIMULTANEOUS EQUATIONS

Algebraic Techniques

25 MINUTES

INTERMEDIATE TEST

Part A Multiple Choice

1 At what point do the lines $x = 3$ and $y = 4$ intersect?

A (3, 4) **B** (4, 3) **C** (–3, –4) **D** (–4, –3) (1 mark)

2 The simultaneous solution of the equations $2x + y = 5$ and $x - 3y = 13$ is: *Hint 1*

A $x = 4, y = 3$ **B** $x = 4, y = -3$ **C** $x = -4, y = 3$ **D** $x = -4, y = -3$ (1 mark)

3 For what value of y do the equations $y = 4x + 7$ and $y = -2x - 5$ have a simultaneous solution?

A –1 **B** –2 **C** 1 **D** 2 (1 mark)

4 Which two equations have a simultaneous solution where $x = 0$?

A $y = 3 - \frac{1}{2}x$ and $y = x - 6$ **B** $y = -\frac{1}{2}x - 6$ and $y = 3 - \frac{1}{2}x$

C $y = x - 6$ and $y = -\frac{1}{2}x - 6$ **D** $y = x + 3$ and $y = -\frac{1}{2}x - 6$ (1 mark)

Part B Short Answer

5 **a** Draw a set of axes and graph the lines $y = x + 3$ and $y = 7 - x$. (2 marks)

b Write down the simultaneous solution of the two equations in part **a**. *Hint 2* (2 marks)

6 Solve the simultaneous equations $7x - 2y = 5$ and $y = 3x$ by substitution. *Hint 3* (2 marks)

7 Solve by elimination $7a + 2b = 16$ and $3a - 2b = 24$. *Hint 4* (2 marks)

8 **a** Make y the subject of the equation $2x + y - 5 = 0$. *Hint 5* (1 mark)

b Solve the equations $2x + y - 5 = 0$ and $3x - 4y + 42 = 0$ by substitution. (2 marks)

9 Solve by elimination $9m + 4p = 41$ and $5m + 2p = 23$. *Hint 6* (2 marks)

10 Solve simultaneously: $2x - y + 8 = 0$ and $3x - y - 2 = 0$. *Hint 7* (2 marks)

11 Find the simultaneous solution of the equations $5a + 3b = 8$ and $3a - 4b = 28$. (3 marks)

12 Madeline bought a total of 24 cans of dog food and cat food at her local store. The dog food cost \$1.40 per can and the cat food was 90c per can. Altogether Madeline spent \$29.10 on the pet food. How many cans of each type of food did she buy? *Hint 8* (4 marks)

Hint 1: There is no need to solve the equations simultaneously. Substitute to see which solution satisfies both equations.
Hint 2: What are the coordinates of the point of intersection?
Hint 3: Replace y with $3x$ in the first equation.
Hint 4: Add the two equations together so that b is eliminated.
Hint 5: The subject of an equation is the value before the equals sign. Rearrange the equation so that it reads $y = \ldots$
Hint 6: First multiply the second equation by 2. Make sure that you multiply every term in the equation.
Hint 7: It doesn't matter which method you use.
Hint 8: Let x be the number of cans of dog food and y be the number of cans of pet food. Form two different equations from the given information.

Your Feedback

$\frac{\square}{26} \times 100\% = \square\ \%$

PAGES 131–132

PAGE 176

SIMULTANEOUS EQUATIONS

Algebraic Techniques

40 MINUTES

ADVANCED TEST

1 Use the substitution method to solve simultaneously:

a $y = 5x - 6$ and $y = 3x + 2$ (2 marks)

b $y = 7 - 2x$ and $x + 3y = 11$ (2 marks)

c $3p - 5q = 23$ and $4p + q = 23$ (2 marks)

2 Use the elimination method to solve simultaneously:

a $a + b = 4$ and $a - b = -10$ (2 marks)

b $8x - 3y = 30$ and $5x - 3y = 3$ (2 marks)

c $9m + 5n - 3 = 0$ and $9m - 7n + 69 = 0$ (2 marks)

d $4p + 3q = 27$ and $3p + q = 14$ (2 marks)

e $9a - 4b = -1$ and $5a + 2b = -9$ (2 marks)

f $6k - 5m = 19$ and $7k - 3m = 42$ (2 marks)

3 a Graph the lines $y = 2x - 4$, $y = -2x + 8$, $y = \frac{1}{2}x - 1$, $y = 5 - x$ and $y = -3$ on the same set of axes. (3 marks)

b Write down the simultaneous solution of $y = \frac{1}{2}x - 1$ and:

i $y = 2x - 4$ (1 mark)

ii $y = 5 - x$ (1 mark)

iii $y = -3$ (1 mark)

c What is significant about the point (3, 2)? (1 mark)

4 a For what value(s) of x does $x = x^2$? (1 mark)

b What is the simultaneous solution of $y = x$ and $y = x^2$? (1 mark)

c In how many places will the line $y = x$ intersect with the parabola $y = x^2$? (1 mark)

d In how many places will the line $y = 4$ intersect with the parabola $y = x^2$? (1 mark)

e What is the simultaneous solution of $y = 4$ and $y = x^2$? (1 mark)

5 Solve simultaneously:

a $y = x^2 + 5x$ and $y = x + 12$ (3 marks)

b $y = x^2 + x - 10$ and $y = 2x - 4$ (3 marks)

6 Solve $y = x^2 + 8x + 3$ and $y = x^2 - 2x - 7$ simultaneously. (2 marks)

7 a Give a reason why a graphical means of solving two equations simultaneously might be the best option. (1 mark)

b Give a reason why an algebraic means of solving two equations simultaneously might be the best option. (1 mark)

Your Feedback

$\frac{\square}{40} \times 100\% = \square\ \%$

PAGES 132–133

PAGE 176

PARALLEL AND PERPENDICULAR LINES

Linear and Non-linear Relationships

STUDY NOTES

1 The **gradient** (or slope) **of a line** is the measure of its **steepness**. A line with **positive gradient** leans to the right (/) and a line with **negative gradient** leans to the left (\). The symbol ***m* is used for gradient**. The gradient is given by $m = \frac{\text{rise}}{\text{run}}$ or the **gradient formula:** $\boldsymbol{m = \frac{y_2 - y_1}{x_2 - x_1}}$.

2 The ***y*-intercept** is where a line meets the *y*-axis. It is the **value of *y* when *x* = 0**. The symbol *b* is used for the *y*-intercept.

3 Any **equation of the form $y = mx + b$** is the **equation of a straight line**.

4 In the equation $y = mx + b$, *m*, the coefficient of *x* (the number by which *x* is multiplied) gives the **gradient**. For example, the line $y = 3x + 4$ has gradient 3. The line $y = -x + 2$ has gradient –1.

5 The constant term, *b*, gives the ***y*-intercept**. For example, The line $y = 3x + 4$ has *y*-intercept 4. The line $y = 2x - 1$ has *y*-intercept –1. The line $y = -x$ has *y*-intercept 0. (It passes through the origin (0, 0).)

6 We say $y = mx + b$ is the **gradient/intercept form** of the equation of a straight line.

7 Given the gradient and *y*-intercept, any **line can be graphed**. The *y*-intercept gives us one point on the line. Use the gradient, $\frac{\text{rise}}{\text{run}}$, to find a second point. For example, if $y = \frac{2}{3}x - 4$, the *y*-intercept is –4 so one point on the line is (0, –4). The gradient is $\frac{2}{3}$ so run across 3 units and rise up 2 units to find another point on the line.

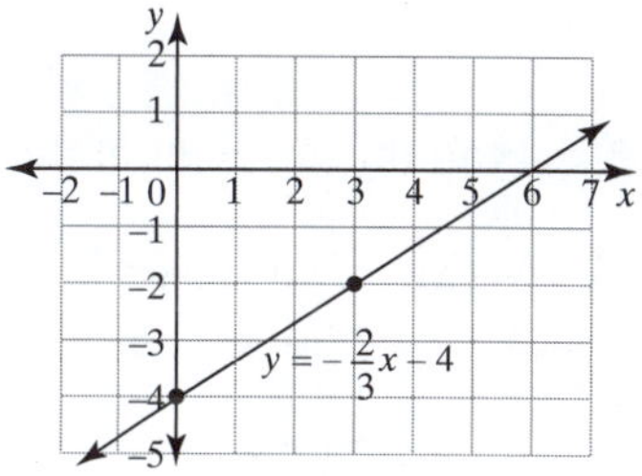

8 The gradient and *y*-intercept can be **read from a graph**. The **equation of the line** can then be formed. For example, the gradient of this line is –1. The *y*-intercept is 2. So the equation of the line is $y = -x + 2$ ($y = -1x + 2$).

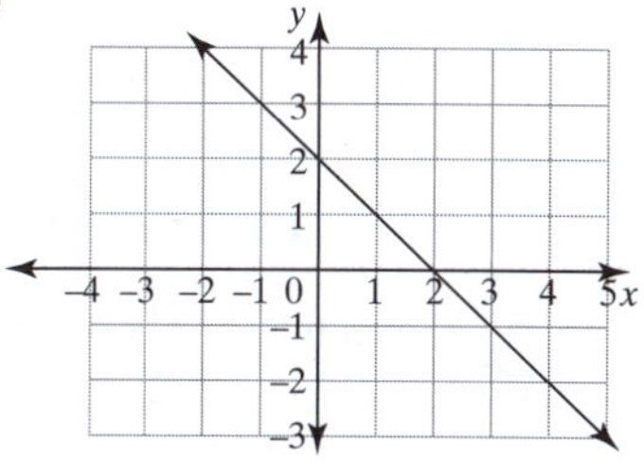

9 **Two lines are parallel** if their gradients are equal. For two lines with gradients m_1 and m_2, if $m_1 = m_2$ the lines are parallel and if the lines are parallel $m_1 = m_2$. To show that two lines are parallel show that their gradients are equal.

10 **Two lines are perpendicular** if they meet at right angles. If two lines are perpendicular their gradients are negative reciprocals. For two lines with gradients m_1 and m_2, the lines are perpendicular if $m_1m_2 = -1$, and if $m_1m_2 = -1$ the lines are perpendicular. (Note: if $m_1m_2 = -1$ then $m_2 = \frac{-1}{m_1}$). To show that two lines are perpendicular, show that their gradients multiply to give negative one.

Checklist

Can you:

1 *Understand what is meant by gradient?* ☐
2 *Understand what is meant by the y-intercept?* ☐
3 *Understand that y = mx + b is the equation of a straight line in gradient/intercept form?* ☐
4 *Understand that in the equation y = mx + b, m is the gradient and b is the y-intercept?* ☐
5 *Graph lines using the gradient and y-intercept?* ☐
6 *Find the gradient, y-intercept and equation of a line from a graph?* ☐
7 *Understand that parallel lines have the same gradient?* ☐
8 *Understand what it means for two lines to be perpendicular?* ☐
9 *Understand that the gradients of perpendicular lines are negative reciprocals?* ☐

PARALLEL AND PERPENDICULAR LINES

Linear and Non-linear Relationships

SKILLS CHECK

1 State whether the gradient of each line is positive or negative:

a
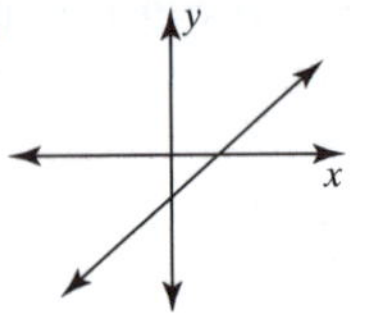

b
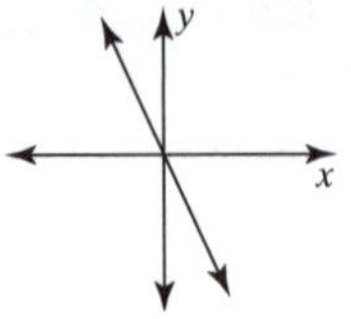

c
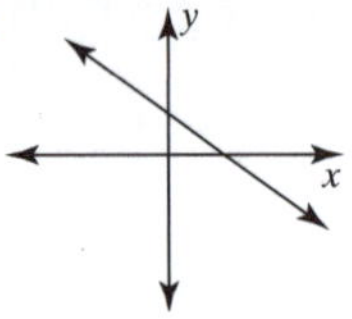

d
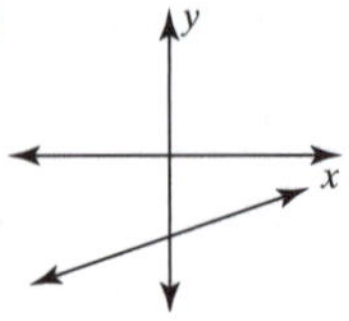

2 Find the gradient of each of these lines:

a
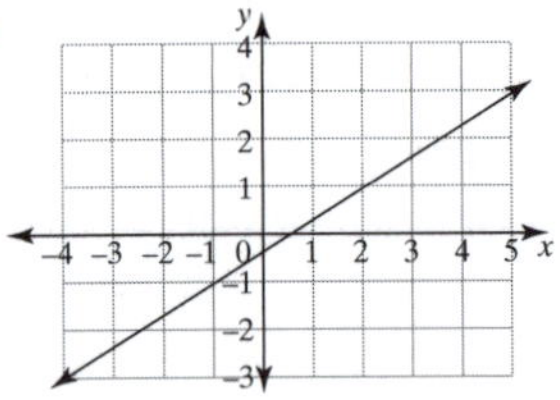

b
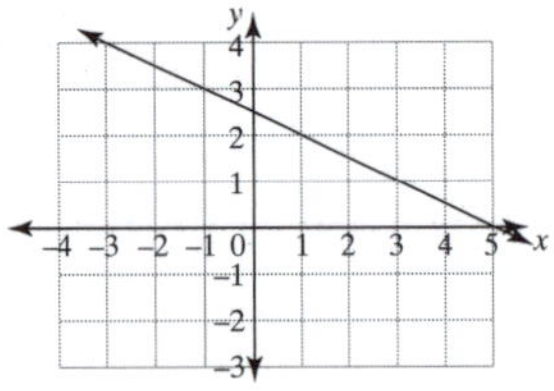

c
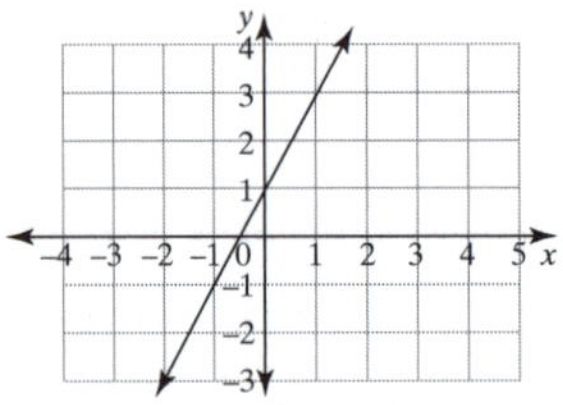

d
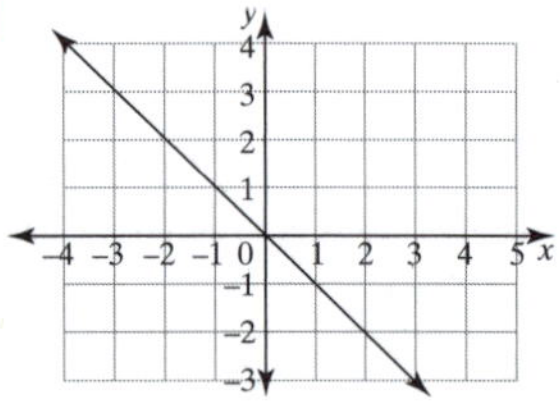

3 Find the gradient of the line joining (1, 2) to (5, 0).

4 Write down the gradient (m) and y-intercept (b) for each line:

a $y = 2x + 3$ b $y = 4x - 1$ c $y = -6x$ d $y = x - 5$

e $y = \frac{1}{3}x - 2$ f $y = \frac{x}{4}$ g $y = \frac{-2x}{3} + \frac{1}{2}$ h $y = 5 - 2x$

5 Write down the equation of the line with gradient m and y-intercept b if:

a $m = 3, b = 2$ b $m = -2, b = 0$ c $m = \frac{1}{2}, b = -\frac{1}{3}$

6 Complete:

a If the gradients of two lines are equal then the lines are ______________________.

b If two lines are parallel then their gradients are ______________________.

7 The gradients of two lines are m_1 and m_2. Complete the following statements.

a If $m_1 = m_2$ the two lines are ______________________.

b If $m_1 m_2 = -1$ the two lines are ______________________.

8 A line l has gradient $-\frac{1}{2}$. Find the gradient of any line which is:

a parallel to l b perpendicular to l

9 The two lines are parallel. True or false?

a $y = 3x - 5$, $y = 3x + 8$

b $y = \frac{1}{2}x - 4$, $y = \frac{x}{2} + 7$

c $y = 7 - x$, $y = -x + 2$

PAGES 133–134

Answers **1 a** positive **b** negative **c** negative **d** positive **2 a** $\frac{2}{3}$ **b** $-\frac{1}{2}$ **c** 2 **d** −1 **3** $-\frac{1}{2}$ **4 a** 2, 3 **b** 4, −1 **c** −6, 0 **d** 1, −5 **e** $\frac{1}{3}$, −2 **f** $\frac{1}{4}$, 0 **g** $-\frac{2}{3}$, $\frac{1}{2}$ **h** −2, 5 **5 a** $y = 3x + 2$ **b** $y = -2x$ **c** $y = \frac{1}{2}x - \frac{1}{3}$ **6 a** parallel **b** equal **7 a** parallel **b** perpendicular **8 a** $-\frac{1}{2}$ **b** 2 **9 a** true **b** true **c** true

PARALLEL AND PERPENDICULAR LINES

Linear and Non-linear Relationships

25 MINUTES

INTERMEDIATE TEST

Part A Multiple Choice

1 Which line is parallel to $y = 3x - 4$?

A $y = 2x - 4$ **B** $y = -3x + 4$ **C** $y = \frac{-1}{3}x - 4$ **D** $y = 3x + 7$ (1 mark)

2 Any line perpendicular to the line $y = 2x - 5$ has gradient:

A 2 **B** –2 **C** $\frac{1}{2}$ **D** $-\frac{1}{2}$ (1 mark)

3 The line which is parallel to the line $y = 2 - x$ and which passes through the origin has equation:

A $y = 2x$ **B** $y = -x$ **C** $x = 2$ **D** $x + y = 2$ (1 mark)

Part B Short Answer

4 Consider the line shown in the diagram.

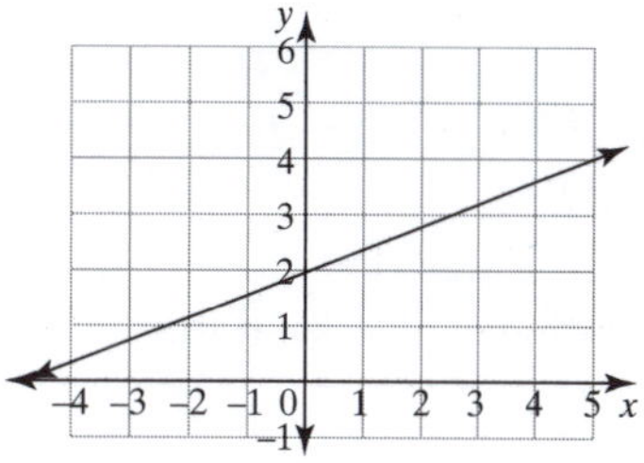

a What is the gradient of the line? (1 mark)

b What is the y-intercept? (1 mark)

c Copy the diagram, and then draw the line $y = \frac{2}{5}x + 4$. *Hint 1* (2 marks)

5 A, B, C and D are the points (–2, 1), (5, 2), (9, –2) and (2, –3) respectively. They form the vertices of a quadrilateral.

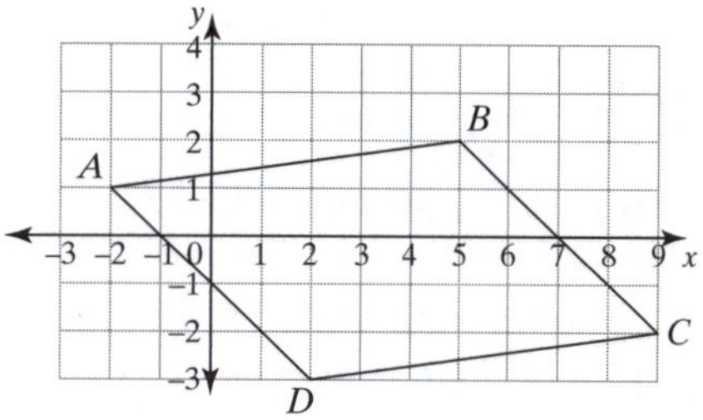

a Show that AB is parallel to DC. *Hint 2* (2 marks)

b Show that AD is parallel to BC. (2 marks)

c What type of quadrilateral is $ABCD$? (1 mark)

6 $P(-8, 5)$, $Q(10, 6)$ and $R(-2, -3)$ are the vertices of a triangle. Is ΔPQR a right-angled triangle? Justify your answer. *Hint 3* (4 marks)

7 The equation of line l is $y = -4x + 3$.

a Line k is parallel to line l and has y-intercept –1. What is the equation of line k? (2 marks)

b Line m is perpendicular to line l and meets it on the y-axis. What is the equation of line m? *Hint 4* (3 marks)

8 A is the point (7, –8), B is the point (1, 1) and C is the point (–3, 7).

a Find the gradient of AB. (1 mark)

b Find the gradient of BC. (1 mark)

c What conclusion can be made about points A, B and C? Justify your answer. (2 marks)

Hint 1: What do you know about the two lines?
Hint 2: Find the gradient of both lines.
Hint 3: Draw a diagram and then investigate whether any sides are perpendicular.
Hint 4: Lines l and m have the same y-intercept.

Your Feedback

$\frac{\square}{25} \times 100\% = \square\%$

PAGES 134–135

PAGE 176

PARALLEL AND PERPENDICULAR LINES

Linear and Non-linear Relationships

35 MINUTES

ADVANCED TEST

1 Determine whether each pair of lines is parallel, perpendicular or neither:

a $y = 3x - 2$; $x + 3y - 4 = 0$ (2 marks)

b $2x + y - 3 = 0$; $x + 2y + 5 = 0$ (2 marks)

c $y = \frac{3}{4}x - 5$; $3x - 4y - 2 = 0$ (2 marks)

2 Line l has equation $y = 7 - x$. What is the equation of the line:

a parallel to l that passes through (2, 1)? (2 marks)

b perpendicular to line l that intersects it on the y-axis? (2 marks)

3 P is the point (1, 8) and Q is the point (–4, 5).

a Find the gradient of the line joining P and Q. (1 mark)

b R is the point on the x-axis such that $PQ \perp QR$. What are the coordinates of R? (2 marks)

4 $S(-10, -5)$, $T(-1, 7)$ and $U(15, -5)$ are the vertices of a triangle.

a Find the gradient of ST. (1 mark)

b Show that ΔSTU is right-angled at T. (2 marks)

c Write down the coordinates of V such that $STUV$ is a parallelogram. (1 mark)

d Find the area of $STUV$. (2 marks)

5 $A(-2, 4)$, $B(6, 2)$ and $C(-4, 2)$ are vertices of a triangle.

a What is the midpoint of AB? (1 mark)

b What is the gradient of AB? (1 mark)

c What is the equation of the perpendicular bisector of AB? (1 mark)

d What is the equation of the perpendicular bisector of AC? (2 marks)

e What is the point of intersection of the lines found in parts **c** and **d**? (2 marks)

f What is the equation of the perpendicular bisector of BC? (1 mark)

g What do you notice about the three perpendicular bisectors? (1 mark)

6 a Rearrange the equation $ax + by + c = 0$ into gradient/intercept form. (1 mark)

b What is the gradient of any line parallel to $ax + by + c = 0$? (1 mark)

c What is the gradient of any line parallel to $5x - 3y + 7 = 0$? (1 mark)

d Write down the equation, in general form, of the line through the origin parallel to the line $5x - 3y + 7 = 0$? (1 mark)

e What is the gradient of any line perpendicular to $ax + by + c = 0$? (1 mark)

f What is the gradient of any line perpendicular to $4x + 2y - 3 = 0$? (1 mark)

g Write down, in general form, the equation of the line through the origin perpendicular to $4x + 2y - 3 = 0$ (1 mark)

Your Feedback

$\frac{\square}{35} \times 100\% = \square\%$

PAGES 135–136

PAGE 176

FURTHER GRAPHS

Linear and Non-linear Relationships

STUDY NOTES

1 The **equation $y = mx + b$** is the **gradient/intercept form** of an equation of a **straight line** (with gradient m and y-intercept b). Any equation of the form $Ax + By + C = 0$ is the equation of a straight line in general form. These equations can be changed from one to the other.

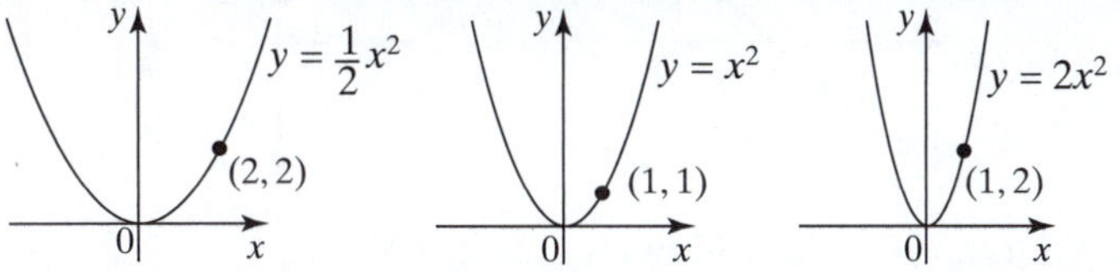

2 A **parabola** is a special type of curve. The graph of $y = ax^2 + c$ $(a \neq 0)$ is a parabola. If $a > 0$, the parabola opens upwards (∪) but if $a < 0$ the parabola opens downwards (∩).

3 The **turning point** of a parabola is called the **vertex**.

4 The parabola $y = ax^2$ has its **vertex at the origin**. The value of a determines the width of the parabola. The larger the value of a, the thinner the parabola.

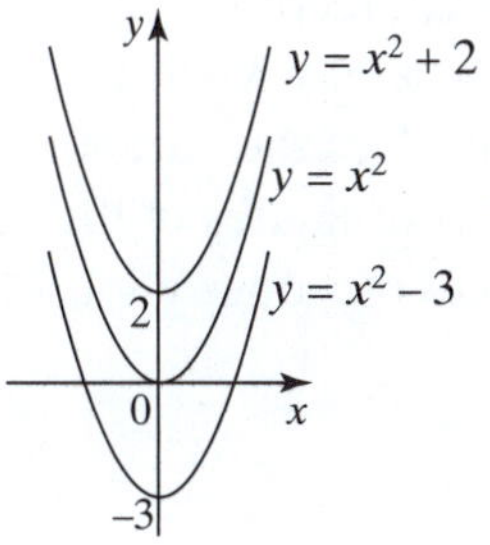

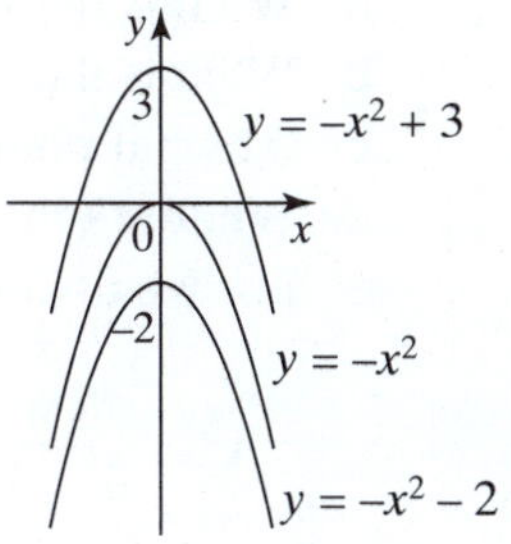

5 The **constant, c,** in the equation $y = ax^2 + c$ has the effect of moving the parabola up or down the y-axis. The vertex of the parabola $y = ax^2 + c$ is at the point $(0, c)$.

6 The graph of $y = ax^2 + bx + c$ (where a, b and c are constants and $a \neq 0$) is also a **parabola**.

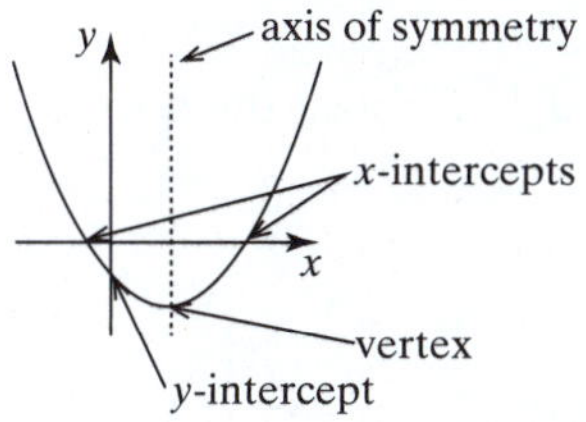

7 The **x-intercepts** of a parabola are the points where the curve meets the x-axis. A parabola may meet the x-axis in 0, 1 or 2 points. The x-intercepts are found when $y = 0$. The y-intercept is the point where the curve meets the y-axis. The **y-intercept** occurs when $x = 0$. If the equation of the parabola is of the form $y = ax^2 + bx + c$ then the y-intercept will be c.

8 The graph of $y = ax^3 + c$ is a **cubic curve**. The value of a determines the width of the curve. c gives the y-intercept.

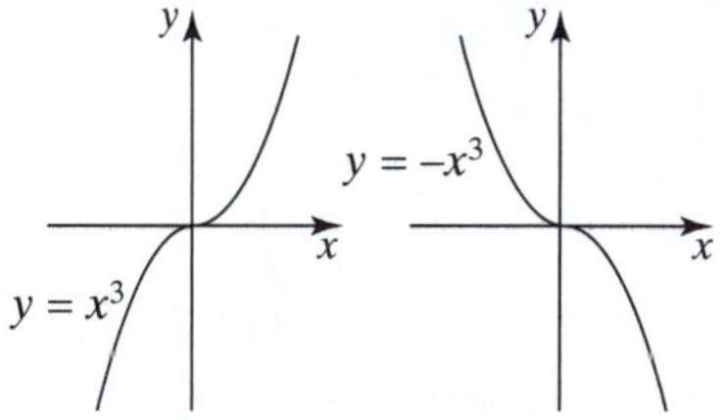

9 Any equation of the form $y = a^x$ is an **exponential curve**. $y = a^x$ has y-intercept 1 for all positive values of a.

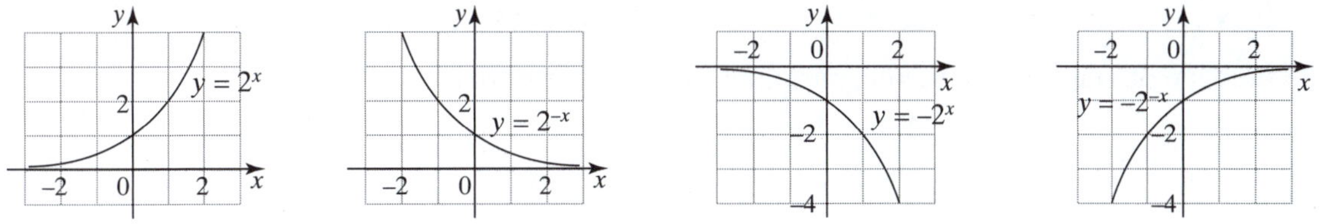

10 A **circle**, **centre the origin**, and **radius** of length r units has equation $x^2 + y^2 = r^2$.

Checklist

Can you:

1. *Identify the type of graph from an equation?* ☐
2. *Graph parabolas, cubic curves, exponential curves and circles?* ☐
3. *Find the radius of a circle (centre the origin) given its equation?* ☐
4. *Use technology (such as graphics calculators or software applications) to draw graphs?* ☐

FURTHER GRAPHS

Linear and Non-linear Relationships

SKILLS CHECK

1 What name is given to each of these curves?

a

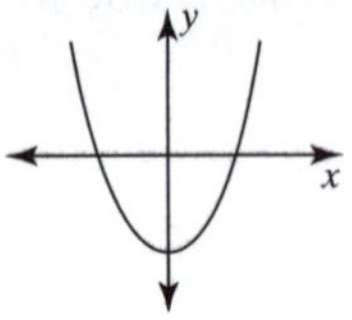

b

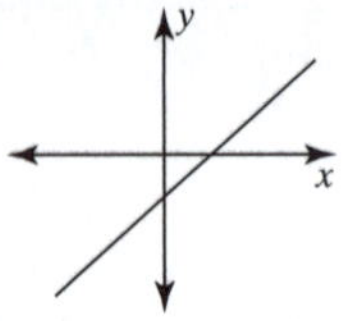

c

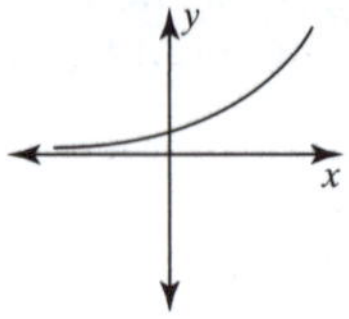

d

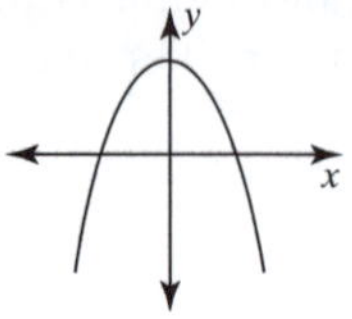

2 Use the parabola shown to answer the following questions.

a What is the y-intercept?

b What is the value of y when $x = 4$?

c At what points does the parabola cut the x-axis?

d What are the coordinates of the vertex (turning point)?

e For what values of x does $y = 13$?

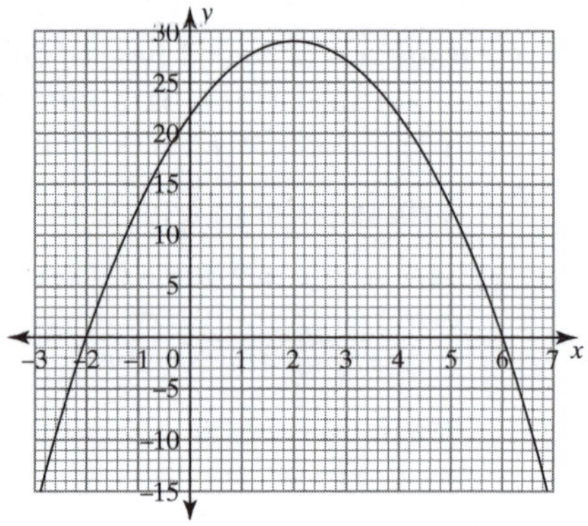

3 Choose the correct equation from the following list to match each curve.

$y = x^2 - 2$, $y = -2^x$, $y = 4 - x^2$, $y = 4^x$, $x^2 + y^2 = 4$, $y = -2x$, $y = 2^{-x}$, $y = \dfrac{x}{4}$

a

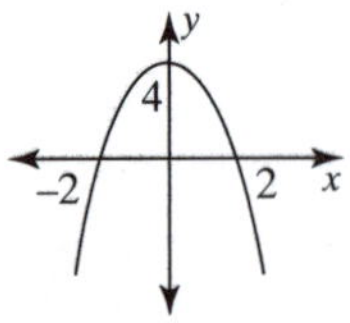

b

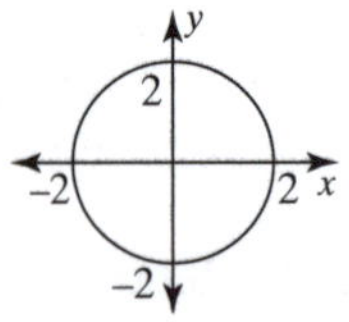

c

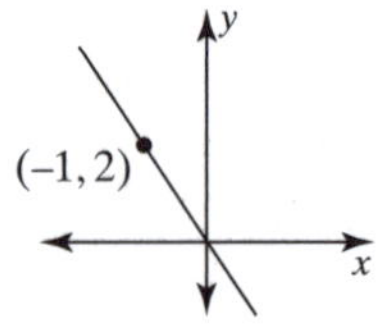

d

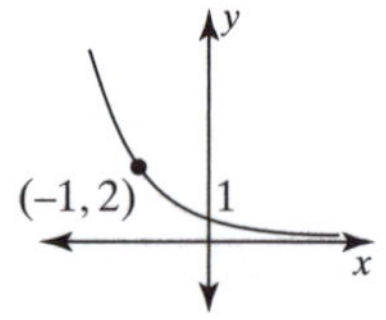

e

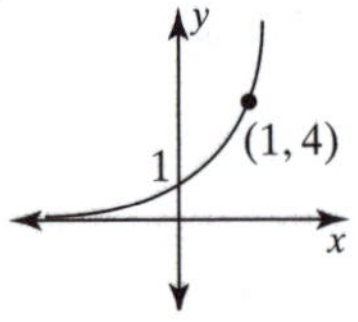

f

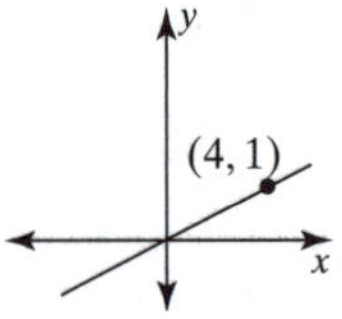

g

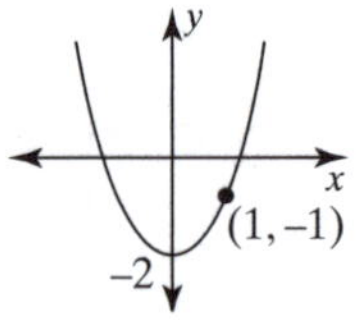

h

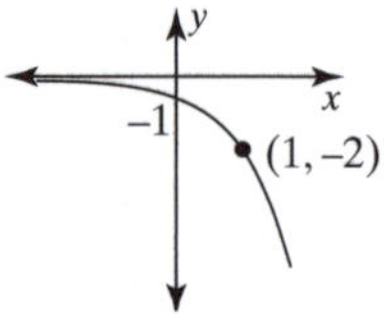

4 Name the type of graph (straight line, parabola, circle, exponential or cubic curve):

a $y = 3 - x$ **b** $y = 3x^2$ **c** $y = 3^x$ **d** $y = x^3$

e $y = x^2 + 3$ **f** $x^2 + y^2 = 3$ **g** $y = \dfrac{x}{3}$ **h** $y = -x^2 + 2$

i $y = 9 - x^2$ **j** $x + y = 9$ **k** $x^2 + y^2 = 9$ **l** $y = 2^{-x}$

PAGES 136–137

Answers **1 a** parabola **b** straight line **c** exponential curve **d** parabola **2 a** 22 **b** 22 **c** (−2, 0) and (6, 0) **d** (2, 29) **e** $x = -1$ and $x = 5$ **3 a** $y = 4 - x^2$ **b** $x^2 + y^2 = 4$ **c** $y = -2x$ **d** $y = 2^{-x}$ **e** $y = 4^x$ **f** $y = \dfrac{x}{4}$ **g** $y = x^2 - 2$ **h** $y = -2^x$ **4 a** straight line **b** parabola **c** exponential curve **d** cubic curve **e** parabola **f** circle **g** straight line **h** parabola **i** parabola **j** straight line **k** circle **l** exponential curve

FURTHER GRAPHS

Linear and Non-linear Relationships

15 MINUTES

INTERMEDIATE TEST

Part A Multiple Choice

1 Which is not the equation of a straight line?

A $x + y = 3$ **B** $x - y = 0$

C $y = \dfrac{x+3}{2}$ **D** $y = \dfrac{3}{x}$ (1 mark)

2 Which point does not lie on the circle $x^2 + y^2 = 50$?

A $(-1, -7)$ **B** $(7, -1)$

C $(2, 6)$ **D** $(5, 5)$ (1 mark)

3 Which diagram could show the graph of $y = x^2 + x$? *Hint 1*

A

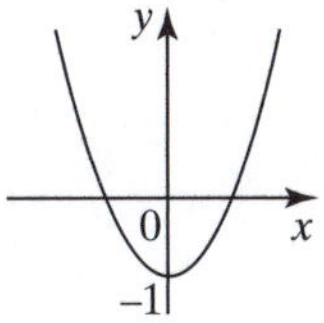

B

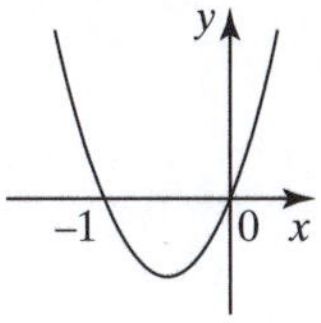

C

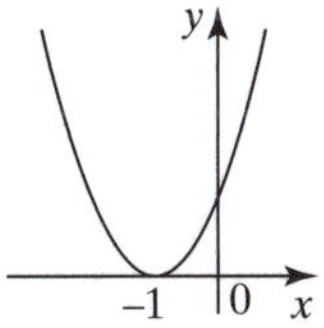

D

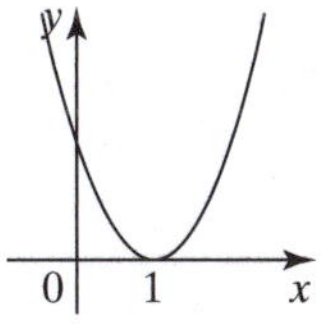

(1 mark)

4 Which is the equation of a parabola?

A $y = (x + 1)(x - 1)$ **B** $y = 2^x$

C $y = x^4$ **D** $y = \dfrac{4}{x^2}$ (1 mark)

Part B Short Answer

5 Kevin used a software program to graph $y = x^3 - 1$ and $y = 1 - x^3$ on the same diagram.

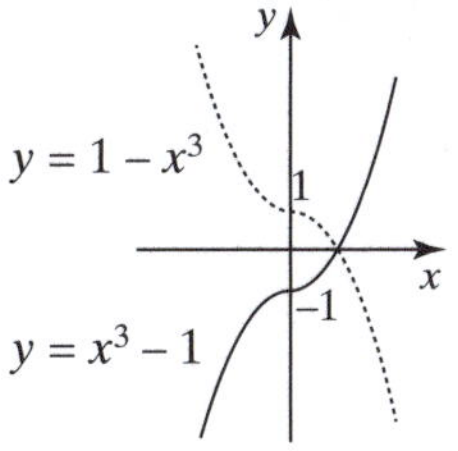

a How many solutions are there to the equation $x^3 - 1 = 1 - x^3$? *Hint 2* (1 mark)

b Solve $x^3 - 1 = 1 - x^3$. (1 mark)

6 A point was plotted on a grid using a software application. The distance from the point $O(0, 0)$ to P was given as 8 units. The circle $x^2 + y^2 = 36$ will then be drawn on the same grid. Will the point P lie inside, outside or on the circle? Justify your answer. (2 marks)

7 A graphics calculator was used to show some exponential graphs. All of the graphs had an equation of the form $y = 2^x + c$ where c is a constant.

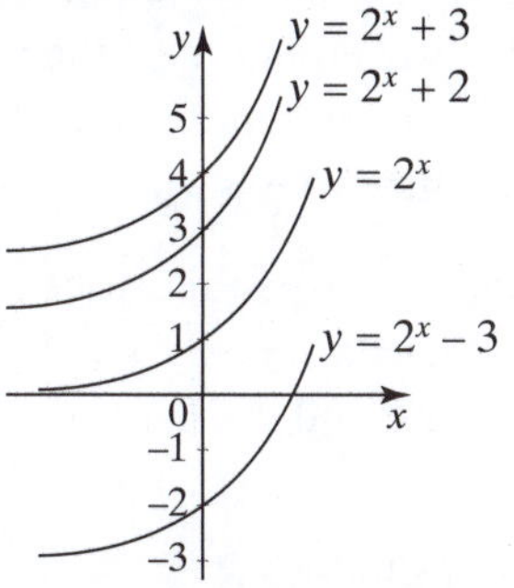

a What effect does the constant have on the graphs? *Hint 3* (1 mark)

b What would the y-intercept be of the graph $y = 2^x + 7$? (1 mark)

c For what value of c will $y = 2^x + c$ pass through the origin? *Hint 4* (1 mark)

8 Stacy used a graphics calculator to graph $y = (x - 2)^2 - 3$ and $y = x^2 - 4x + 1$

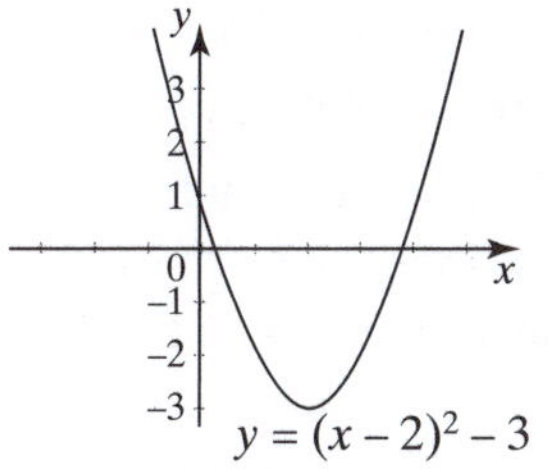

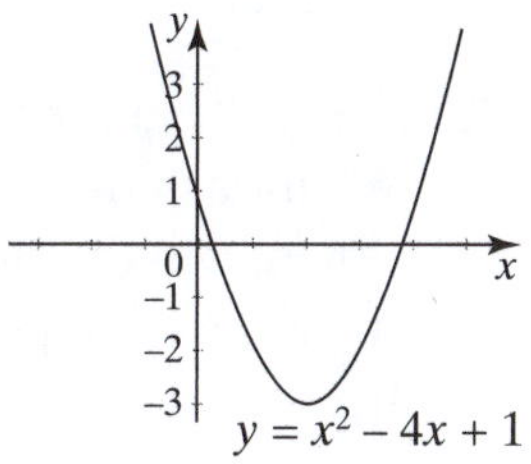

a What similarities can you notice with the two graphs? (1 mark)

b Use algebra to explain the result in part **a**. *Hint 5* (3 marks)

Hint 1: Substitute some of the points that each parabola passes through.

Hint 2: Use the graph.

Hint 3: What happens to the graphs as the number added changes?

Hint 4: The origin is the point (0, 0)

Hint 5: Expand and simplify $(x - 2)^2 - 3$.

Your Feedback

$\dfrac{\square}{15} \times 100\% = \square\%$

PAGES 137–138

PAGE 176

FURTHER GRAPHS

Linear and Non-linear Relationships

35 MINUTES

ADVANCED TEST

1 Consider the curve $y = x^2 - 10x + 21$.

- **a** What name is given to the type of curve? (1 mark)
- **b** What is the y-intercept? (1 mark)
- **c** What are the x-intercepts? (2 marks)
- **d** What is the equation of the axis of symmetry? (1 mark)
- **e** What is the minimum value? (1 mark)
- **f** Sketch the curve. (3 marks)

2 Briefly explain any similarities and differences between the curve $y = x^2$ and each of the following:

- **a** $y = x^2 + 2$ and $y = x^2 - 3$ (1 mark)
- **b** $y = (x + 2)^2$ and $y = (x - 3)^2$ (1 mark)
- **c** $y = -x^2$ (1 mark)
- **d** $y = 2x^2$ and $y = \frac{1}{2}x^2$ (1 mark)

3 Find the turning point and state whether it is a maximum or minimum:

- **a** $y = (x + 2)^2 - 3$ (2 marks)
- **b** $y = 5 - (3 - x)^2$ (2 marks)

4 Describe the curve given by the equation $(x - 1)^2 + (y + 3)^2 = 16$. (2 marks)

5 Describe the symmetrical properties of the curve $y = x^n$ if:

- **a** n is even (1 mark)
- **b** n is odd (1 mark)

6 Consider the curve $y = (x - 2)(x - 4)(x + 1)$.

- **a** Where will the curve cross the y-axis? (1 mark)
- **b** Where will the curve meet the x-axis? (1 mark)
- **c** If x has a large positive value, what can you say about the value of y? (1 mark)
- **d** If x has a large negative value, what can you say about the value of y? (1 mark)
- **e** Using the information found, sketch the curve $y = (x - 2)(x - 4)(x + 1)$. (3 marks)

7

- **a** Solve simultaneously $x^2 + y^2 = 25$ and $y = 7 - x$. (2 marks)
- **b** Sketch $x^2 + y^2 = 25$ and $y = 7 - x$ on the same diagram, clearly showing the points of intersection. (2 marks)

8 The diagram shows the graphs of $y = 2^x$ and $y = 2^x - 3$.

- **a** What is the value of a? (1 mark)
- **b** Find the value of b. (1 mark)
- **c** Use your calculator to find the value of c to three decimal places. (1 mark)

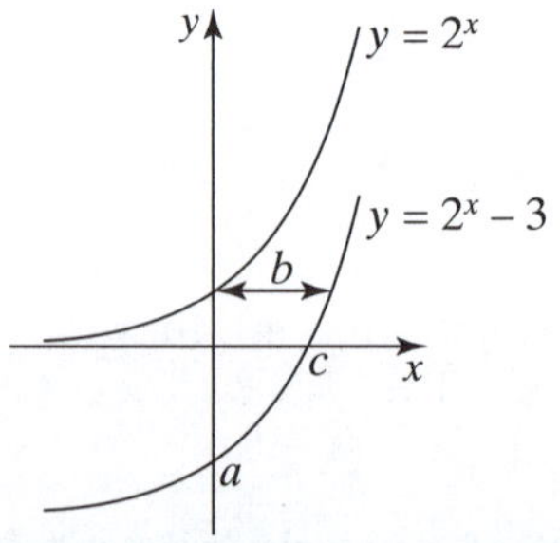

Your Feedback

$\frac{\square}{35} \times 100\% = \square\%$

PAGES 138–139

PAGE 176

SURFACE AREA AND VOLUME

Using Units of Measurement

STUDY NOTES

1 The **perimeter** of a shape is the distance around its boundary. The perimeter can be found by adding all the lengths that make up the boundary.

2 The **circumference** of a circle is the distance around its boundary. The circumference can be found by either of the formulas $C = \pi d$, where d is the diameter, or $C = 2\pi r$ where r is the radius. (The diameter is twice the radius.)

3 An arc of a circle is part of the circumference. The **length of an arc** can be found by finding a fraction of the length of the circumference. For example, if the angle at the centre of a sector of a circle is 57°, then the arc length of the sector is $\frac{57}{360}$ of the length of the circumference of the circle. If finding the **perimeter of a sector**, the arc length must be added to the two radii.

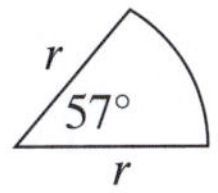

4 **Area** is measured in **square units**, (m^2, cm^2, km^2, ha). You must know the formulas for finding the **area of rectangles** ($A = lb$), **triangles** ($A = \frac{1}{2}bh$), **parallelograms** ($A = bh$), **rhombuses** and **kites** ($A = \frac{1}{2}xy$), **trapeziums** ($A = \frac{1}{2}h(a + b)$) and **circles** ($A = \pi r^2$). The **area of a composite shape** can be found by adding the areas of each of the comprising parts.

5 Volume is measured in cubic units (m^3, cm^3, km^3). The **volume** of any **right prism** can be found by multiplying the area of its base by the perpendicular height ($V = Ah$). Similarly, the **volume of a cylinder** is given by $V = \pi r^2 h$. The **capacity** of a solid is directly related to its volume. An object with volume of 1 m^3 will hold 1000 L. A volume of 1 cm^3 will hold 1 mL The volume of **composite solids** can be found by first finding the volume of the separate component solids. These volumes will then be added together.

6 The **surface area** of a solid is the sum of the areas of all of its faces. For example, the surface area of a rectangular prism is given by $A = 2(lb + lh + bh)$ because the rectangular prism has 6 faces. The area of 2 of those faces is found by multiplying the length by the breadth. For another 2 faces, the area is the length by the height and for the remaining faces, the breadth by the height.

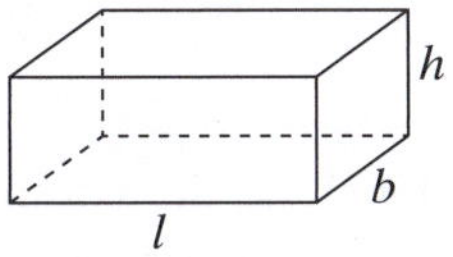

7 Every **prism** has two identical faces (the cross-section after which the prism is named) and a number of rectangular faces. The total area of those rectangular faces can be found by multiplying the perimeter of the cross-section by the height of the prism. The surface area of the prism is then that total area plus twice the area of the cross-section.

8 A closed cylinder is made up of a curved surface and two circles, top and bottom. The **surface area of a closed cylinder** is given $A = 2\pi r^2 + 2\pi rh$, where r is the radius and h the height. If the cylinder is open at the top, the surface area will only include one circle ($A = \pi r^2 + 2\pi rh$) and, if the cylinder is completely open, the surface area is that of the curved surface only ($A = 2\pi rh$).

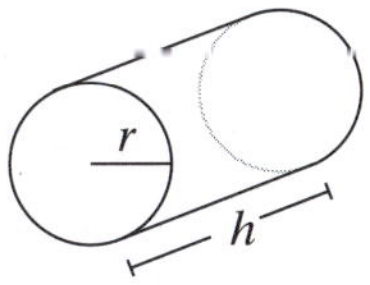

9 If two **similar figures** have matching **intervals** in the ratio $1:k$ then the matching **areas** are in the ratio $1:k^2$ and the matching **volumes** are in the ratio $1:k^3$.

Checklist

Can you:

1 *Understand what is meant by perimeter and find the perimeter of simple and composite shapes?* ☐
2 *Find the circumference of circles, arc lengths and the perimeter of sectors?* ☐
3 *Find the area of triangles, quadrilaterals, circles and composite figures?* ☐
4 *Find the volume of prisms, cylinders and composite solids?* ☐
5 *Understand the relationship between capacity and volume and find the capacity of prisms, cylinders and composite solids?* ☐
6 *Find the surface area of prisms and cylinders?* ☐

SURFACE AREA AND VOLUME

Using Units of Measurement

SKILLS CHECK

1 Find the perimeter of a regular hexagon of side length 26 cm.

2 Find the perimeter of:

a

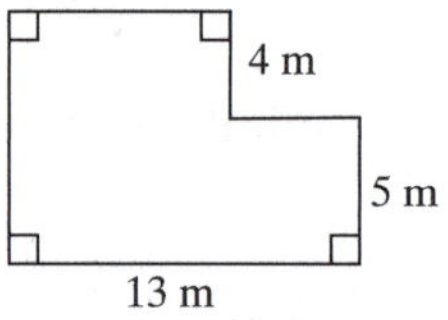

b

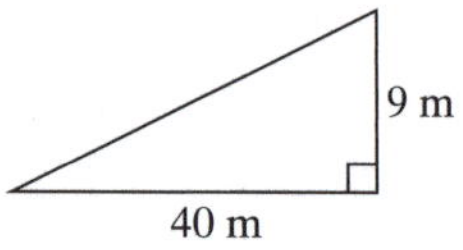

3 Find the circumference (correct to 1 decimal place), of a circle of radius 19 cm.

4 Find the area of:

a a rectangle 3 m long and 70 cm wide

b a circle of diameter 16 m

c a rhombus with diagonals 23 cm long and 17 cm long

5 Find the area of each shape.

a

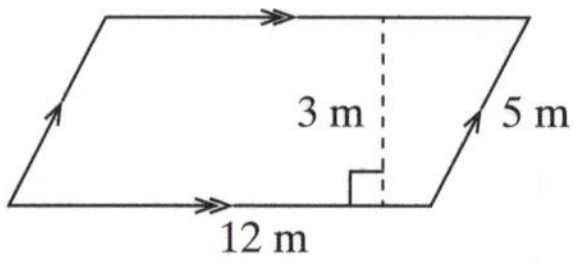

b

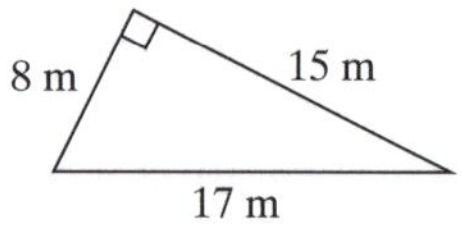

c

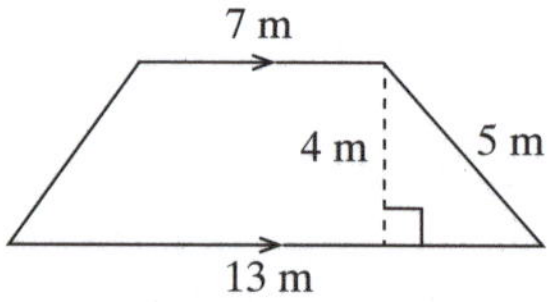

6 Find the volume of:

a an octagonal prism with base area 65 cm^2 and height 7 cm

b a rectangular prism 8 m long, 5 m wide and 3 m high

c a cylinder with radius 0.6 m and height 1 m

7 Find the surface area of:

a a cube of side length 14 cm

b a rectangular prism with length 6 m, breadth 4 m and height 3 m

c a closed cylinder with diameter 20 cm and height 15 cm

8 For this triangular prism, find:

a the volume

b the surface area

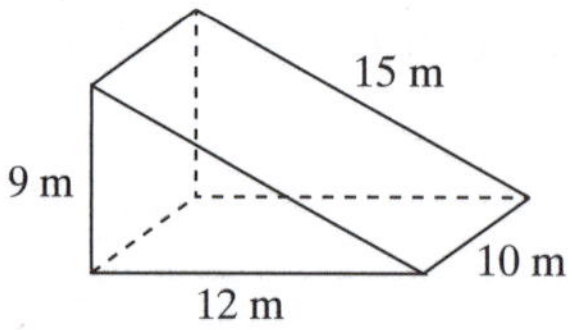

PAGES 139–140

Answers **1** 156 cm **2 a** 44 m **b** 90 m **3** 119.4 cm **4 a** 2.1 m^2 **b** 201.1 m^2 **c** 195.5 cm^2 **5 a** 36 m^2 **b** 60 m^2 **c** 40 m^2
6 a 455 cm^3 **b** 120 m^3 **c** 1.13 m^3 **7 a** 1176 cm^2 **b** 108 m^2 **c** 1570.8 cm^2 **8 a** 540 m^3 **b** 468 m^2

SURFACE AREA AND VOLUME

Using Units of Measurement

25 MINUTES

INTERMEDIATE TEST

Part A Multiple Choice

1 A rectangular prism has length 6 cm, breadth 5 cm and height 4 cm. The surface area of this prism is:

A 120 cm² **B** 74 cm² **C** 148 cm² **D** 60 cm² (1 mark)

2 If A is the area of the shaded face, in which diagram is the volume found by multiplying A by h?

A

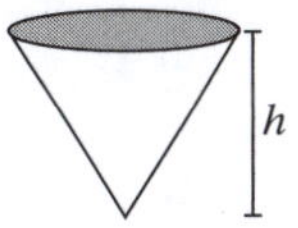

B

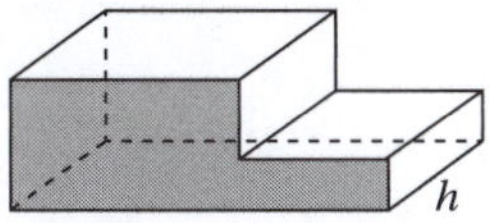

C

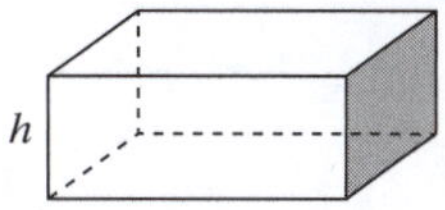

D

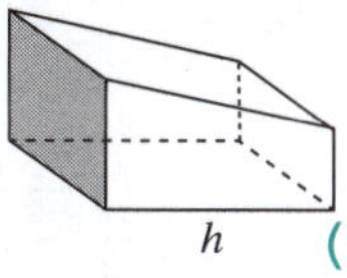

(1 mark)

3 The volume of a cylinder of radius 50 cm and height 20 cm is closest to:

A 157 080 cm³ **B** 628 320 cm³ **C** 209 440 cm³ **D** 52 360 cm³ (1 mark)

4 The surface area of a cube is 13.5 m². Which is closest to its volume?

A 50 m³ **B** 34 m³ **C** 11 m³ **D** 3 m³ (1 mark)

Part B Short Answer

5 **a** Find the perpendicular height of the shaded face. *Hint 1* (1 mark)

b Find the volume. (2 marks)

c Find the surface area. (2 marks)

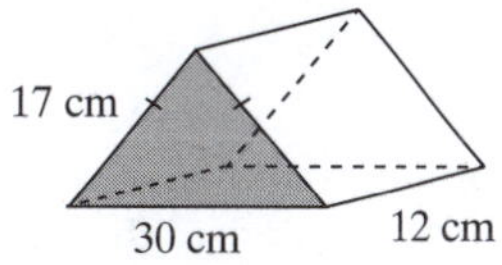

6 A water trough has a semicircular cross-section.

a Find the volume of the trough. Give the answer in square metres, correct to two decimal places. *Hint 2* (2 marks)

b Given that 1m³ = 1000 L, find the approximate capacity of the trough in litres. (1 mark)

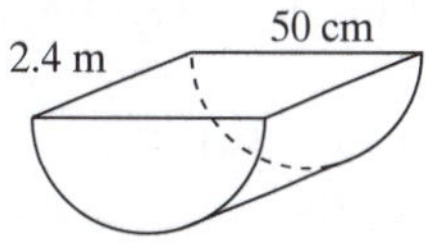

7 A swimming pool is 15 m long and 7 m wide. The pool is 1 m deep at one end and 2.4 m deep at the other as shown in the diagram (which is not to scale).

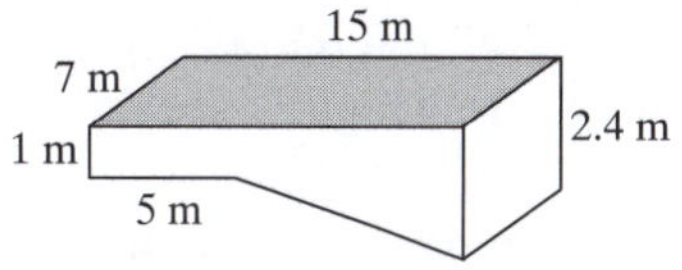

a What is the area of the cross-section of the pool? *Hint 3* (2 marks)

b The pool is to be filled from a hose. Water flows from the hose at a rate of 5 litres per second. How long, to the nearest half-hour, will it take to fill the swimming pool? (3 marks)

8 A swimming pool is in the shape of a trapezoidal prism as shown in the diagram. The four walls and the bottom of the pool are to be painted. They will need two coats of paint.

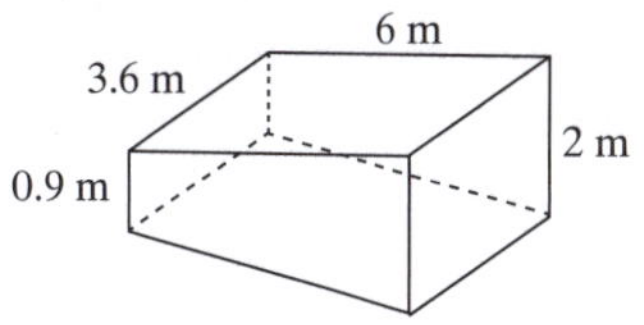

a What is the area of each trapezoidal wall? (1 mark)

b What is the total area of the four walls? (2 marks)

c What is the area of the bottom of the pool? *Hint 4* (2 marks)

d The manufacturer of the paint says that each tin of paint should cover an area between 19 m² and 24 m². How many tins of paint should be bought for the job? *Hint 5* (3 marks)

Hint 1: Use Pythagoras' Theorem. The bottom measurement of the right-angled triangle will not be 30 cm.

Hint 2: The diameter of the semicircle is given in cm but the length of the trough is given in metres.

Hint 3: Divide the area into a rectangle and trapezium.

Hint 4: The bottom is a rectangle but you will first need to find its length.

Hint 5: There must be enough paint for two coats and you must be sure that there is enough.

Your Feedback

$\frac{\square}{25} \times 100\% = \square\ \%$

PAGES 140–141

PAGE 176

SURFACE AREA AND VOLUME

Using Units of Measurement

40 MINUTES

ADVANCED TEST

1 Find the volume of each solid in cubic metres to two decimal places.

a

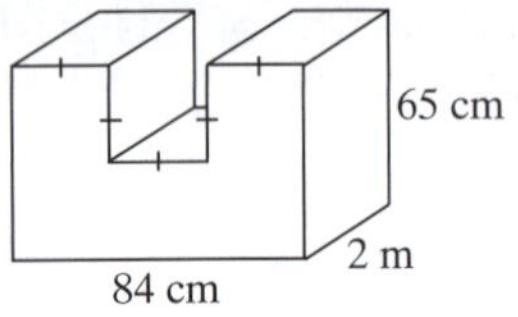

b

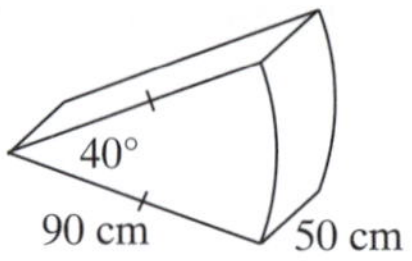

c

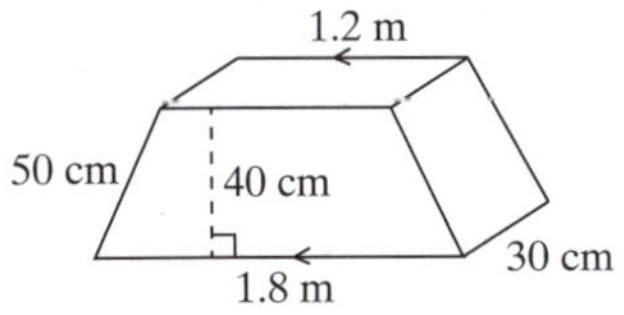

d

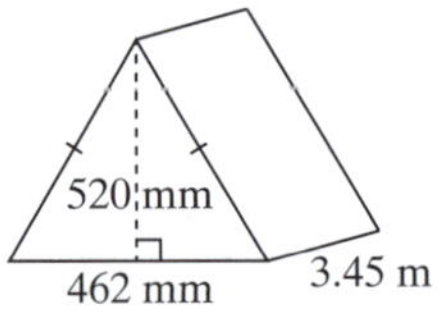

(2 marks each)

2 Find the surface area, in square metres, of each solid in question 1. (2 marks each)

3 An octagonal prism has perpendicular height 19 cm. The perimeter of the cross-section is 96 cm and its area is 697 cm^2. Find, for the prism, the:

a volume (1 mark)

b surface area (1 mark)

4 A cross-section of a prism is made up of a triangle with a rectangular hole in it as shown in the diagram. The perpendicular height of the prism is 23.45 m. Find:

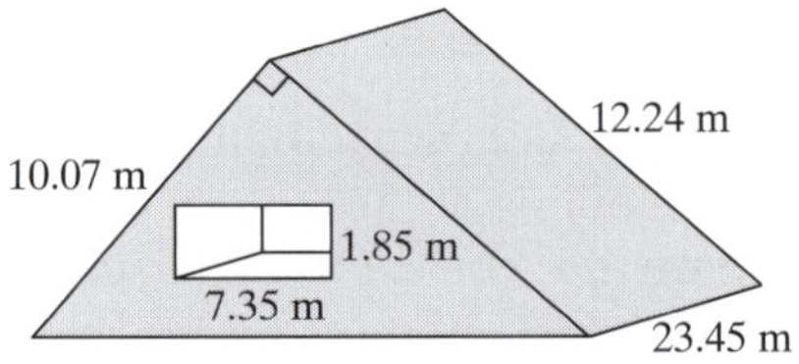

a the volume of the prism (2 marks)

b the area of all the surfaces (including those of the hole) (3 marks)

5 The diagram shows the cross-section of a solid which has a perpendicular height of 1.4 m. The cross-section is made up of a rectangle and semicircle. Find:

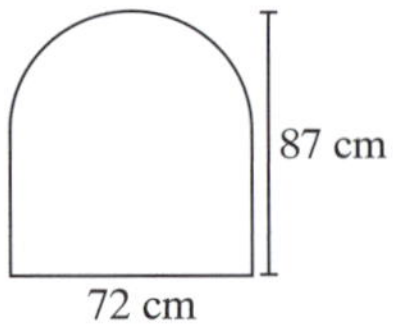

a the volume of the solid (in m^3) (3 marks)

b the surface area in square metres (3 marks)

6 The diagram shows a square inscribed in a circle. This forms the cross-section of a solid of height 6 cm.

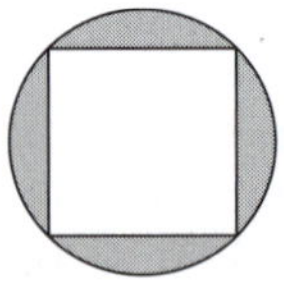

a Explain why the shaded area is given by $A = r^2(\pi - 2)$ units^2 where r is the radius of the circle. (2 marks)

b Find the shaded area, in terms of π, given that the diameter is 8 cm. (1 mark)

c Hence find the volume of the solid. Give the answer to one decimal place. (1 mark)

d Show that, when the diameter is 8 cm, the exterior surface area is $(80\pi - 64)$ cm^2. (2 marks)

7 The surface area of a closed cylinder of height 15 cm is 2287 cm^2.

a If r is the radius of the cylinder explain why solving $r^2 + 15r - 364 = 0$ will give an approximate value for the radius. (2 marks)

b What is the approximate radius of the cylinder? (2 marks)

c Find the approximate volume of the cylinder. (1 mark)

Your Feedback

$\frac{\square}{40} \times 100\% = \square\%$

PAGES 141–143

PAGE 176

ANGLE PROPERTIES

Geometric Reasoning

STUDY NOTES

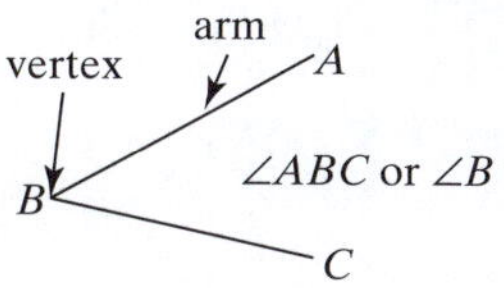

1 **Angles** are usually **named** by three points, the middle point being the **vertex** of the angle, and the other two points lying one on each arm. If there is no chance of confusion one point can be given.

2 A **right angle** measures 90°. Angles in a right angle add to 90°. A **straight angle** measures 180°. Angles in a straight line add to 180°. A **complete revolution** measures 360°. Angles in a revolution add to 360°.

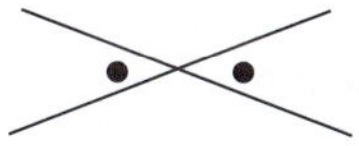

3 **Vertically opposite** angles are formed when two straight lines intersect. Vertically opposite angles are equal.

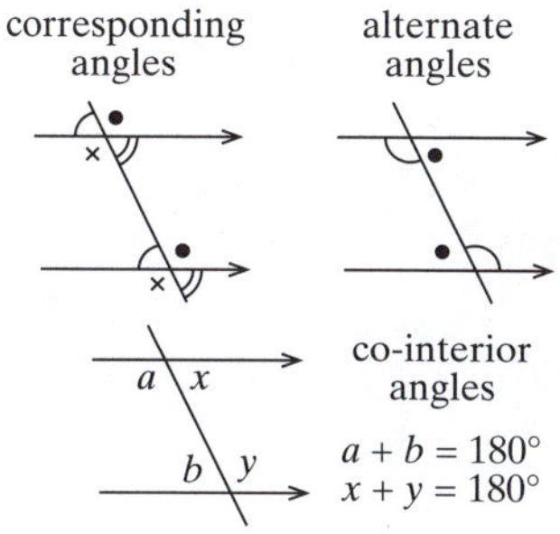

4 When a pair of lines is cut by a transversal, three types of pairs of angles are formed. **Corresponding angles** are in the same position in relation to each of the lines. **Alternate angles** are on alternate sides of the transversal, inside the pair of lines. **Co-interior angles** are inside the lines on the same side of the transversal. When the lines are parallel the corresponding angles are equal, the alternate angles are equal and the co-interior angles add to 180°.

5 The **angle sum of a triangle** is 180°. This means that for any triangle the angles always add up to 180°. The **angle sum of a quadrilateral** is 360°.

6 When a side of a triangle is extended an exterior angle is formed. The **exterior angle of a triangle** is equal to the sum of the interior opposite angles.

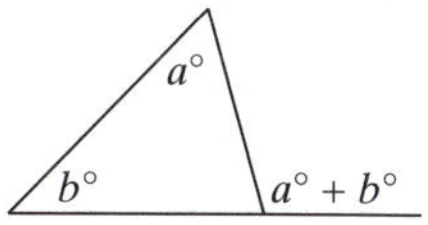

7 A **polygon** is a many-sided plane figure. Some polygons have special names. These include the pentagon (5 sides), hexagon (6 sides), heptagon (7 sides), octagon (8 sides), decagon (10 sides) and dodecagon (12 sides).

8 A **regular polygon** has all its sides equal and all its angles equal.

9 The **angle sum of a polygon** with n sides is $(n - 2) \times 180°$. By joining one vertex of a polygon to every other vertex (except the two it already adjoins), the polygon will be divided into a number of triangles. The number of triangles will always be 2 less than the number of sides of the polygon. Because the angle sum of a triangle is 180°, the angle sum of a polygon with n sides will be $(n - 2) \times 180°$.

10 A convex polygon is one where every angle is less than 180°. If a side of a convex polygon is extended an exterior angle is formed. The angle sum of the **exterior angles of a convex polygon** is 360°. (The sum of all the exterior angles that face the same way as you rotate around the polygon.)

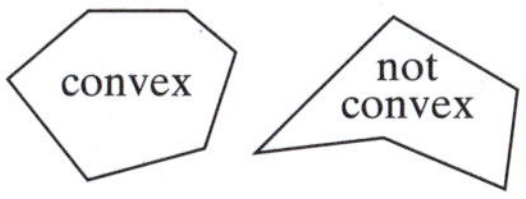

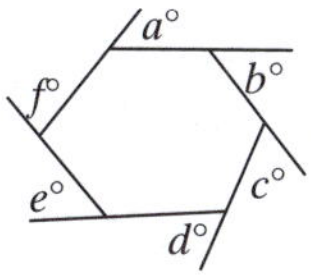

$a + b + c + d + e + f = 360$

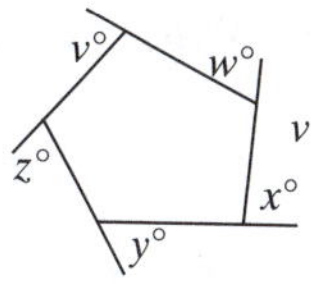

$v + w + x + y + z = 360$

Checklist

Can you:

1 *Correctly name and identify angles?* ☐

2 *Find the size of angles in a right angle, straight angle or complete revolution?* ☐

3 *Understand what is meant by vertically opposite, corresponding, alternate and co-interior angles?* ☐

4 *Use the angle sum of a triangle and angle sum of a quadrilateral?* ☐

5 *Understand what is meant by the exterior angle of a triangle and use it to find angles?* ☐

6 *Understand what is meant by a polygon and a regular polygon?* ☐

7 *Know the names of the common polygons and their number of sides?* ☐

8 *Find the angle sum of a polygon and the sum of the exterior angles?* ☐

ANGLE PROPERTIES

Geometric Reasoning

SKILLS CHECK

1 Find the value of x. (Give reasons.)

a
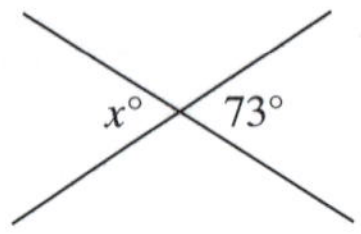

b
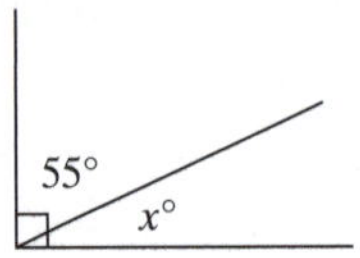

c
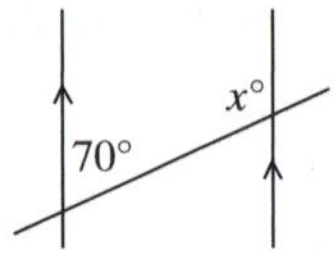

d
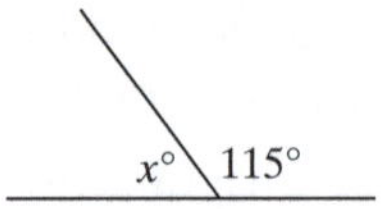

e
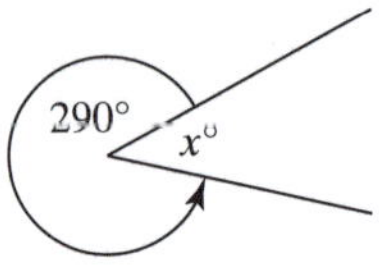

f
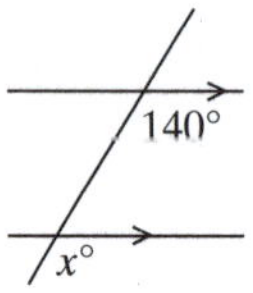

g
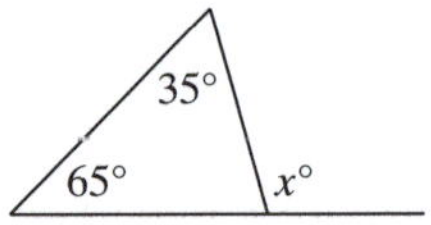

h
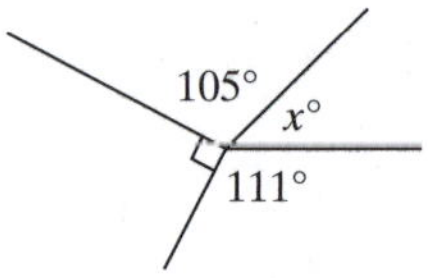

i
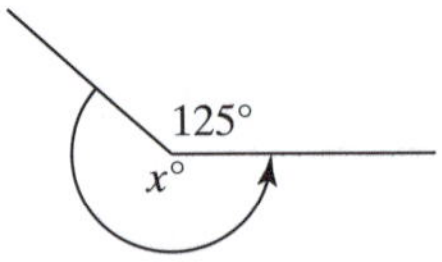

j
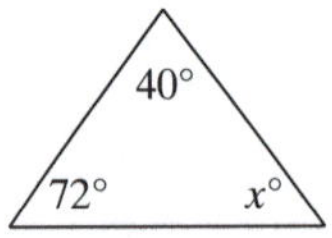

k
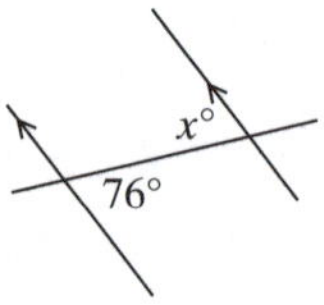

l
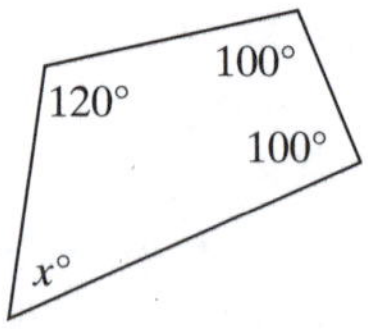

2 Find the value of x:

a
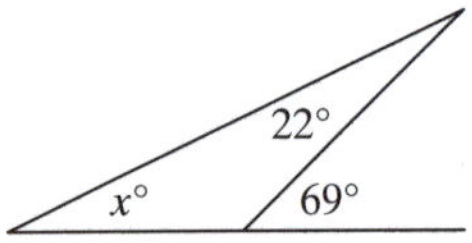

b
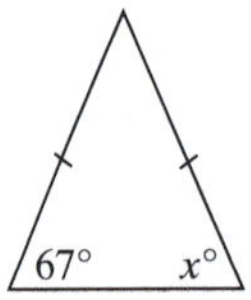

c
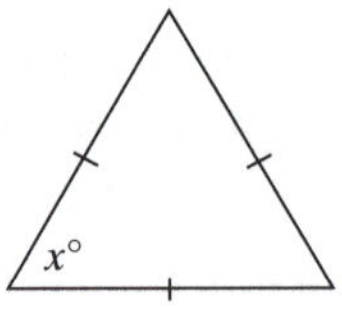

d
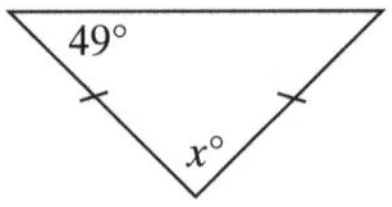

e
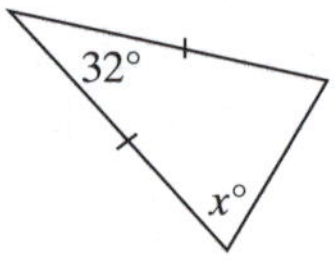

3 Find the value of x:

a
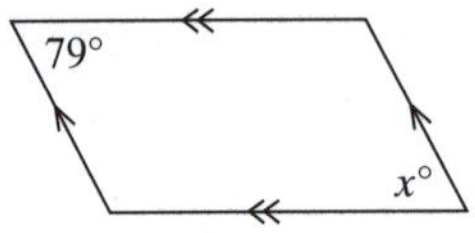

b
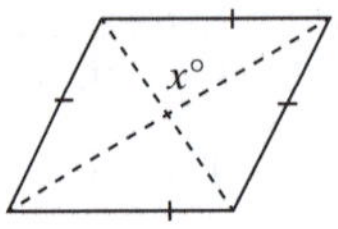

c
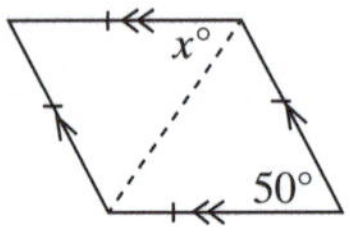

d
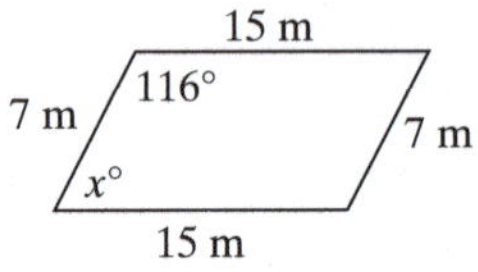

e
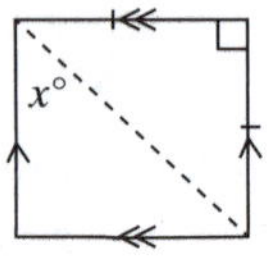

4 What name is given to a polygon with:

a 6 sides? **b** 10 sides? **c** 12 sides?

5 How many sides has the given polygon?

a pentagon **b** octagon **c** heptagon

6 This polygon has been divided into 5 triangles. What is its angle sum?

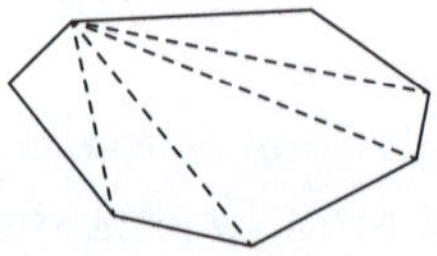

PAGES 143–144

Answers **1 a** 73 **b** 35 **c** 110 **d** 65 **e** 70 **f** 140 **g** 100 **h** 54 **i** 235 **j** 68 **k** 76 **l** 40 **2 a** 47 **b** 67 **c** 60 **d** 82 **e** 74 **3 a** 79 **b** 90 **c** 65 **d** 64 **e** 45 **4 a** hexagon **b** decagon **c** dodecagon **5 a** 5 **b** 8 **c** 7 **6** 900°

ANGLE PROPERTIES
Geometric Reasoning

25 MINUTES
INTERMEDIATE TEST

Part A Multiple Choice

1 $x = ?$

A 54   B 63
C 72   D 84

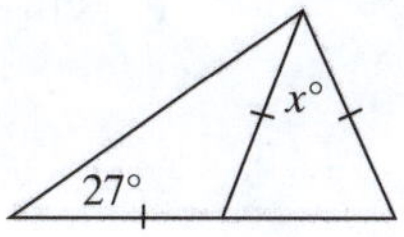

(1 mark)

2 $x = ?$

A 20   B 30
C 40   D 50

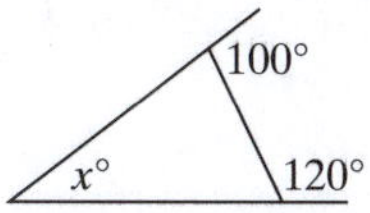

(1 mark)

3 $x = ?$

A 123   B 131
C 141   D 143

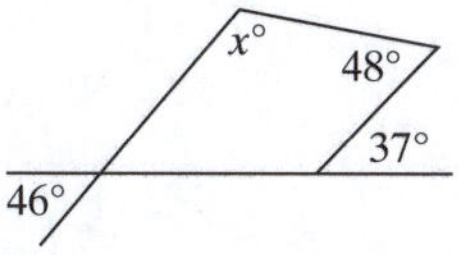

(1 mark)

4 B is a point on AC. $BCDE$ is a parallelogram. $AB = DC$.
$\angle DEB = 108°$. What is the value of x? *Hint 1*

A 36   B 54
C 63   D 72

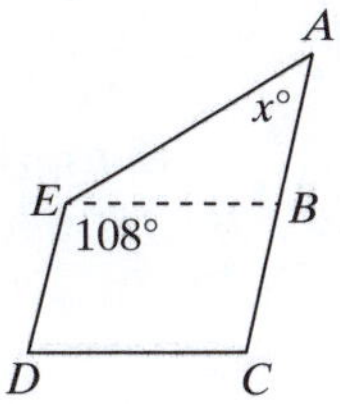

(1 mark)

Part B Short Answer

5 Find the value of the given pronumeral, giving reasons:

a x *Hint 2* (1 mark)
b y (1 mark)
c z (1 mark)

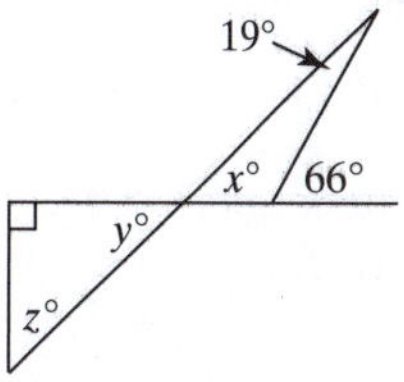

6 $ABCDE$ is a pentagon.
$\angle BCD = \angle CDE = 90°$.
$\angle ABC = \angle AED = 125°$.

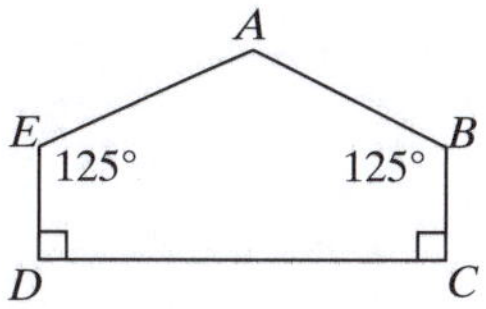

a What is the angle sum of a pentagon? (2 marks)
b What is the size of $\angle EAB$? *Hint 3* (2 marks)

7 Find the size of each interior angle of a regular octagon. (2 marks)

8 a Find the size of each exterior angle of a regular decagon. (2 marks)
b Use the answer to part **a** to find the size of each interior angle of a regular decagon. *Hint 4* (2 marks)
c Use the answer to part **b** to find the angle sum of any decagon. *Hint 5* (1 mark)

9 $ABCD$ is a parallelogram.
$\angle ABC = 72°$. $CB = CD$.
$ADEF$ is a square.
Find the size of $\angle EFB$. *Hint 6* (3 marks)

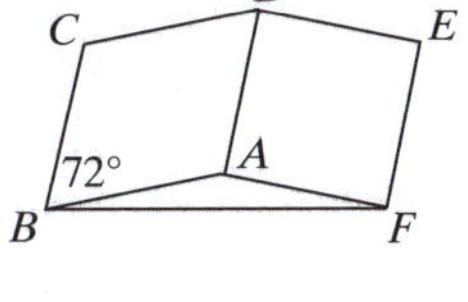

10 $PQ = PR = RS$ $\quad \angle QPR = 36°$

a Show that $\angle PSR = \angle QPR$. *Hint 7* (2 marks)
b What type of triangle is ΔPSQ? Justify your answer. (2 marks)

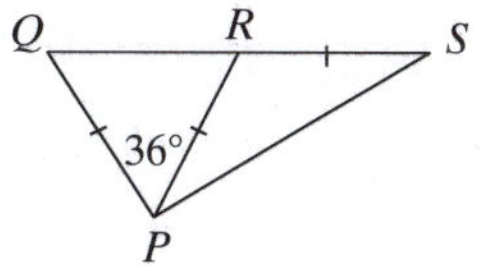

Hint 1: Mark all the given information on your diagram.
Hint 2: Use the exterior angle.
Hint 3: The angles must add to the answer in part **a**.
Hint 4: Angles in a straight line.
Hint 5: How many angles does a decagon have?
Hint 6: What type of quadrilateral is *ABCD*? Show all the equal sides on your diagram.
Hint 7: Find the size of $\angle PSR$.

Your Feedback

$\frac{\square}{25} \times 100\% = \square\ \%$

PAGES 144–145
PAGE 176

ANGLE PROPERTIES
Geometric Reasoning

30 MINUTES
ADVANCED TEST

1 Find the value of x in each diagram, giving reasons.

a
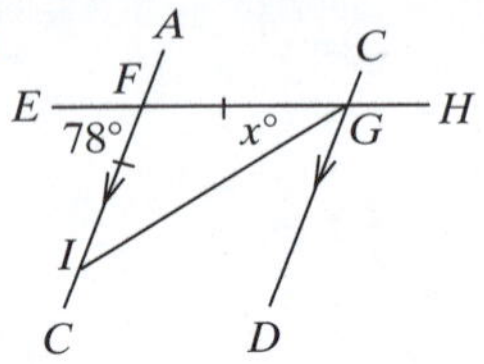

b
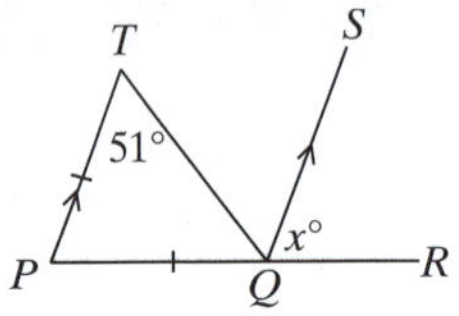

c
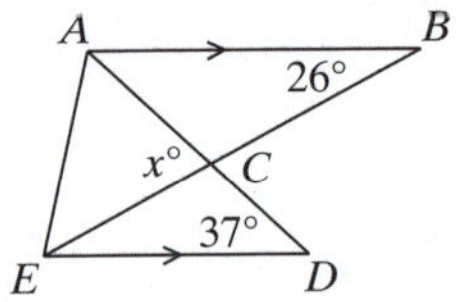

d
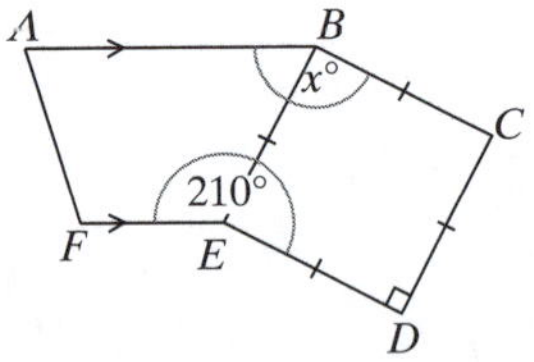

(2 marks each)

2 $ABCD$ is a trapezium with $AB \parallel DC$. $AB = BC$.

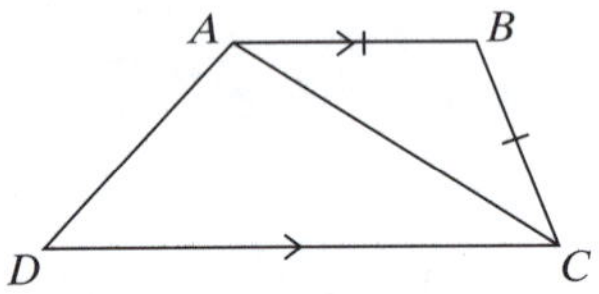

Show that AC bisects $\angle BCD$. (1 mark)

3 ΔADC is right-angled at D. $\angle BDA = \angle ACD$

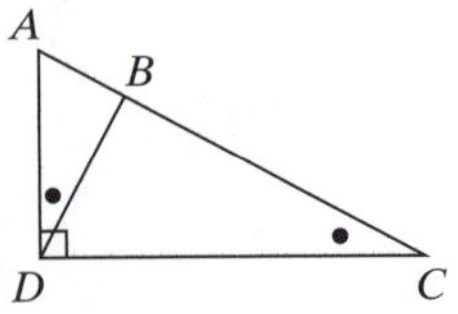

Show that BD is perpendicular to AC. (2 marks)

4 $\angle RST = \angle RTQ$

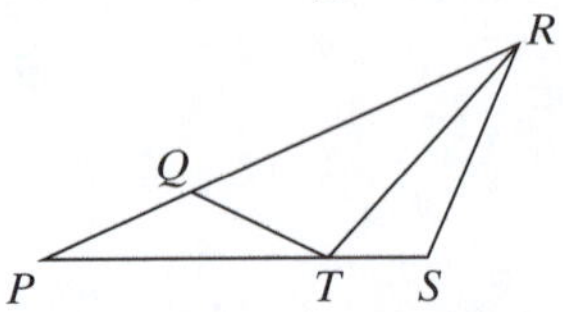

Show that $\angle QTP = \angle SRT$. (2 marks)

5 $ABCDEF$ is a regular hexagon. BC and ED both produced meet at X.

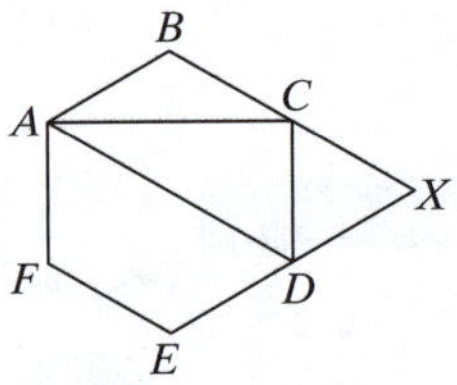

a What is the size of $\angle ABC$? (1 mark)

b Show that $\angle ACD = 90°$. (2 marks)

c Show that C is the midpoint of BX. (2 marks)

6 $AB = BD = DC$

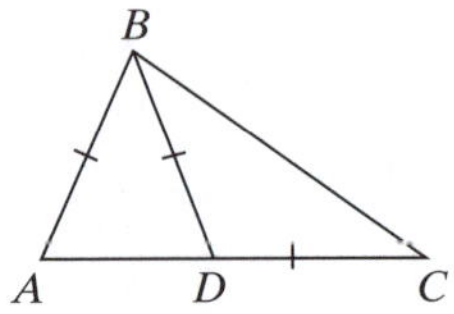

Show that $\angle BAD = 2\angle BCA$. (2 marks)

7 $PQ = QS$ and $QT = QR$
$\angle RTS = 45°$. Let $\angle QPS = \theta$.

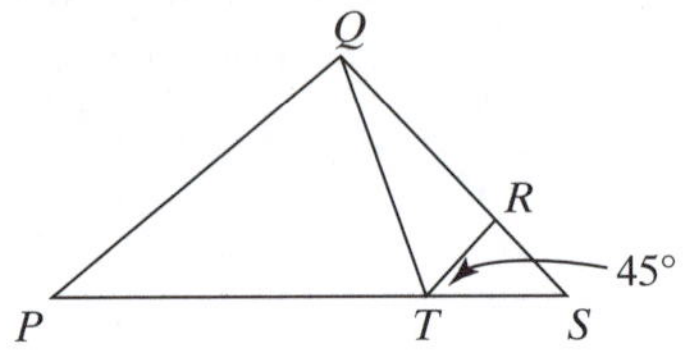

a Find $\angle QRT$ in terms of θ. (1 mark)

b Find $\angle TQR$ in terms of θ. (2 marks)

c Show that ΔPQT is right-angled. (1 mark)

8 $ABCDE$ is a pentagon with all its sides equal. $AE = EB$. $\angle EDC = 90°$. X is a point on BC such that XE bisects $\angle AED$.

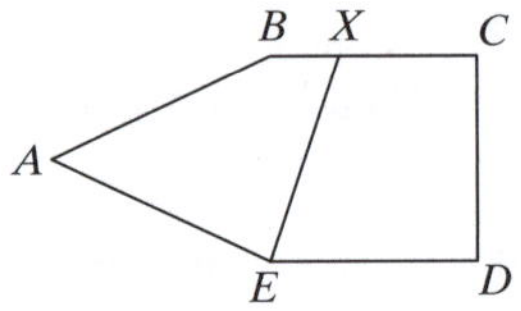

a Emil said that $ABCDE$ is a regular pentagon as all the sides are equal. Briefly explain why Emil is wrong. (1 mark)

b What is the size of $\angle BCD$? Justify your answer. (1 mark)

c What is the size of $\angle EAB$? Justify your answer. (1 mark)

d What is the size of $\angle AEX$? (1 mark)

e Y is a point on AB. $\angle YEB = \angle XEB$. What is the size of $\angle EYB$? (2 marks)

Your Feedback

$\frac{\square}{30} \times 100\% = \square\%$

PAGES 145–146

PAGE 176

PROOFS INVOLVING CONGRUENT TRIANGLES

Geometric Reasoning

STUDY NOTES

1 **Congruent triangles** are exactly the same shape and exactly the same size. The symbol ≡ means 'is congruent to'.

2 There are **four tests** for **congruent triangles:**
- three sides of one triangle are respectively equal to three sides of the other triangle **SSS**
- two sides and the included angle of one triangle are respectively equal to two sides and the included angle of another triangle **SAS**
- two angles and one side of one triangle are respectively equal to two angles and the corresponding side of the other triangle **AAS**
- the hypotenuse and one side of one right-angled triangle are respectively equal to the hypotenuse and one side of another right-angled triangle **RHS**.

3 If two triangles are congruent, then the **corresponding angles are equal** and the **corresponding sides are equal**. Note: we use the term corresponding to mean the matching angles. The angles in the same position in each figure (when they face the same way). This is also what we mean when we talk about corresponding angles on parallel lines. By talking about corresponding angles in triangles we don't mean that there are any parallel lines.

4 A **scalene triangle** has no sides equal in length. All of the angles are different sizes. An **isosceles triangle** has two sides equal. The angles opposite the equal sides are equal. An **equilateral** triangle has all three sides equal. Each angle measures 60°.

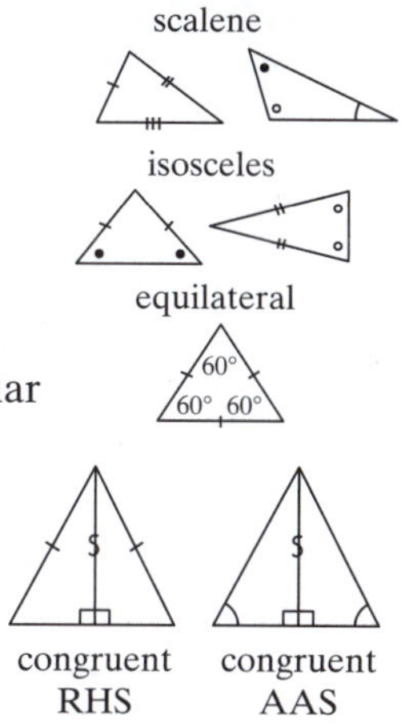

5 There are two **theorems** concerning isosceles triangles:
- if two sides of a triangle are equal the angles opposite the equal sides are equal
- if two angles of a triangle are equal the sides opposite the equal angles are equal.

These theorems can easily be proven. For example, by constructing the perpendicular from the top of the triangle to the base, the two resulting triangles can be proven congruent and hence the corresponding sides and angles are equal.

6 A quadrilateral is a **four-sided figure**. The following **definitions** give the minimum amount of information to identify certain special quadrilaterals.
- A **trapezium** is a quadrilateral with at least one pair of opposite sides parallel.
- A **parallelogram** is a quadrilateral with both pairs of opposite sides parallel.
- A **rhombus** is a parallelogram with two adjacent sides equal.
- A **rectangle** is a parallelogram with one angle a right angle.
- A **square** is a rectangle with two adjacent sides equal.

7 Applying congruence tests allows us to prove results about the **special quadrilaterals**, e.g. $ABCD$ is a rectangle. By proving triangles ADC and BCD congruent we can prove that the diagonals are equal.

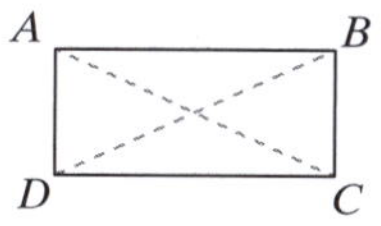

Checklist

Can you:

1 *Understand what is meant by congruent triangles?* ☐
2 *Know the four tests for congruent triangles?* ☐
3 *Prove two triangles congruent?* ☐
4 *Understand that if two triangles are congruent the matching sides and matching angles are equal?* ☐
5 *Understand and prove results about triangles and special quadrilaterals?* ☐
6 *Solve geometric problems?* ☐

PROOFS INVOLVING CONGRUENT TRIANGLES

Geometric Reasoning

SKILLS CHECK

1 **a** Which test shows these two triangles are congruent?

b Complete: $\Delta ABC \equiv \Delta$______

c Which angle corresponds to:

i $\angle ABC$? **ii** $\angle CAB$? **iii** $\angle BCA$?

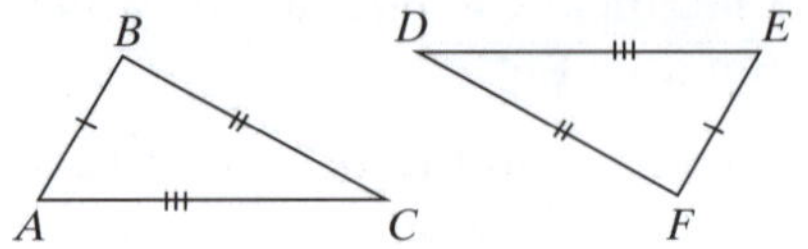

2 $\Delta ABC \equiv \Delta DFE$

a Which test shows that the triangles are congruent?

b Which side corresponds to AC?

c Which side corresponds to AB?

d Which angle corresponds to $\angle BCA$?

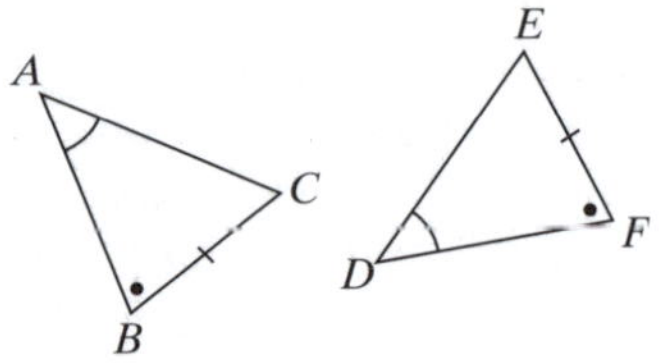

3 Determine whether the given pair of triangles is necessarily congruent. If they are congruent which test shows this?

a ΔABC and ΔEDF

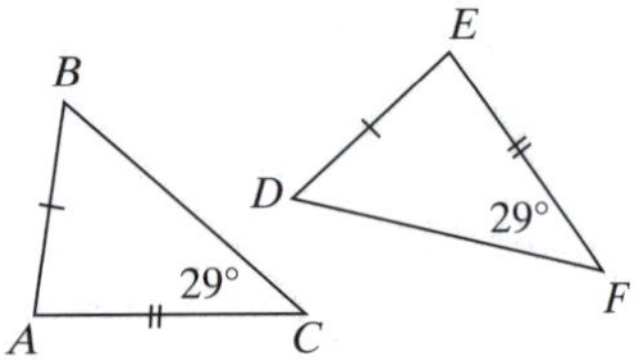

b ΔGHI and ΔGJI

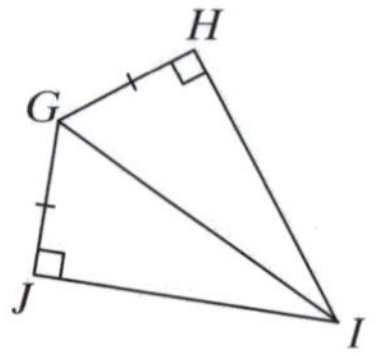

c ΔKOL and ΔNML (KM and NO are straight lines)

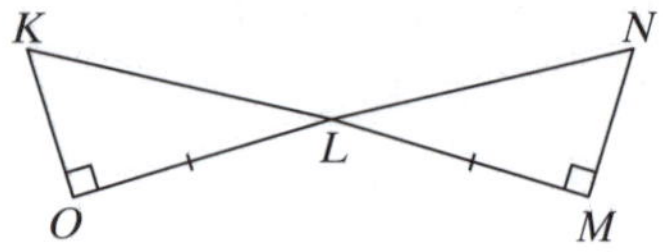

4 What type of triangle has:

a all sides equal in length? **b** no sides equal in length? **c** two sides equal in length?

5 Complete:

a The sum of the interior angles of a triangle is ______.

b Each angle of an equilateral triangle is equal to ______.

c The exterior angle of a triangle is equal to ______.

d If two sides of a triangle are equal then the angles opposite those sides are ______.

e If two angles of a triangle are equal then the sides opposite those angles are ______.

6 Find the value of x:

a

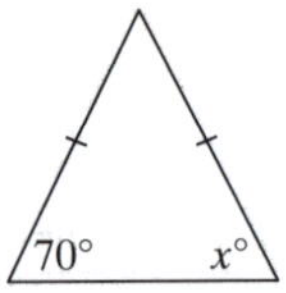

b

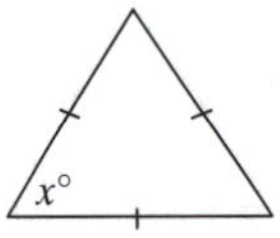

c

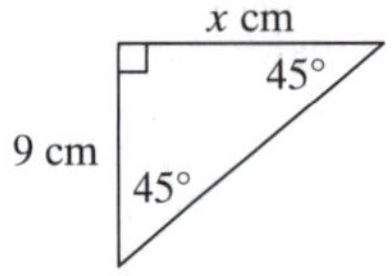

d

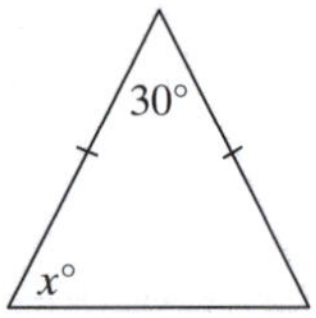

7 Write down the name of the special quadrilateral given by each definition.

a A parallelogram with 2 adjacent sides equal.

b A rectangle with 2 adjacent sides equal.

c A quadrilateral with at least one pair of opposite sides parallel.

d A parallelogram with one angle a right angle.

e A quadrilateral with both pairs of opposite sides parallel.

PAGE 147

Answers **1 a** SSS **b** EFD **c i** $\angle EFD$ **ii** $\angle DEF$ **iii** $\angle FDE$ **2 a** AAS **b** DE **c** DF **d** $\angle FED$ **3 a** not necessarily congruent **b** RHS **c** AAS **4 a** equilateral **b** scalene **c** isosceles **5 a** 180° **b** 60° **c** the sum of the interior opposite angles **d** equal **e** equal **6 a** 70 **b** 60 **c** 9 **d** 75 **7 a** rhombus **b** square **c** trapezium **d** rectangle **e** parallelogram

PROOFS INVOLVING CONGRUENT TRIANGLES

Geometric Reasoning

25 MINUTES

INTERMEDIATE TEST

Part A Multiple Choice

1 $\Delta ABC \equiv \Delta FDE$.
What is the size of $\angle FED$?

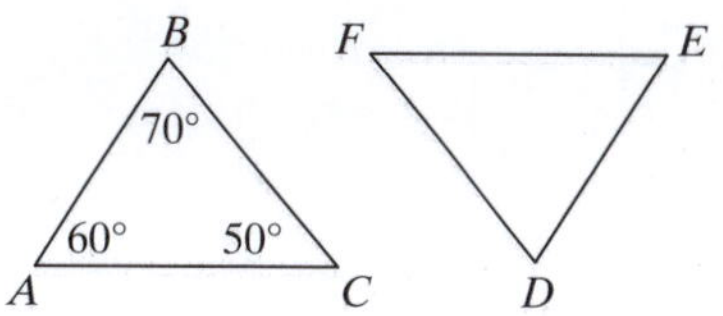

A 50° **B** 60° **C** 70°
D There is not enough information to determine the size. (1 mark)

2 AB and CD are straight lines intersecting at E. $AB = CD$ and $AE = DE$. Which test would be used to prove the two triangles congruent?

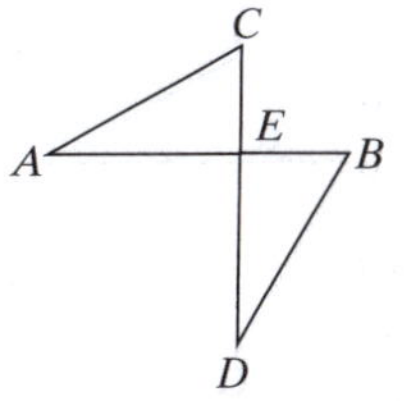

A AAS **B** SAS
C SSS **D** RHS (1 mark)

3 Which statement is correct?
I All similar triangles are congruent. II All congruent triangles are similar.
A I only **B** II only **C** both I and II **D** neither I nor II (1 mark)

Part B Short Answer

4 $ABCD$ is a quadrilateral with AB parallel to DC and AD parallel to BC. AC is a diagonal of the quadrilateral.

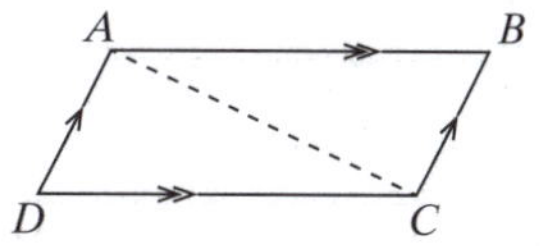

a What type of quadrilateral is $ABCD$? (1 mark)
b Why does $\angle BAC = \angle DCA$? (1 mark)
c Why does $\angle BCA = \angle DAC$? (1 mark)
d Why does $AC = CA$? (1 mark)
e Which test proves triangle ABC congruent to triangle CDA? *Hint 1* (1 mark)
f What does this prove about angles ABC and CDA? Justify your answer. (2 marks)
g What property have you just proven? *Hint 2* (1 mark)

5 In ΔABC, $AB = AC$. P is a point on BC such that AP bisects $\angle BAC$.

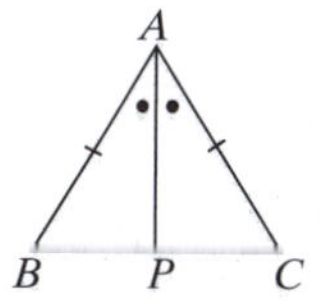

a By proving that ΔABP is congruent to ΔACP, show that $\angle ABP = \angle ACP$. *Hint 3* (3 marks)
b What property have you proven? (1 mark)

6 $PQRS$ is a quadrilateral in which the diagonals PR and QS meet at T. $PQ = PS$ and PR bisects $\angle QPS$.

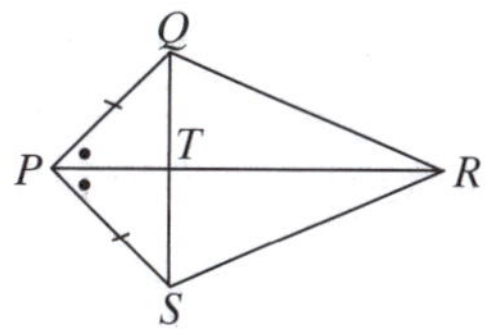

a Prove that $\Delta PQT \equiv \Delta PST$. *Hint 4* (3 marks)
b Prove that PR and QS meet at right angles. (2 marks)

7 S is the midpoint of TR.

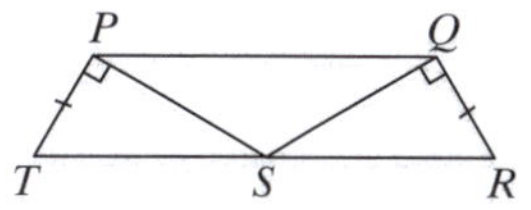

a Show that triangles TPS and RQS are congruent. *Hint 5* (3 marks)
b What type of triangle is ΔPQS? Justify your answer. *Hint 6* (2 marks)

Hint 1: There is no need for a formal proof. Use the answers to parts **b**, **c** and **d**.
Hint 2: What can we say about angles *ABC* and *CDA* in terms of the parallelogram?
Hint 3: You are trying to prove the base angles equal. You cannot use the fact that they are equal as one of the reasons why the triangles are congruent.
Hint 4: *PT* is a common side.
Hint 5: If *S* is the midpoint of *TR*, what does this mean about *TS* and *RS*?
Hint 6: What information do we know about the sides *PS* and *QS* of the triangle?

Your Feedback

$\frac{\square}{25} \times 100\% = \square\%$

PAGE 148
PAGE 176

PROOFS INVOLVING CONGRUENT TRIANGLES

Geometric Reasoning

30 MINUTES

ADVANCED TEST

1 $BD = FD$; $\angle BCF = \angle BEF$

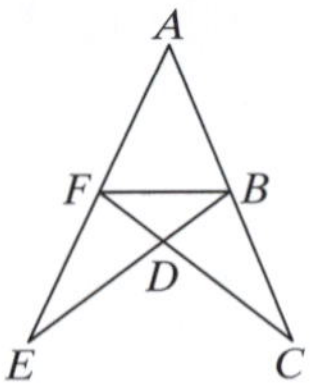

Show that $\Delta FBE \equiv \Delta BFC$. (2 marks)

2 ΔAEC is right-angled at E.
BD is perpendicular to EC.
D is the midpoint of EC.

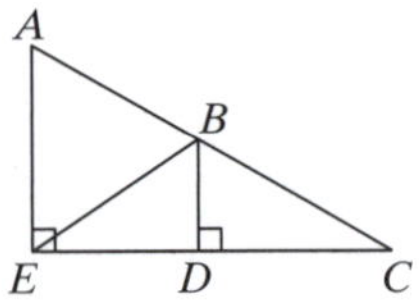

a Show that $\Delta EBD \equiv \Delta CBD$. (2 marks)
b Show that ΔABE is isosceles. (2 marks)

3 AB and DC intersect at E.
$AE = BE$ and $DE = CE$

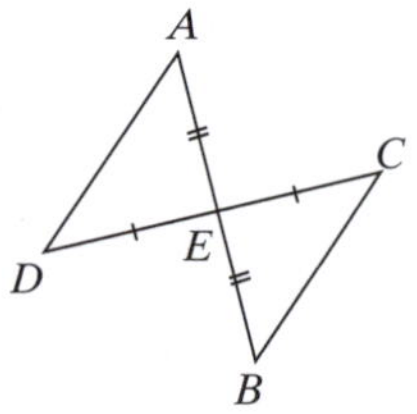

a Prove that $\Delta AED \equiv \Delta BEC$. (2 marks)
b Prove that AD is parallel to CB. (1 mark)

4 $PQRS$ is a quadrilateral.
$PQ = PS$ and $QR = SR$

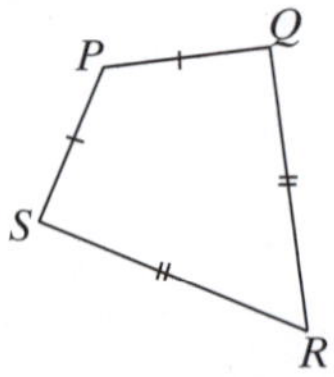

a Show that $\Delta PQR \equiv \Delta PSR$. (2 marks)
b Show that diagonal PR bisects the angles of the quadrilateral through which it passes. (2 marks)

5 $PQRST$ is a regular pentagon.
X is the foot of the perpendicular from S to TR.

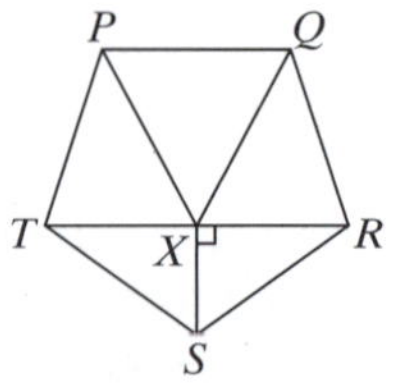

a Prove that triangles TXS and RXS are congruent. (2 marks)
b Find the size of $\angle TSX$. (2 marks)
c Show that $\angle PTX = \angle QRX$. (1 mark)
d Prove that $\Delta PTX \equiv \Delta QRX$. (2 marks)
e Show that ΔPQX is isosceles. (1 mark)
f Show that PQ is parallel to TR. (2 marks)

6 $ABCD$ is a trapezium with AB parallel to DC.
$\angle ABC = \angle ADC$

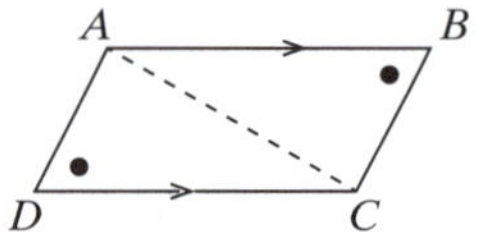

a Prove that $\Delta ABC \equiv \Delta CDA$. (2 marks)
b Prove that AD is parallel to BC. (1 mark)

7 $PQ = RS = 70.9$ m
$RQ = 40.1$ m
$QT = 25.9$ m
$RT \perp PS$

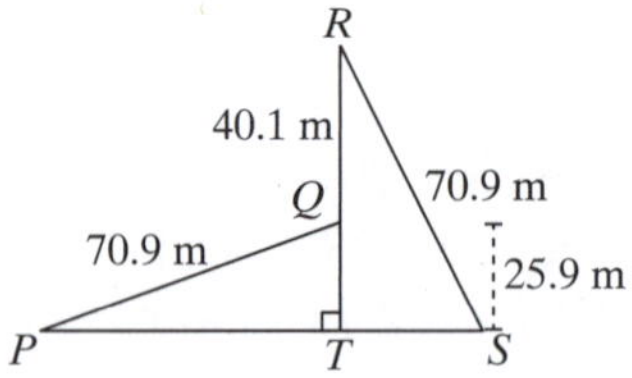

Use congruent triangles to show that $\angle TPQ = \angle TRS$. (4 marks)

Your Feedback

$\frac{\square}{30} \times 100\% = \square\%$

PAGES 148–149
PAGE 176

LOGICAL REASONING

Geometric Reasoning

STUDY NOTES

1 All four-sided figures are quadrilaterals. Don't forget that many of the **special quadrilaterals** fit into more than one category. If a quadrilateral has one pair of opposite sides parallel it is a trapezium, so every parallelogram, rhombus, rectangle or square is also a trapezium because they all have at least one pair of opposite sides parallel. All rhombuses are parallelograms and all rectangles are parallelograms. A square is both a rectangle and a rhombus (and a parallelogram and a trapezium).

2 **Similar figures** are exactly the same shape but not necessarily the same size. The symbol ||| means 'is similar to'. Similar figures are related by a **scale factor**. If two triangles are related by a scale factor of 2, then every side of the larger triangle is twice as long as the corresponding side of the smaller triangle. If the second figure is larger than the first, it is an **enlargement** of the first. If the second figure is smaller than the first it is a **reduction**.

3 In any pair of similar triangles, the **corresponding angles are equal** and the **corresponding sides are in the same ratio**. There are four tests for similar triangles:

- If the **three sides of one triangle are proportional to the three sides of another triangle** then the two triangles are similar.
- If **two sides of one triangle are proportional to two sides of another triangle and the included angles are equal**, then the triangles are similar.
- If **two angles of one triangle are equal to two angles of another triangle** then the triangles are similar.
- If **the hypotenuse and one side of a right-angled triangle are proportional to the hypotenuse and a side of another right-angled triangle** then the two triangles are similar.

4 If two triangles are similar, we can use the **proportions** to find the size of an unknown side.

5 If two similar figures are related by a scale factor of one, they are **congruent**. Every pair of congruent figures is also similar. The four tests for congruent triangles (SSS, SAS, AAS and RHS), are special cases of the four tests for similar triangles.

6 When naming two similar figures, or two congruent figures, the **vertices should be listed in the same order**, matching corresponding angles. If $\Delta ABC \ ||| \ \Delta DEF$, for example, then $\angle A = \angle D$, $\angle B = \angle E$ and $\angle C = \angle F$.

7 When **proving two triangles congruent**, name the triangles being considered, show the three pairs of equal sides or angles (giving reasons) and then show the conclusion. A similar process is used to prove triangles similar.

8 If **solving a geometric problem**, draw a large clear **diagram** and mark on it all the given information. Make sure that you are **logical** in your approach. Take time to stop and think and **plan** how the proof will go before starting to write things down. If you become stuck or are unsure how to continue, check that you have used every piece of **given information**. Many proofs follow from pairs of triangles being congruent or similar, so remember to look for those. Remember, if triangles are congruent then the corresponding sides are equal and the corresponding angles are equal. If the triangles are similar then the corresponding sides are in the same ratio and the corresponding angles are equal.

Checklist

Can you:

1 *Identify the special quadrilaterals and know how they are related?* ☐
2 *Understand what it means for two figures to be similar?* ☐
3 *Know the four tests for similar triangles?* ☐
4 *Understand what it means for two figures to be congruent?* ☐
5 *Correctly name pairs of similar figures and pairs of congruent figures?* ☐
6 *Know how to write a formal proof involving congruent or similar triangles?* ☐
7 *Solve geometrical problems?* ☐

LOGICAL REASONING

Geometric Reasoning

SKILLS CHECK

1 Complete, by naming the vertices in their correct order.

a $\Delta ABC \equiv \Delta$ ______

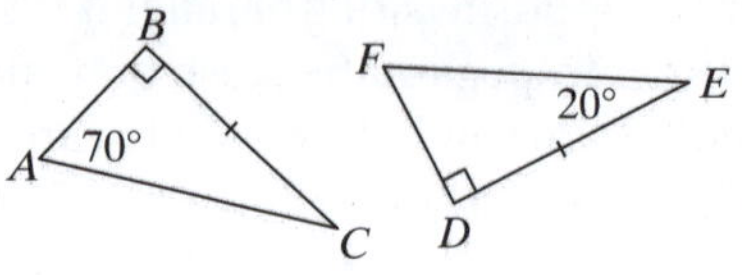

b $\Delta XYZ \,|||\, \Delta$ ______

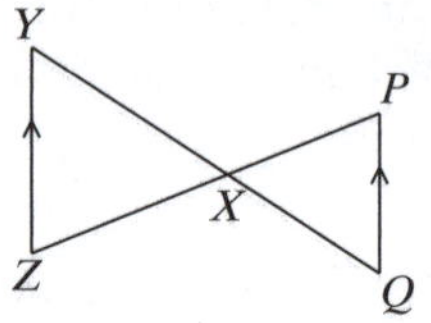

2 Determine whether the pair of triangles is congruent, similar (but not congruent), or neither congruent nor similar. If the triangles are congruent or similar, briefly explain why this is so.

a

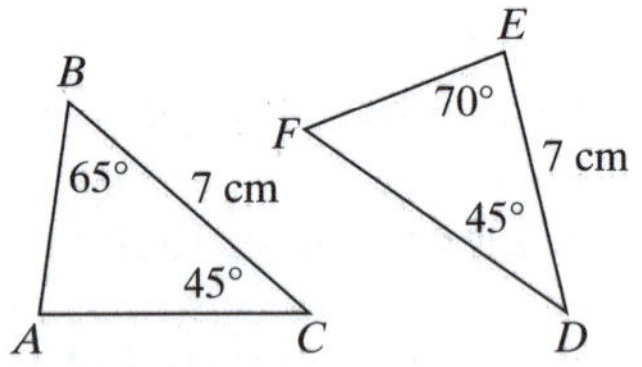

b

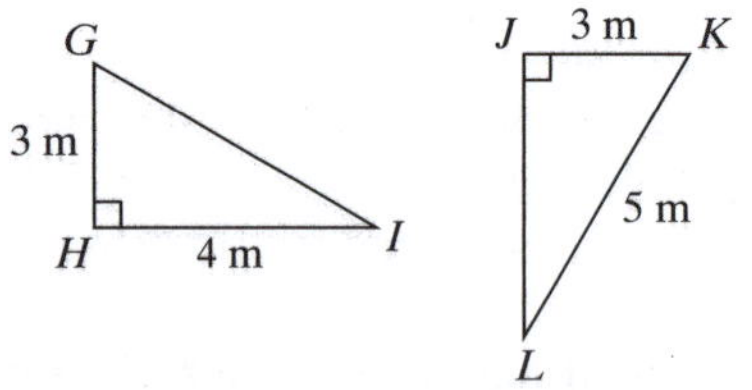

c

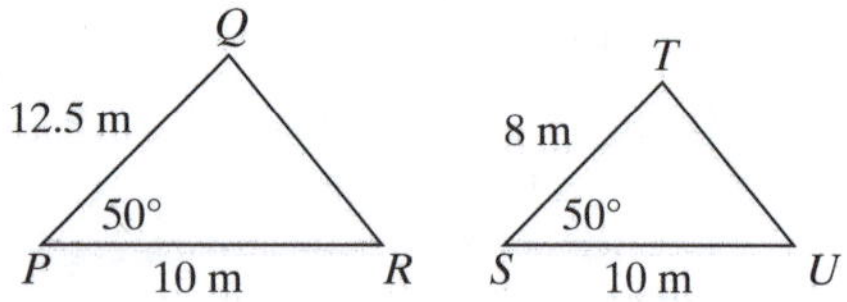

3 Find the value of each pronumeral:

a

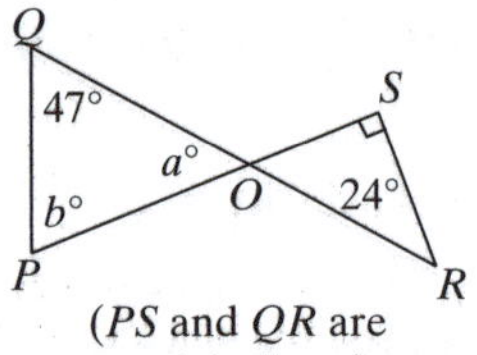

(*PS* and *QR* are straight lines.)

b

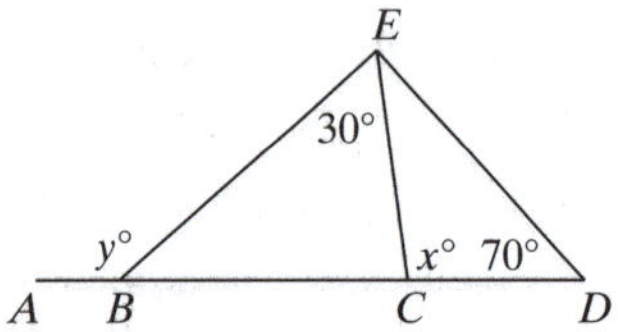

$BE = DE$
(*AD* is a straight line.)

c

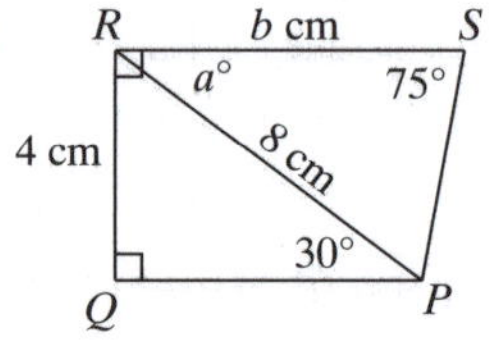

d

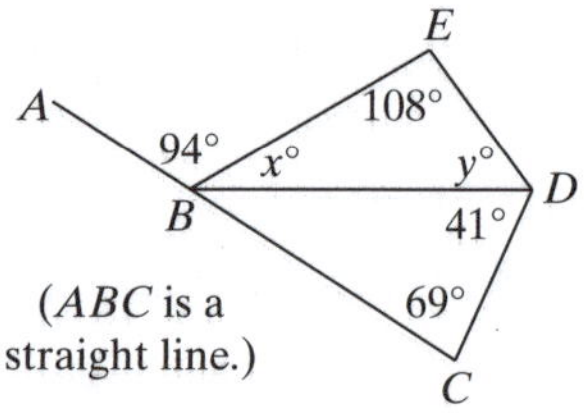

(*ABC* is a straight line.)

4 Determine whether the given statement is true or false.

a Every square is a rhombus.

b Every rhombus is a square.

c Every rectangle is a parallelogram.

d Every rhombus is a parallelogram.

e Every parallelogram is a trapezium.

f Every trapezium is a parallelogram.

g Every square is a parallelogram.

h Every square is a rectangle.

5 Find the value of x:

a

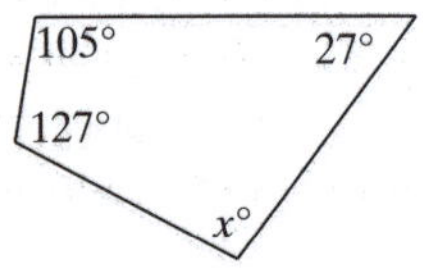

b

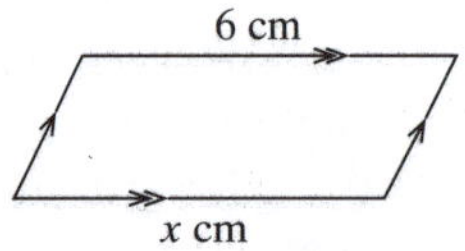

c

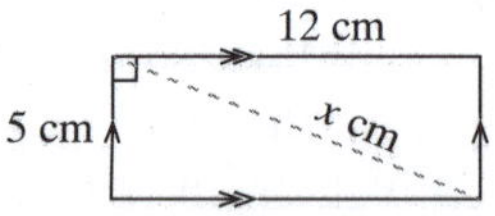

d

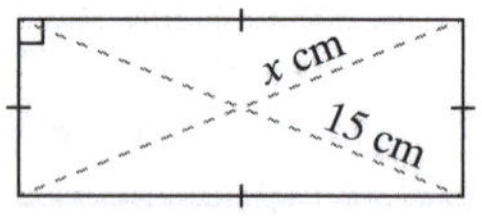

6 Find the value of each pronumeral:

a

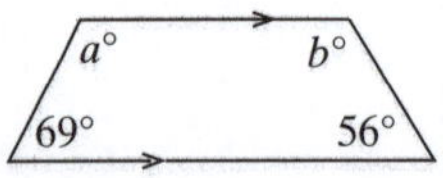

b

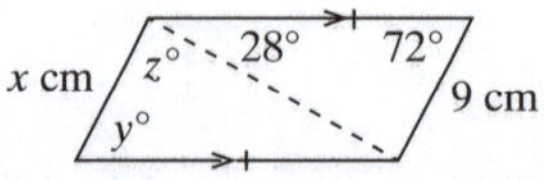

PAGES 149–150

Answers **1 a** *FDE* **b** *XQP* **2 a** similar **b** congruent **c** similar **3 a** $a = 66, b = 67$ **b** $x = 100, y = 110$ **c** $a = 30, b = 8$ **d** $x = 16, y = 56$ **4 a** T **b** F **c** T **d** T **e** T **f** F **g** T **h** T **5 a** 101 **b** 6 **c** 13 **d** 15 **6 a** $a = 111, b = 124$ **b** $x = 9, y = 72, z = 80$

LOGICAL REASONING

Geometric Reasoning

25 MINUTES

INTERMEDIATE TEST

Part A Multiple Choice

1 If $a° + b° = c°$ and $b° + c° = d°$ then $a° = ?$

A $d° - 2b°$ **B** $d° - 2c°$

C $2b° - d°$ **D** $2c° + d°$ (1 mark)

2 Which statement is correct?

A every trapezium is a parallelogram

B every rhombus is a trapezium

C every parallelogram is a rectangle

D every rectangle is a rhombus (1 mark)

3

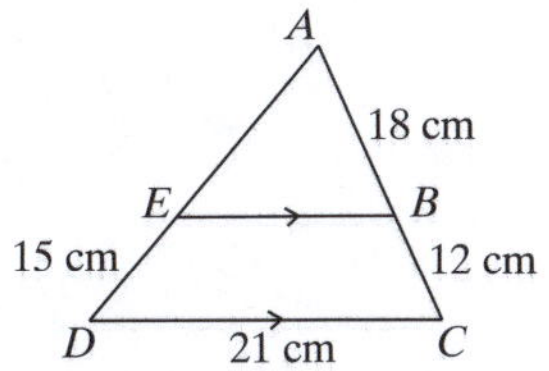

Which is correct?

A $EB = 12.6$ cm and $AE = 21$ cm

B $EB = 14$ cm and $AE = 21$ cm

C $EB = 12.6$ cm and $AE = 22.5$ cm

D $EB = 14$ cm and $AE = 22.5$ cm (1 mark)

Part B Short Answer

4 In the diagram, ABC is a straight line. BE is parallel to CD.

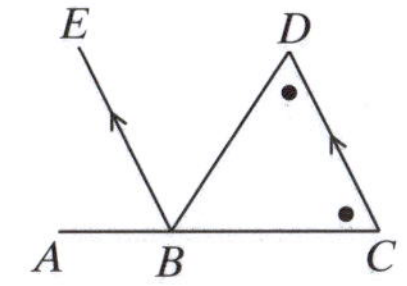

$\angle BDC = \angle BCD$. Prove that BE bisects $\angle ABD$. Hint 1 (2 marks)

5 $\angle BAC = 20°$, $\angle ACD = 123°$ and $\angle BDE = 77°$.

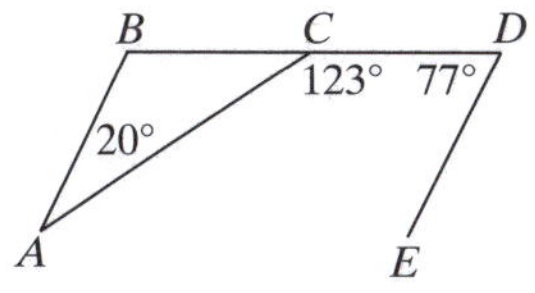

Show that AB is parallel to ED. Hint 2 (2 marks)

6 $PQRS$ is a rhombus. $PR = 42$ cm and $QS = 56$ cm.

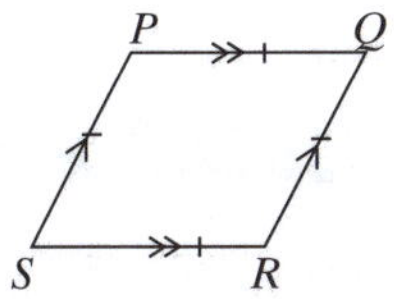

Find the perimeter of the rhombus. Hint 3 (3 marks)

7 AC is a straight line.
$AD = BD = DC$.
$\angle ADB = \angle CDB$.

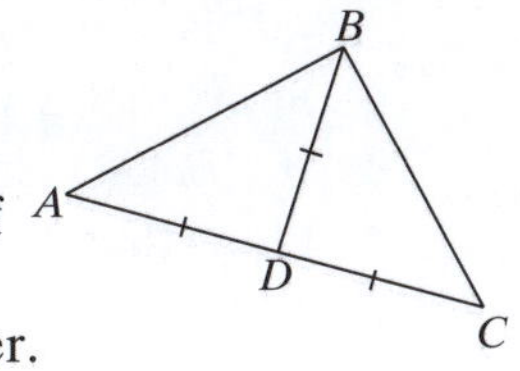

a What is the size of angle ADB? Justify your answer. Hint 4 (2 marks)

b What is the size of $\angle BAD$? Justify your answer. (2 marks)

c What special type of triangle is ΔABC? Justify your answer. Hint 5 (2 marks)

8 $VX = VW$, $VZ = ZX$, $YX = YW$, $\angle XVZ = 20°$.

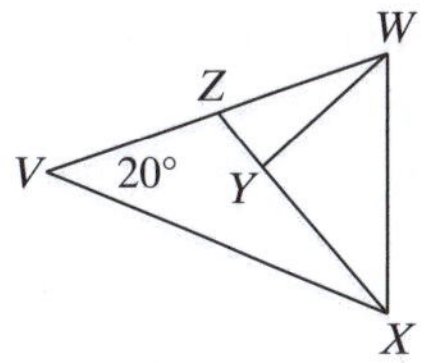

a Find the size of $\angle YXW$. (3 marks)

b Show that $YW = XW$. Hint 6 (2 marks)

9 $\angle WTU = \angle UVW = x°$. $\angle TUV = \angle VWT = y°$

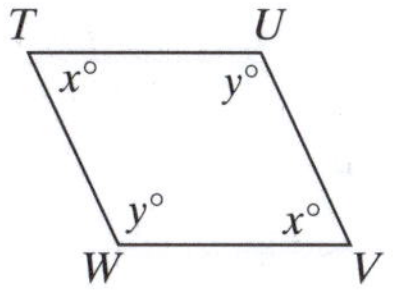

a Show that $x + y = 180$. Hint 7 (2 marks)

b Briefly explain why TU must be parallel to WV and why TW is parallel to UV. Hint 8 (2 marks)

Hint 1: Use alternate and corresponding angles.
Hint 2: Use co-interior angles.
Hint 3: The diagonals bisect each other at right angles. Form a right-angled triangle and find the hypotenuse.
Hint 4: You have been given all the necessary information. It should not be a complicated proof.
Hint 5: There are two special things about the triangle.
Hint 6: You know that $YW = YX$. What type of triangle are you looking for?
Hint 7: Use the sum of the angles of a quadrilateral.
Hint 8: What information do you have about the co-interior angles?

Your Feedback

$\frac{\square}{25} \times 100\% = \square\%$

PAGE 151
PAGE 176

LOGICAL REASONING

Geometric Reasoning

30 MINUTES

ADVANCED TEST

1 *ABCD* is a square. *X* is a point on diagonal *BD*.

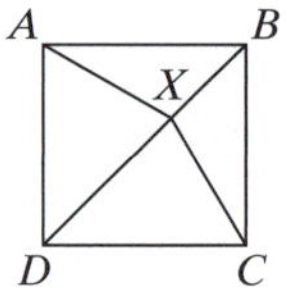

a Show that $\Delta ABX \equiv \Delta CBX$. (2 marks)

b If $\angle BAX = 22°30'$, find the size of $\angle DXC$. (2 marks)

c Show that ΔDXC is isosceles. (1 mark)

2 *ABCD* and *EFBD* are rectangles. $AB = 4$ cm and $BC = 3$ cm

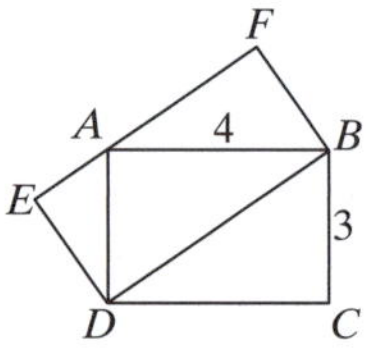

a Show that $\angle FAB = \angle CDB$. (1 mark)

b Show that $\Delta AFB \parallel\!\mid \Delta DCB$. (2 marks)

c Find the length of *FB* and hence show that both rectangles have the same area. (3 marks)

3 Prove that the diagonals of a rectangle are equal. (3 marks)

4 ΔACB is right-angled at *C*. $DE = CE$. $\angle ACD = \angle EDB$. Let $\angle DCE = \alpha$. Show that *E* is the midpoint of *BC*. (3 marks)

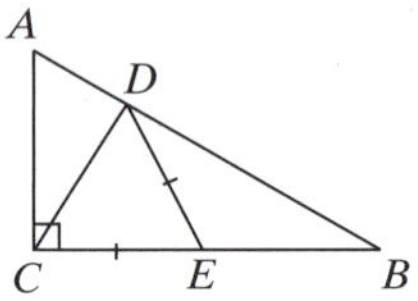

5 *B* is a point on *AC* and *D* is a point on *CE* such that $\angle ABE = \angle ADE = \alpha$. Show that $\angle AFE + \angle BCD = 2\alpha$. (2 marks)

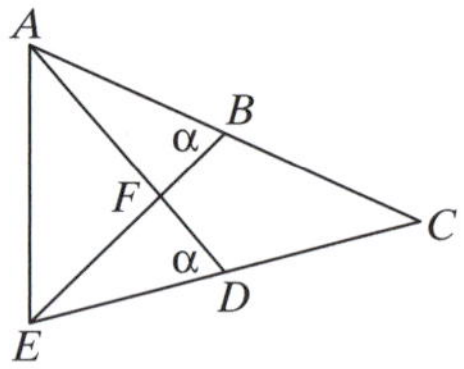

6 *ABCD* is a trapezium with $AD \parallel BC$, *X* is the midpoint of *AD*. $XB \perp BC$. *AC* and *BX* intersect at *Y*. Show that $\angle XYD = \angle BYC$. (3 marks)

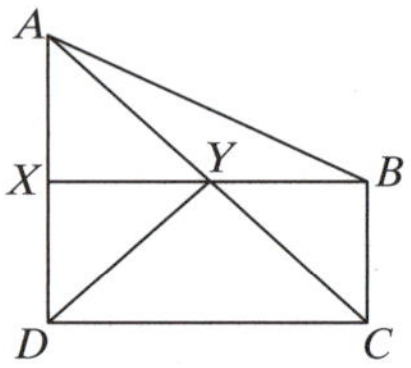

7

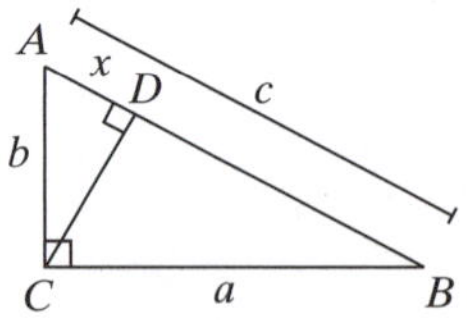

a Prove that $\Delta ADC \parallel\!\mid \Delta ACB$. (2 marks)

b Let $AB = c$, $AC = b$, $BC = a$ and $AD = x$. Show that $b^2 = cx$. (2 marks)

c Prove that $\Delta BDC \parallel\!\mid \Delta BCA$ (2 marks)

d Show that $\dfrac{a}{c} = \dfrac{c - x}{a}$. (1 mark)

e Show that $c^2 = a^2 + b^2$. (1 mark)

Your Feedback

$\dfrac{\square}{30} \times 100\% = \square\,\%$

PAGES 152–153

PAGE 176

ANGLES OF ELEVATION AND DEPRESSION, AND BEARINGS

Trigonometry

STUDY NOTES

1 The **ratios of the sides** of a right-angled triangle are the same whenever the angles are equal.
For example, for every angle of 30°, the opposite divided by the hypotenuse will always be 0.5.

$$\textbf{Tangent: } \frac{\text{opposite}}{\text{adjacent}} \qquad \textbf{Sine: } \frac{\text{opposite}}{\text{hypotenuse}} \qquad \textbf{Cosine: } \frac{\text{adjacent}}{\text{hypotenuse}}.$$

These ratios are abbreviated to tan, sin and cos.

2 An angle can be measured in whole degrees or in **degrees and minutes**. There are 60 minutes in a degree and 60 seconds in a minute. When rounding off to the nearest minute, the number must be rounded up if there are 30 or more seconds.

3 An **angle of elevation** is always measured upwards from the horizontal.

4 An **angle of depression** is always measured downwards from the horizontal.

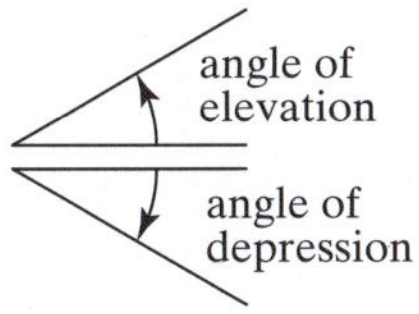

5 It may be necessary to use your knowledge of angles associated with **parallel lines** to find the size of required angles. When two parallel lines are cut by a transversal, the corresponding angles formed are equal, the alternate angles are equal and the co-interior angles add to 180°.

6 Make sure you are familiar with the sixteen points of the **compass**.
The four main points are North, East, South and West. The four points NE, SE, SW and NW are found between those main points.
NE is 'halfway' between North and East for example. The remaining eight directions are found between those points. For example, NNE is between north and north-east and WSW is between west and south-west.

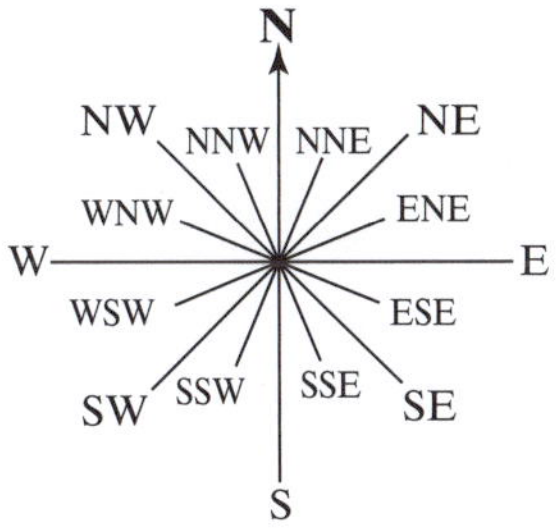

7 True **bearings** are measured from **north** in a **clockwise** direction. Three digits are always used to express a bearing.
For example:
The bearing of Q from P is 070°.
The bearing of A from B is 212°.
The bearing of Y from X is 293° (because 360° – 67° = 293° and bearings are measured in a clockwise direction)

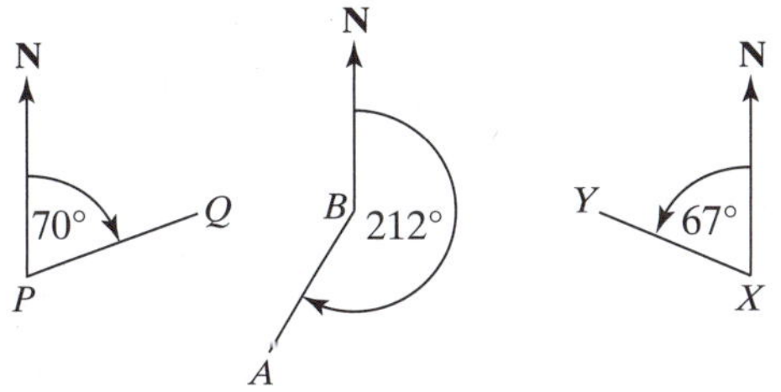

8 Always **draw a diagram** if one is not supplied with the question. Draw the diagram roughly to scale and use the actual (approximate) directions.

9 When **solving problems** involving trigonometry, place all the given information on your diagram and add any other freely available information. Use trigonometry to answer the question and give the answer as a sentence.

Checklist

Can you:

1 *Use trigonometry to find the length of a side or the size of an angle in a right-angled triangle?* ☐
2 *Understand and use degrees and minutes and round off to the nearest minute?* ☐
3 *Understand and use angles of elevation and depression?* ☐
4 *Understand and use the sixteen points of the compass?* ☐
5 *Understand and use bearings?* ☐
6 *Draw diagrams from given information?* ☐
7 *Solve problems involving right-angled triangles?* ☐

ANGLES OF ELEVATION AND DEPRESSION, AND BEARINGS

Trigonometry

SKILLS CHECK

1 Find, correct to four decimal places:

a $\sin 29°37'$ b $\tan 56°21'$ c $\cos 72°40'$

2 Find θ to the nearest minute if:

a $\cos\theta = 0.35$ b $\tan\theta = 1.7$ c $\sin\theta = 0.2916$

3 Find the value of x to one decimal place:

a
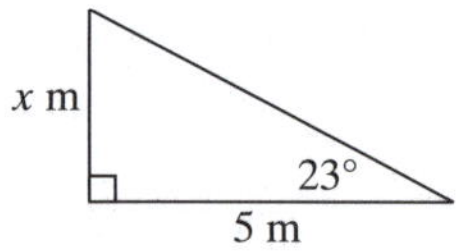

b
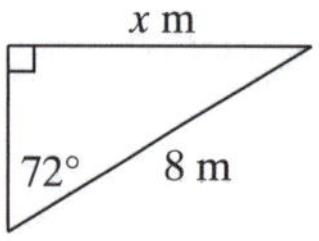

c
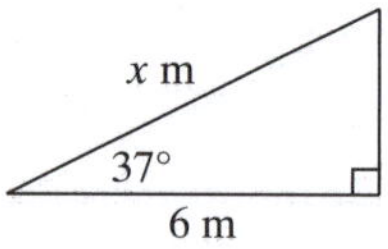

4 Find the value of θ, to the nearest degree:

a
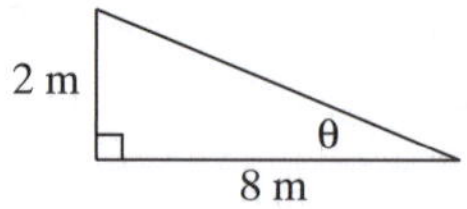

b
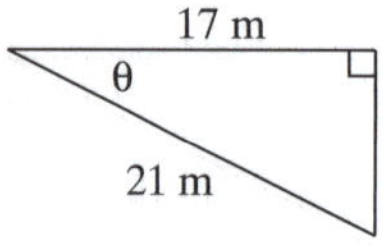

c
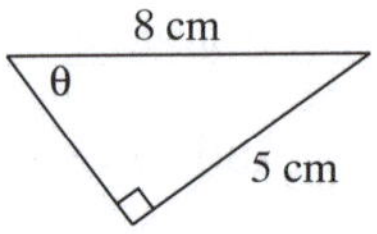

5 Choose the correct word to complete the statement:
Angles of elevation and depression are always measured from the ____________ (horizontal/vertical).

6 Which direction is:

a P? b Q? c R?
d T? e U? f V?

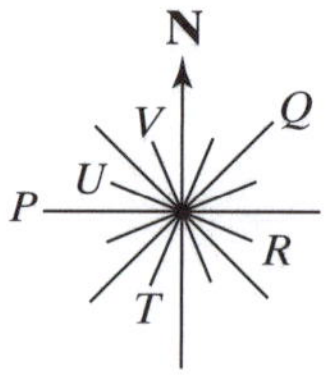

7 Find the bearing of Q from P:

a
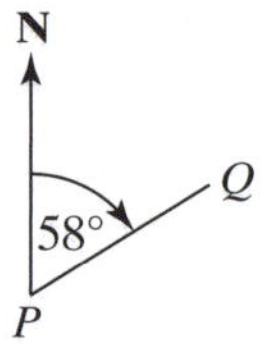

b
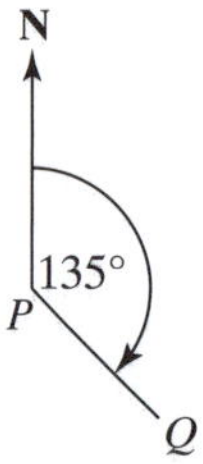

c
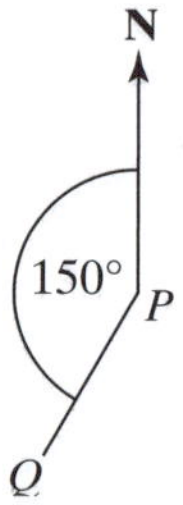

d
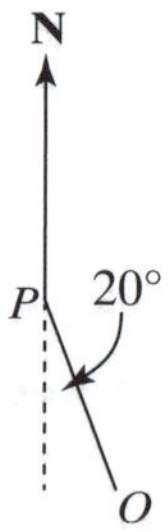

e
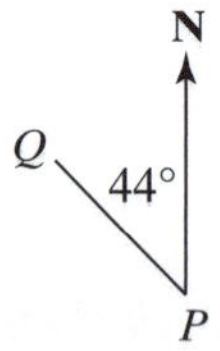

f
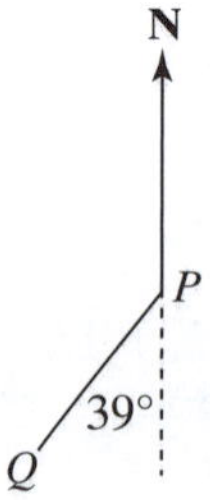

g
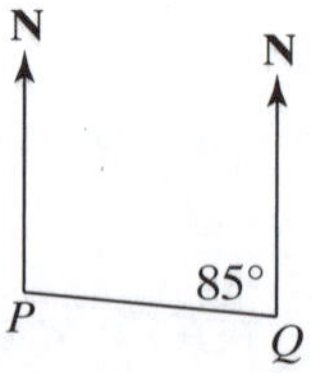

h
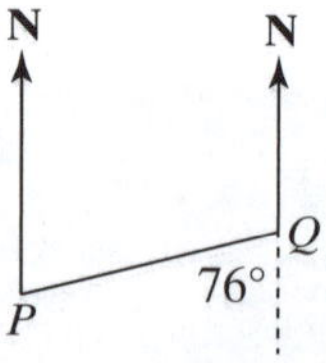

PAGES 153–154

Answers **1 a** 0.4942 **b** 1.5023 **c** 0.2979 **2 a** 69°31′ **b** 59°32′ **c** 16°57′ **3 a** 2.1 **b** 7.6 **c** 7.5 **4 a** 14° **b** 36° **c** 39°
5 horizontal **6 a** W **b** NE **c** ESE **d** SSW **e** WNW **f** NNW **7 a** 058° **b** 135° **c** 210° **d** 160° **e** 316° **f** 219° **g** 095° **h** 076°

ANGLES OF ELEVATION AND DEPRESSION, AND BEARINGS

Trigonometry

20 MINUTES

INTERMEDIATE TEST

Part A Multiple Choice

1 Which diagram best fits the given information? A helicopter leaves P and flies 23 km south-east to Q. From Q if flies due west to R. From R it then flies straight back to P, a distance of 35 km.

A
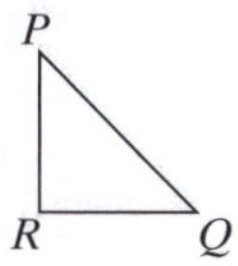

B
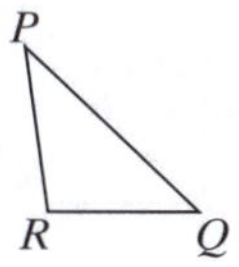

C
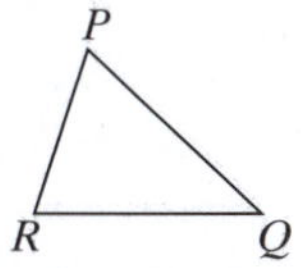

D
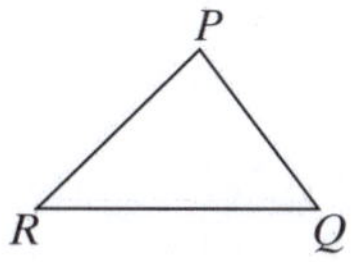

(1 mark)

2 G is south-east of H. What is the bearing of H from G? *Hint 1*

A 135° **B** 215°
C 225° **D** 315° (1 mark)

3 A is due west of C. $\angle ABC = 110°$ and $\angle BCA = 30°$. What is the bearing of B from A?

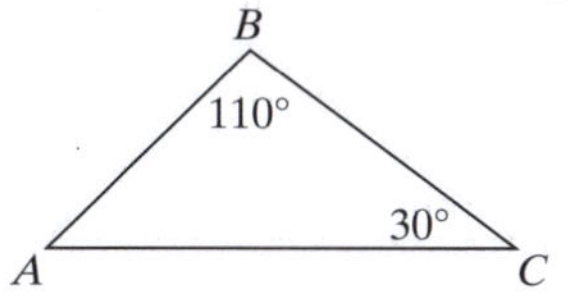

A 040° **B** 045° **C** 050°
D There is insufficient information to determine the bearing. (1 mark)

4 Y is due south of X and north west of Z. The bearing of Z from X is 160°. What is the size of $\angle XZY$? *Hint 2*

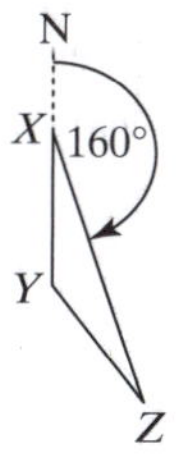

A 20°
B 25°
C 35°
D 45° (1 mark)

Part B Short Answer

5 Find the angle of elevation of the sun, to the nearest degree, when a 3 m high pole casts a shadow (on level ground) which is 4.5 m long.

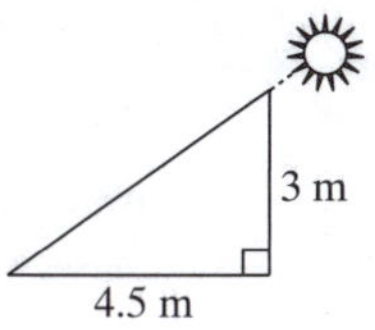

(3 marks)

6 From the top of a 40 m high cliff, the angle of depression of a boat is 29°. How far is the boat from the base of the cliff? (Give the answer to the nearest metre.) *Hint 3* (3 marks)

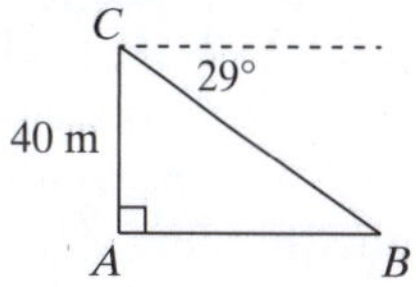

7 K is due west of J and L is due south of J. K is 64 km from L on a bearing of 304°. How far south of J is L? *Hint 4* (3 marks)

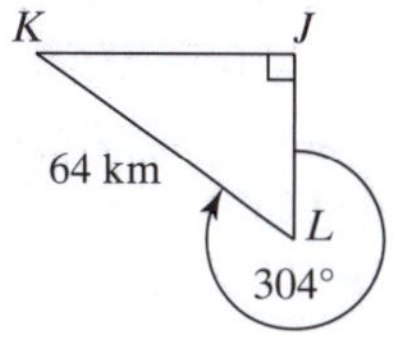

8 Richmount is 68 km from Penton on a bearing of 145°. The bearing of Quarrytown from Penton is 055° and Quarrytown is 95 km from Richmount.

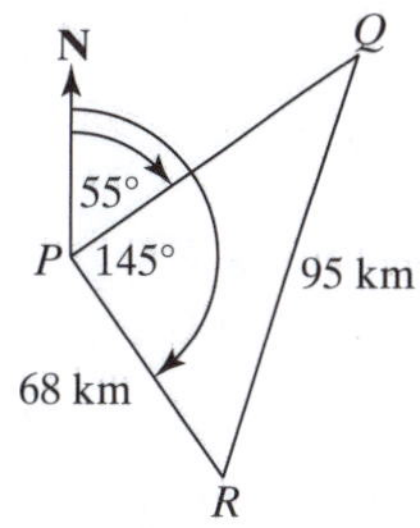

a Find the size of $\angle QPR$. *Hint 5* (1 mark)
b Find the size of $\angle PQR$ to the nearest whole degree. (3 marks)
c Find the bearing of R from Q. *Hint 6* (3 marks)

Hint 1: Draw a diagram.
Hint 2: If Y is north-west of Z, what is the size of $\angle XYZ$?
Hint 3: What is the size of $\angle ACB$?
Hint 4: First find the size of angle JLK.
Hint 5: There is no need to use trigonometry. Use the bearings.
Hint 6: Use the answer to part **b** and the bearing of Q from P.

Your Feedback

$\frac{\square}{20} \times 100\% = \square\ \%$

PAGES 154–155

PAGE 176

1 a If P is south-east of Q, what is the bearing of Q from P? (1 mark)

b Point R is 70 m due south of Q and south-west of P. Show this information on a diagram. (1 mark)

c How far is it from P to R to the nearest metre? (2 marks)

2 From the top of a building 18 m high, the angle of depression of the foot of a neighbouring building is 38° and the angle of elevation of the top of the building is 24°.

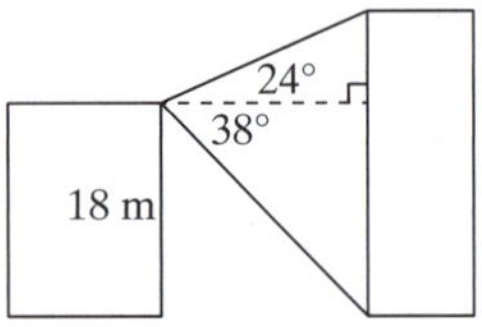

a How far apart are the buildings to the nearest metre? (2 marks)

b How tall is the second building? (2 marks)

3 A plane flies 316 km on a bearing of 130° from A to B. From B it flies 827 km on a bearing of 040° to C.

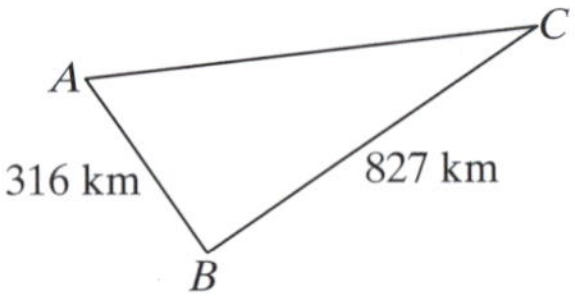

a Show that $\angle ABC = 90°$. (1 mark)

b Find the size of $\angle CAB$ to the nearest degree. (2 marks)

c Find the distance from A to C. (1 mark)

d On what bearing should the plane fly to return directly to A? (1 mark)

4 From a boat 36 m from the base of a cliff, the angle of elevation of the top of the cliff is 54°.

a How high is the cliff? (2 marks)

b What will the angle of elevation be when the boat is twice as far from the base of the cliff? (2 marks)

5

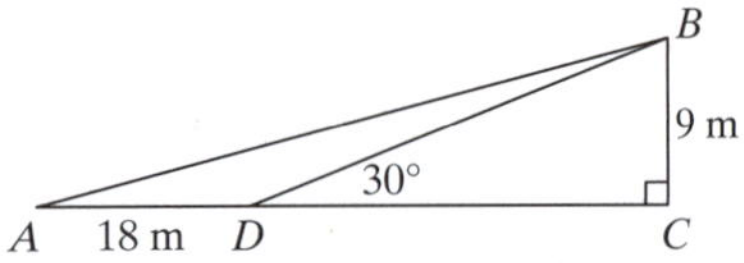

a Find the length of BD. (1 mark)

b Explain why $\angle BAD = 15°$. (1 mark)

c Find the length of AB to the nearest 10 cm. (2 marks)

6 A green marker is 78 m from the base of a lookout, on level ground. From the top of the lookout, the angle of depression of the green marker is 27°. A red marker lies between the green marker and the lookout. The angle of depression of the red marker is 41°.

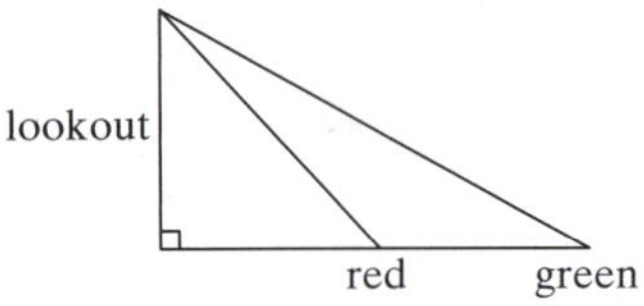

a How high is the lookout? (2 marks)

b How far apart are the red and green markers? (2 marks)

7 B is 250 km from A on a bearing of 195° and C is 67 km from A on a bearing of 285°.

a Show this information on a diagram. (1 mark)

b What is the size of $\angle BAC$? (1 mark)

c Show that B is almost exactly due south of C. (3 marks)

Your Feedback

$\frac{\square}{30} \times 100\% = \square\ \%$

PAGES 155–156

PAGE 176

COMPOUND EVENTS

Chance

STUDY NOTES

1 In probability we consider the likelihood of certain events. The set of all possible outcomes is called the **sample space**.

2 $P(A)$ is the **probability of an event** $\boldsymbol{A}$. $P(A) = \dfrac{\text{number of outcomes that give } A}{\text{number of outcomes in the sample space}}$.

3 The probability of an event is often expressed as a **fraction**, but it could also be given as a decimal or percentage. If the probability of an event is **0**, the event is **impossible**. If the probability of an event is **1**, the event is **certain**. Otherwise the probability of an event is always between 0 and 1. $\boldsymbol{0 \le P(A) \le 1}$. The **sum** of the probabilities of **all possible outcomes** must always be **1**.

4 If two events are such that if one doesn't happen then the other must, the events are **complementary.** The probability of an event is one minus the probability of the complementary event. For example, if the probability that it will rain tomorrow is $\frac{1}{3}$, the probability that it will not rain tomorrow is $1 - \frac{1}{3} = \frac{2}{3}$.

5 A **tree diagram** is simply a tool used to help determine the sample space and find probabilities. For example, if drawing a tree diagram to show the possible outcomes if a coin is tossed three times, the first coin could either be a head or a tail. If the first coin was a head, the second coin could either be a head or a tail and if the first coin was a tail, the second coin can either be a head or tail. For each of these possibilities, the third toss could again produce either a head or a tail.

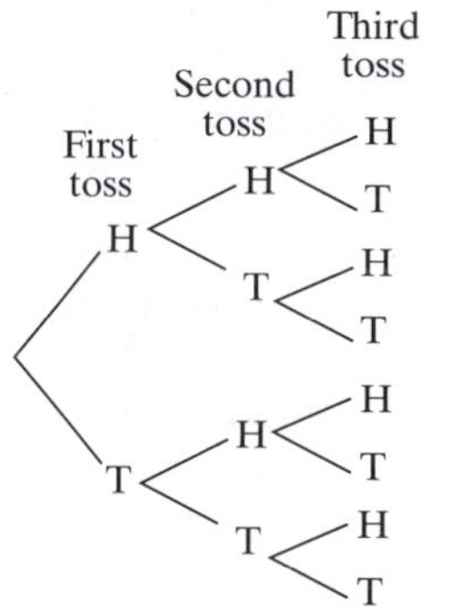

6 Two events are **independent** if the likelihood of one occurring does not depend on the likelihood of the other. For example, the results when tossing two coins are independent. Whether the result is a head or a tail on the first coin, will not affect whether it is a head or a tail on the second coin.

7 Two events are **dependent** if the likelihood of one occurring is influenced by the likelihood of the other.

8 It is important to know when balls are chosen from a bag for example, whether it is **with or without replacement**. That is whether or not the first ball is replaced before the second ball is drawn. If the first ball is replaced the two events are independent but if it is not replaced the second event is dependent on the first.

9 If two events A and B are independent then $\boldsymbol{P(A \text{ and } B) = P(A) \times P(B)}$.

10 In **conditional statements** care must be taken as to the exact meaning of the condition and its implications must be carefully considered.

Checklist

Can you:

1 *Understand what is meant by sample space and list the sample space for events?* ☐

2 *Find the probability of simple events and understand that probability must always be between 0 and 1 and understand what it means if the probability of an event is 0 or 1?* ☐

3 *Understand what is meant by complementary events and find the probability of a complementary event?* ☐

4 *Understand the difference between independent and dependent events and the need to determine whether events are independent or dependent when calculating probabilities?* ☐

5 *Read and interpret conditional statements?* ☐

COMPOUND EVENTS

Chance

SKILLS CHECK

1 A bag holds 7 blue, 5 red and 4 white counters. If one counter is selected at random, what is the probability that it is:

a blue? **b** red? **c** white? **d** not blue?

2 A card is chosen at random from a normal pack of 52 playing cards. What is the probability that the card is:

a red? **b** a club? **c** an ace? **d** the queen of hearts? **e** a red seven?

3 The letters A, B, C and D are written on four cards, one on each card. If one card is chosen at random, what is the probability that the card reads:

a A? **b** D? **c** E? **d** B or C? **e** A, B, C or D?

4 There is a 92% chance that a seed will germinate. What is the probability that the seed will not germinate?

5 What is the probability that a card drawn from a pack of playing cards is:

a not red? **b** not a club? **c** not the ace of spades?

6 State whether the two events are dependent or independent.

a The two numbers resulting when two dice are thrown together.

b The colour of the jelly beans when choosing and eating 2 jelly beans from a bag holding 7 red and 3 black jelly beans.

7 A tree diagram has been drawn to show the possible results when three coins are tossed. What is the probability that:

a all the coins show heads?

b all the coins show tails?

c exactly two coins show heads?

d exactly one coin shows heads?

e at least one coin shows a head?

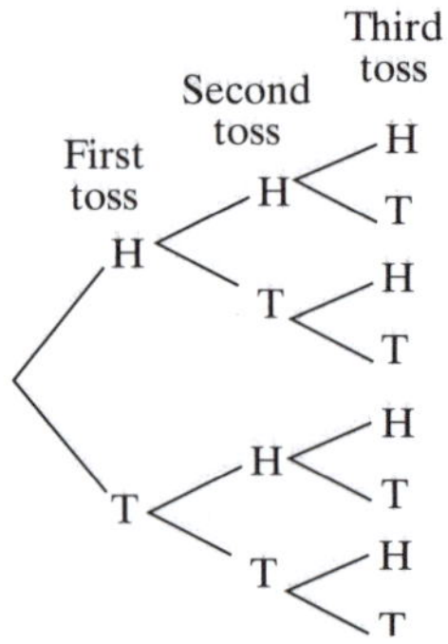

8 A coin is tossed and a die is rolled at the same time. What is the probability of getting a head and a 5?

9 A bag holds 5 yellow and 4 white golf tees. One tee is chosen at random from the bag.

a What is the probability that it is white?

b A second tee is chosen at random. What is the probability that it is also white (the first tee not having been replaced)?

c If the first tee was replaced, what is the probability that the second tee chosen is white?

d If two tees were taken at the same time, what is the probability that both were white?

10 **a** What is the probability of throwing a 4 on a die?

b Jo threw a die and got a number greater than 2. What is the probability that Jo threw a 4?

c Jenna threw a die and got 4. She then threw the die again. What is the probability that she got a 4 again?

PAGES 156–157

Answers **1** **a** $\frac{7}{16}$ **b** $\frac{5}{16}$ **c** $\frac{1}{4}$ **d** $\frac{9}{16}$ **2** **a** $\frac{1}{2}$ **b** $\frac{1}{4}$ **c** $\frac{1}{13}$ **d** $\frac{1}{52}$ **e** $\frac{1}{26}$ **3** **a** $\frac{1}{4}$ **b** $\frac{1}{4}$ **c** 0 **d** $\frac{1}{2}$ **e** 1 **4** 8% **5** **a** $\frac{1}{2}$ **b** $\frac{3}{4}$ **c** $\frac{51}{52}$ **6** **a** independent **b** dependent **7** **a** $\frac{1}{8}$ **b** $\frac{1}{8}$ **c** $\frac{3}{8}$ **d** $\frac{3}{8}$ **e** $\frac{7}{8}$ **8** $\frac{1}{12}$ **9** **a** $\frac{4}{9}$ **b** $\frac{3}{8}$ **c** $\frac{4}{9}$ **d** $\frac{1}{6}$ **10** **a** $\frac{1}{6}$ **b** $\frac{1}{4}$ **c** $\frac{1}{6}$

COMPOUND EVENTS

Chance

20 MINUTES

INTERMEDIATE TEST

Part A Multiple Choice

1 A card is chosen at random from a standard pack of 52 playing cards. The probability that the card is a red five is:

A $\frac{5}{52}$ B $\frac{1}{26}$ C $\frac{1}{13}$ D $\frac{2}{13}$ (1 mark)

2 The probability of getting a 7 when picking an envelope from a group of numbered envelopes is $\frac{2}{5}$. What is the probability of not picking a 7?

A $\frac{1}{5}$ B $\frac{2}{5}$ C $\frac{3}{5}$ D $\frac{4}{5}$ (1 mark)

3 Two independent events, R and B, are such that $P(R) = \frac{3}{5}$ and $P(B) = \frac{1}{3}$. What is $P(R \text{ and } B)$?

A $\frac{1}{5}$ B $\frac{4}{15}$ C $\frac{14}{15}$ D $\frac{1}{2}$ (1 mark)

4 A coin is tossed three times. What is the probability of getting head, tail, head in that order?

A $\frac{1}{8}$ B $\frac{1}{4}$ C $\frac{3}{8}$ D $\frac{1}{2}$ (1 mark)

Part B Short Answer

5 'If I buy a ticket in a raffle I will either win a prize or not win a prize. Therefore the probability that I will win a prize is $\frac{1}{2}$'. Is this statement true or false? Justify your answer. *Hint 1* (2 marks)

6 A bowl holds 2 green, 3 black, 1 yellow and 4 red jelly beans. Without looking, three people take and eat a jelly bean, one after the other. What is the probability that there are then no black jelly beans left in the bowl? *Hint 2* (2 marks)

7 Caleb buys 5 tickets in a raffle in which 100 tickets are sold. There are three prizes.

a What is the probability that Caleb wins first prize? (1 mark)

b What is the probability that Caleb wins all three prizes? (3 marks)

8 'A family of three children is much more likely to consist of a boy and two girls than of three girls, so because there are already two girls in the family the next baby will probably be a boy.' Is this statement true? Justify your answer. *Hint 3* (2 marks)

9 Kelly and Chris play a game. They each roll a die and the numbers on the uppermost faces are multiplied together to give the score. Kelly wins if the score is an odd number. Chris wins if the score is an even number greater than 16.

a Who has the best chance of winning in the first roll of the dice? Justify your answer. *Hint 4* (3 marks)

b If neither wins in the first roll of the dice, a second roll takes place. What is the probability that Chris wins on a second roll? *Hint 5* (2 marks)

c What is the probability that no-one has won after 3 rolls of the dice? (1 mark)

Hint 1: Is a person equally likely to win or not win a prize?
Hint 2: Find the probability that three blacks are chosen, one after another, without replacement.
Hint 3: Are the events dependent or independent?
Hint 4: Draw up a table to show the sample space.
Hint 5: Consider the need to have a second roll of the dice.

Your Feedback

$\frac{\square}{20} \times 100\% = \square\ \%$

PAGE 157

PAGE 176

COMPOUND EVENTS
Chance

30 MINUTES

ADVANCED TEST

1 There are 18 marshmallows in a packet, 12 are white and the rest are pink. Daisy eats three marshmallows from the packet. What is the probability that:

- **a** the first one is pink? (1 mark)
- **b** the third one is pink? (1 mark)
- **c** all are pink? (1 mark)
- **d** at least one is pink? (2 marks)

2 A bag holds 1 yellow, 1 blue and 2 red counters. Two counters are drawn from the bag, one after the other, without replacement.

- **a** Draw a tree diagram to show the possible results. (2 marks)
- **b** What is the probability that the counters are both red? (1 mark)
- **c** Briefly explain why the probability that the counters are both the same colour is the same as the probability that the counters are both red. (1 mark)
- **d** The first counter drawn was red. Explain how that makes a difference to the answer to part **b**. (1 mark)
- **e** If the first counter had been replaced, explain how that makes a difference to the answer to part **b**. (1 mark)

3 A fair die is rolled three times.

- **a** What is the probability of getting at least one 6? (2 marks)
- **b** What is the probability of the same number occurring each time? (1 mark)
- **c** What is the probability of rolling 5 each time? (1 mark)
- **d** The number 5 did occur each time. Don said, 'If the die is rolled again it is more likely to show 5 than any other number.' Ron said, 'If the die is rolled again it is now less likely to show 5 than any other number.' Who is right? Justify your answer. (1 mark)

4 John buys 10 tickets in a raffle in which 500 tickets are sold. There are three prizes: first worth \$200, second worth \$100 and third worth \$50. The tickets cost \$1 each. What is the probability, as a percentage, that John:

- **a** wins second prize? (1 mark)
- **b** wins all three prizes? (1 mark)
- **c** makes a profit? (2 marks)

5 Some students were asked whether they had represented their school in netball, tennis or hockey. The results are shown in the Venn diagram.

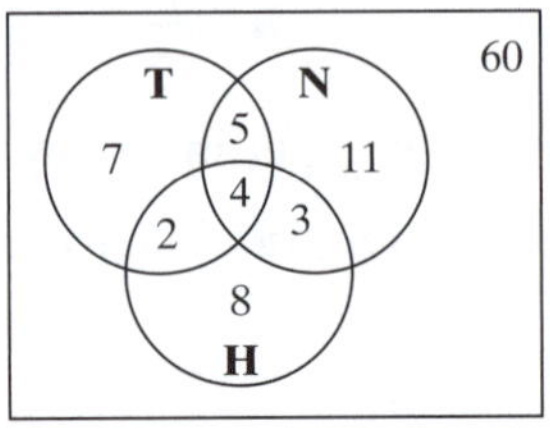

- **a** What is the probability that a randomly chosen student from the group represented their school in netball? (1 mark)
- **b** What is the probability that a randomly chosen student from the group who represented the school in netball also represented the school in tennis? (1 mark)
- **c** How could you describe the chance that three randomly chosen students from the group all represented their school in both tennis and hockey but not netball? (1 mark)
- **d** What is the probability that three randomly chosen students from the group all represented the school in hockey? (1 mark)

6 Sally rolls two dice and multiplies the result on the uppermost faces for a score.

- **a** Draw up a table to show the possible results. (1 mark)
- **b** In a single roll of the two dice, what is the probability of a score of 4? (1 mark)
- **c** What is the probability that Sally gets a score of 4 in each of three rolls of the two dice? (1 mark)
- **d** Is Sally more likely to get a score less than 10 or greater than 10? Justify your answer. (1 mark)
- **e** In three rolls of the dice, what is the probability that Sally gets a total score of more than 100? (2 marks)

Your Feedback

$\frac{\square}{30} \times 100\% = \square\%$

PAGES 157–159

PAGE 176

QUARTILES AND INTERQUARTILE RANGE

Data Representation and Interpretation

STUDY NOTES

1 Statistics is concerned with the collection, analysis and representation of **data**.

2 The **mean** is commonly known as the average. To find the mean, add all the scores together and divide by the number of scores.

3 The **mode** is the score with the highest frequency. It is the score that happens the most.

4 The **median** is the middle value when the scores are arranged in order. If there is an odd number of scores, there is one middle score. When there is an even number of scores the median is the average of the two middle scores.

5 The mean, mode and median are all measures of **location**. They give us information about where the data is placed.

6 The **upper extreme** is the highest score and the **lower extreme** is the lowest score.

7 The **range** is the difference between the upper and lower extremes.

8 **Quartiles** divide the scores into four equal parts. The **upper quartile** is the value for which one quarter of the scores are above. The **lower quartile** is the value for which one quarter of the scores are below. The median is the middle quartile. The notations Q_1 for lower quartile, Q_2 for median and Q_3 for upper quartile are sometimes used.

9 The **interquartile range** is the difference between the upper quartile and the lower quartile.

10 The range and interquartile range are measures of **spread**. They give us information about how consistent the scores are, whether they are all close together or whether there is a large spread of scores.

11 To **determine the upper or lower quartiles**, or interquartile range, first place all the scores in order (from lowest to highest). Then find the median. The lower quartile could be thought of as the median of the scores below the median. We consider the scores below the median and the lower quartile will be the middle of those (the middle score if there is an odd number and the average of the two middle scores if there is an even number). Similarly the upper quartile is the middle score of those scores that are above the median. For example, if the scores are 2, 4, 5, 8, 9, 11, 12, 14 and 16, then the median is 9. To find the lower quartile consider the scores below the median: 2, 4, 5, 8. There are two middle scores so the lower quartile is the average of those: 4.5. For the upper quartile consider 11, 12, 14 and 16. The average of the two middle scores is 13 so the upper quartile is 13.

12 Because the quartiles divide the data into **quarters**, 25% of scores lie below the lower quartile, 50% lie between the lower and upper quartiles, and another 25% lie above the upper quartile.

13 If a **cumulative frequency polygon** is drawn from data, the median and upper and lower quartiles can easily be read. To find the lower quartile, for example, find the 25% mark on the vertical axis, follow this straight across to the polygon, and then drop straight down to read the score.

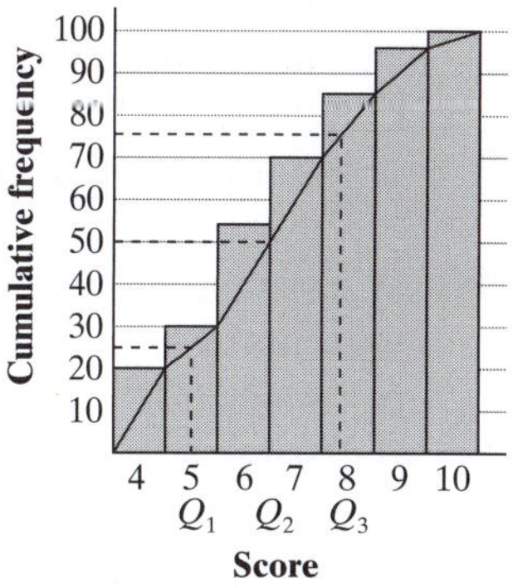

Checklist

Can you:

1 *Find the mean, mode, median and range of a set of scores?* ☐

2 *Determine the upper and lower quartiles and find the interquartile range?* ☐

3 *Understand the percentages of the data that fall within the different quartiles?* ☐

4 *Make judgements and interpret data by finding the range or interquartile range?* ☐

QUARTILES AND INTERQUARTILE RANGE

Data Representation and Interpretation

SKILLS CHECK

1 For the scores 1, 2, 2, 5, 6, 7, 7, 7, 8 find the:
a mean **b** mode **c** median **d** range

2 Find the median of the scores:
a 2, 7, 9, 5, 7, 6, 8, 4
b 58, 63, 62, 37, 58, 67

3 Find the interquartile range if a set of scores has lower quartile 57, median 63 and upper quartile 72.

4 For the scores 73, 75, 79, 82, 85, 86, 87, 87, 88, 91, 94 find the:
a median **b** lower quartile **c** upper quartile

5 For the scores 13, 13, 14, 15, 15, 15, 16, 16, 16, 16, 17, 17, 18, 18, 18, 18, 18, 19, 19, 20 find the:
a lower quartile **b** upper quartile **c** interquartile range

6 Use the cumulative frequency polygon to find:

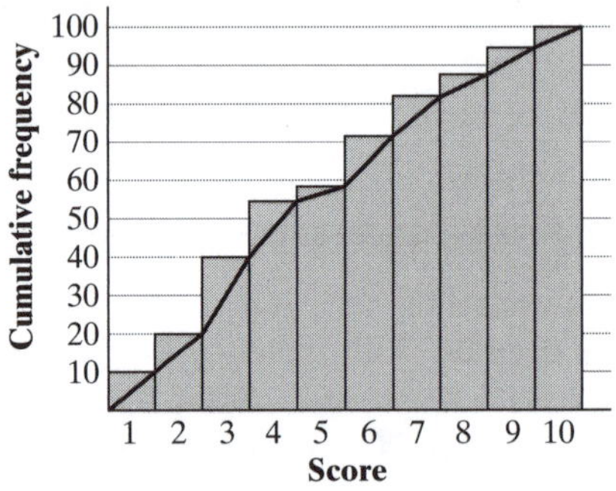

a the median **b** the lower quartile **c** the upper quartile

7 For the scores 78, 56, 32, 25, 48, 52, 63, 50, 61 find:
a upper extreme **b** lower extreme **c** median
d upper quartile **e** lower quartile **f** range
g interquartile range

8 A data set has lower extreme 16, lower quartile 39, median 54, upper quartile 68 and upper extreme 87. Approximately what percentage of the data lies between:
a 16 and 87? **b** 16 and 54? **c** 68 and 87?
d 39 and 54? **e** 39 and 68? **f** 16 and 68?

9 For the scores 20, 13, 9, 12, 16, 14, 17, 17 find:
a range
b interquartile range

PAGES 159–160

Answers **1 a** 5 **b** 7 **c** 6 **d** 7 **2 a** 6.5 **b** 60 **3** 15 **4 a** 86 **b** 79 **c** 88 **5 a** 15 **b** 18 **c** 3 **6 a** 4 **b** 3 **c** 7 **7 a** 78 **b** 25 **c** 52 **d** 62 **e** 40 **f** 53 **g** 22 **8 a** 100% **b** 50% **c** 25% **d** 25% **e** 50% **f** 75% **9 a** 11 **b** 4.5

QUARTILES AND INTERQUARTILE RANGE

Data Representation and Interpretation

20 MINUTES

INTERMEDIATE TEST

Part A Multiple Choice

1 What is the median of the scores 19, 27, 25, 32, 20, 27, 24?

A 24 B 25 C 27 D 32 (1 mark)

2 Consider the scores 8, 10, 9, 10, 5, 12, 6. Which is not correct?

A the median is 9
B the range is 7
C the lower extreme is 6
D the interquartile range is 4 (1 mark)

3 For the set of scores 4, 5, 5, 5, 6, 7, 7, 8, 9, 9, 10, 10 which is the largest?

A mean B median C mode D interquartile range (1 mark)

4 The interquartile range for the scores 3, 5, 9, 10, 12, 13, 14, 16, 19 is:

A 6 B 7 C 8 D 9 (1 mark)

5 For this set of marks, 59, 70, 64, 76, 71, 57, 65, 61, 77, 55, 78, 71, 59, 67, 56, 79, 69, 73, the biggest difference is between:

A lower extreme and lower quartile
B lower quartile and median
C median and upper quartile
D upper quartile and upper extreme (1 mark)

Part B Short Answer

6 Approximately 80 000 students sat for a national examination. The lowest mark was 4% and the highest was 99%. The median was 62%, the upper quartile 78% and the lower quartile was 50%.

a About how many of the students scored more than 78%? *Hint 1* (2 marks)

b About how many of the students scored more than 50%? (2 marks)

7 The scores in an exam were collected and displayed in a cumulative frequency histogram.

a What is the median? (1 mark)

b Between what 2 marks are the middle 50% of scores? *Hint 2* (2 marks)

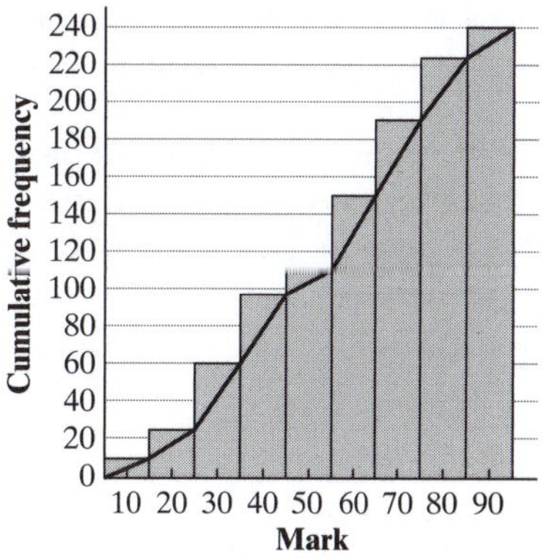

8 Annabel recorded the time (in minutes) to travel home from work over the last 3 weeks. She arranged the times in ascending order: 32, 35, 36, 38, 38, 39, 42, 44, 45, 45, 46, 48, 56, 75, 88

a What is the median? (1 mark)

b What is the range? (1 mark)

c What is the interquartile range? (3 marks)

d Discuss the merits of the range and interquartile range as measures of spread. *Hint 3* (3 marks)

Hint 1: Use the percentages that fall between the different quartiles.
Hint 2: Find the upper and lower quartiles.
Hint 3: Which gives a better picture of the times? What influences the range but not the interquartile range?

Your Feedback

$\frac{\square}{20} \times 100\% = \square\%$

PAGE 160

PAGE 176

QUARTILES AND INTERQUARTILE RANGE

Data Representation and Interpretation

35 MINUTES

ADVANCED TEST

1 These are the percentage marks in an exam for each member of a class.

47 49 52 54 58 60 60 61 63 65
69 70 74 75 77 79 81 85 92

a What is the median? (1 mark)

b What is the lower quartile? (1 mark)

c What is the upper quartile? (1 mark)

d What is the interquartile range? (1 mark)

e Stacey said that the interquartile range should be 50% of the range. Is that the case in this question? Is Stacey correct? Discuss. (1 mark)

2 The heights of a group of students were recorded. The interquartile range was 26 cm and the upper quartile was 154 cm.

a What was the lower quartile? (1 mark)

b About what percentage of the students were taller than 154 cm? (1 mark)

c Megan looked at the above information and said that the median must have been 141 cm. Is Megan correct? Justify your answer. (1 mark)

d Briefly explain any advantage gained from knowing the interquartile range in this case. Would it be beneficial without also knowing the upper quartile? (1 mark)

3 These are the numbers of vehicles parked in a parking lot at various times one day.

7 26 32 37 28 44 51 69 48 43
35 46 59 41 39 34 40 28 31 19

a What is the mean? (1 mark)

b What is the mode? (1 mark)

c What is the median? (1 mark)

d What is the range? (1 mark)

e What is the interquartile range? (1 mark)

f How many scores are less than a position that is $1\frac{1}{2}$ times the interquartile range below the lower quartile? (1 mark)

g Which scores are more than a position that is $1\frac{1}{2}$ times the interquartile range above the upper quartile? (1 mark)

h If any scores found in parts **f** and **g** are ignored, what effect does this have on the mean, mode, median, range and interquartile range? (3 marks)

4 The five-number summary was published after an exam: 37, 48, 62, 79, 96

a What was the range for the exam? (1 mark)

b What was the interquartile range? (1 mark)

c Between what marks did the middle 50% of marks lie? (1 mark)

d What percentage of examinees scored between:

i 37 and 96? (1 mark)

ii 62 and 37? (1 mark)

iii 48 and 96? (1 mark)

e If 32 000 people sat for the exam, about how many scored:

i less than 48? (1 mark)

ii more than 62? (1 mark)

iii less than 79? (1 mark)

5 The frequency distribution table shows times to complete a certain task.

Class centre	Frequency	Cumulative frequency
11	9	
19	12	
27	13	
35	15	
	20	
51	21	
59	17	
67	13	

a One class centre is missing. What should it be? (1 mark)

b Copy and complete the frequency distribution table. (1 mark)

c Draw a cumulative frequency histogram for the data. (2 marks)

d Draw a cumulative frequency polygon (ogive). (1 mark)

e What is the median? (1 mark)

f What is the interquartile range? (1 mark)

Your Feedback

$\frac{\square}{35} \times 100\% = \square\%$

PAGES 160–162

PAGE 176

BOX PLOTS AND SCATTER PLOTS

Data Representation and Interpretation

STUDY NOTES

1 The **median** is the middle value when the scores are arranged in order. The **upper quartile** is the value for which one quarter of the scores are above. The **lower quartile** is the value for which one quarter of the scores are below. The median is the middle quartile. The **interquartile range** is the difference between the upper quartile and the lower quartile.

2 The **five-point summary** (or five-number summary) consists of five values calculated from the data. The first point is the **lower extreme** (the lowest score), the second point is the **lower quartile**, the third point is the **median**, the fourth point is the **upper quartile** and the fifth point is the **upper extreme** or highest score.

3 A **box plot** (sometimes called a box-and-whisker plot), is drawn using the five-point summary.

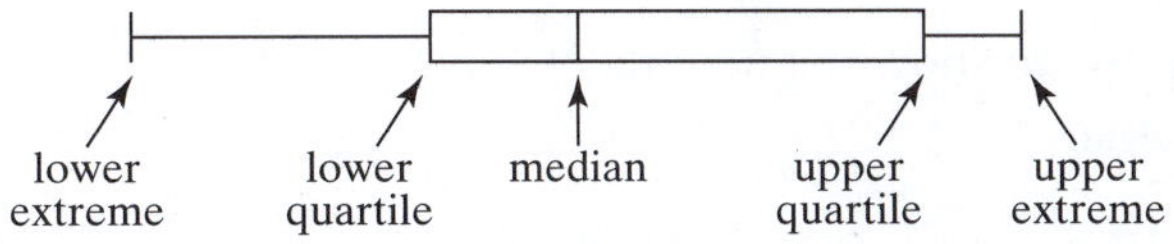

4 The **shape of a distribution** gives information about the data. A distribution may be **symmetrical** or **skewed**. In a symmetrical distribution scores are evenly distributed about the mean. In a skewed distribution more scores are located at one end and very few at the other end. A distribution is **positively skewed** if there are more lower scores than higher scores. A distribution is **negatively skewed** if there are more higher scores.

5 **Sets of data** can be **compared** by considering the similarities and differences in box plots. For example, if two box plots are drawn on the same scale differences in spread can immediately be seen.

6 A **histogram** is like a column graph. In a frequency histogram the scores are shown on the horizontal axis and the frequency on the vertical axis. In a cumulative frequency histogram the cumulative frequency is shown on the vertical axis.

7 A **scatterplot** can be drawn to show relationships between two variables.

8 It is not always necessary to have a precise graph drawn to come to conclusions about data. The **relative position** of a point on the graph can give information about the item at that point. For example, point A on this graph is close to the median age but its price is lower than average.

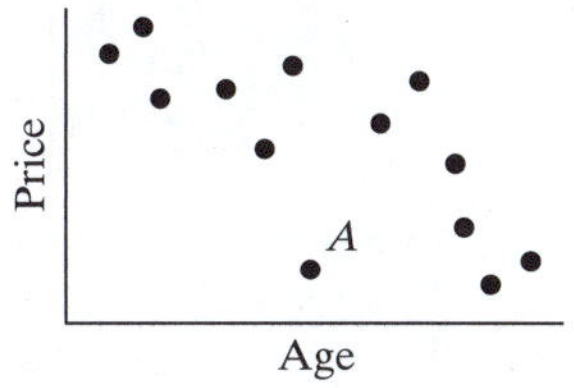

9 If there is a mathematical relationship between the two variables in a scatterplot, then they will form a pattern. The **association** between the variables can be described in terms of its **strength and direction**. The closer the points are to forming a straight line, the stronger the association. If the pattern leans to the right (/), it is positive, and if it leans to the left (\ ), it is negative.

10 If there is an association between the variables, the pattern can be used to make **predictions**.

Checklist

Can you:

1 *Understand what is meant by the five-point summary?* ☐
2 *Determine the upper and lower extremes, upper and lower quartiles and median?* ☐
3 *Construct box plots using the five-point summary?* ☐
4 *Recognise when a distribution is symmetrical or skewed and describe the shape of a distribution?* ☐
5 *Compare box plots with dot plots and histograms?* ☐
6 *Understand what is meant by a scatter plot?* ☐
7 *Determine whether there is an association between variables in a scatter plot and comment on the direction and strength of any association?* ☐

BOX PLOTS AND SCATTER PLOTS

Data Representation and Interpretation

SKILLS CHECK

1 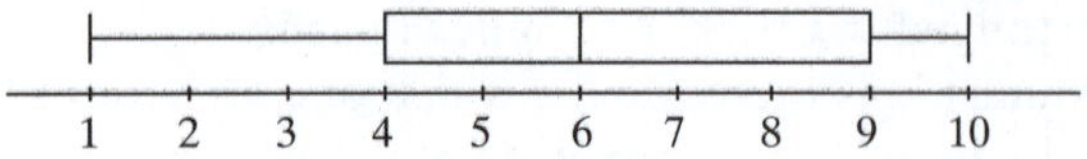

From the box plot write down:

a the median
b the lower extreme
c the upper quartile
d the upper extreme
e the lower quartile
f the interquartile range

2 For a particular set of scores the lower extreme is 2, the lower quartile is 5, the median is 9, the upper quartile 11 and the upper extreme 15. Draw a box plot for the scores.

3 A box plot has been drawn to show the heights of some plants.

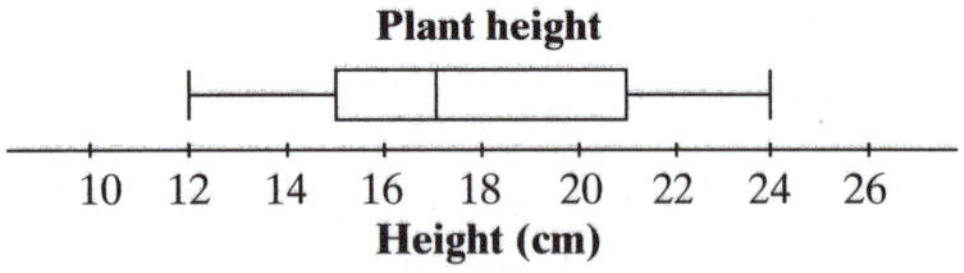

What percentage of plants have a height:

a between 15 cm and 21 cm?
b more than 17 cm?
c less than 21 cm?
d between 17 cm and 21 cm?
e more than 15 cm?
f between 21 cm and 24 cm?

4 For each display state whether the data is symmetrical or skewed. If it is skewed state whether it is positively or negatively skewed.

a 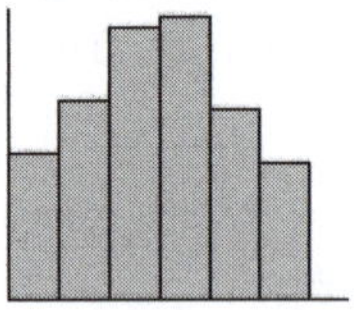
b 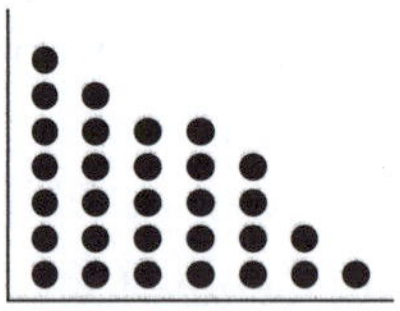
c 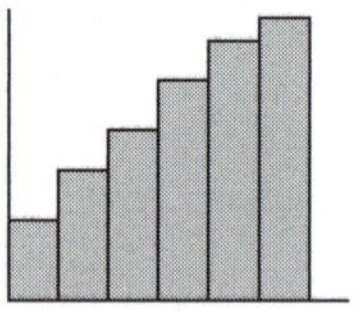
d 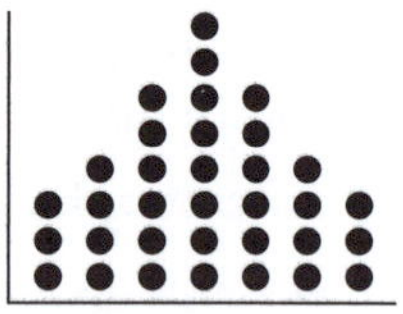

5 A study of cars compared their age with the amount of money that had been spent on repairs. Some of the results were plotted on the graph. Which point would represent:

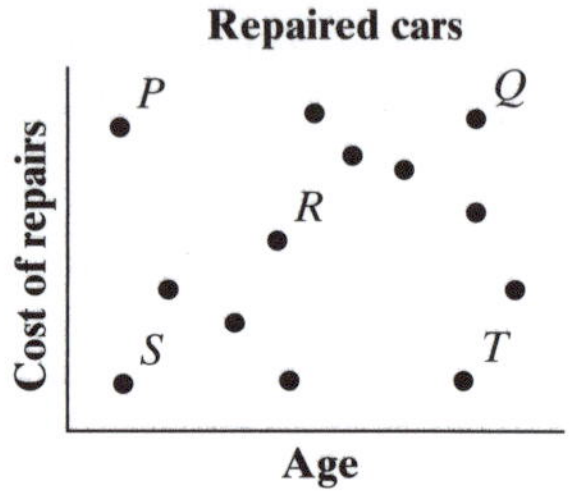

a an old car which had had a lot of repairs?
b a new car which had had very few problems?
c an older car which had few repairs?
d a car of average age which had had an average amount spent on repairs?
e a car which is not very old but which had had a lot of problems?

6 Determine whether or not each scatterplot shows an association between the two variables. If so, is there a positive or negative association and is it strong or weak?

a 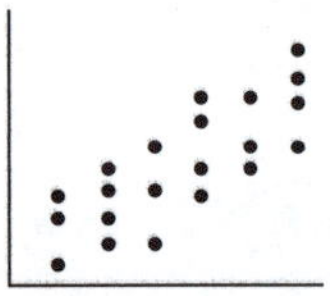
b 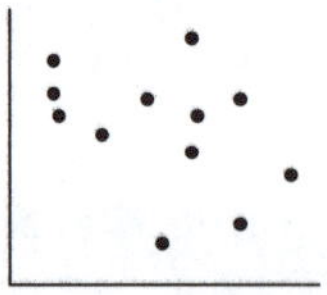
c 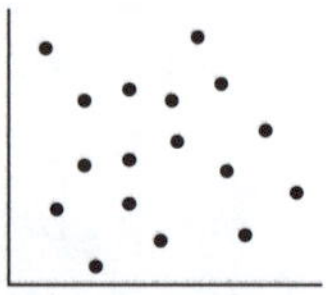
d 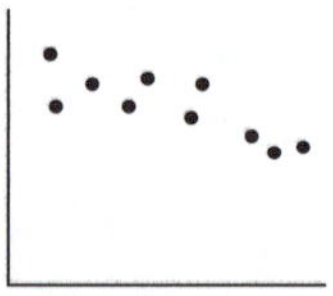

PAGE 162

Answers **1 a** 6 **b** 1 **c** 9 **d** 10 **e** 4 **f** 5 **2** see worked solutions **3 a** 50% **b** 50% **c** 75% **d** 25% **e** 75% **f** 25%
4 a symmetrical **b** positively skewed **c** negatively skewed **d** symmetrical **5 a** *Q* **b** *S* **c** *T* **d** *R* **e** *P* **6 a** strong positive
b weak negative **c** no association **d** strong negative

BOX PLOTS AND SCATTER PLOTS

Data Representation and Interpretation

INTERMEDIATE TEST

Part A Multiple Choice

1 The fuel consumption of some cars (in litres per 100 km), was measured and the results are shown in the box plot. What is the interquartile range?

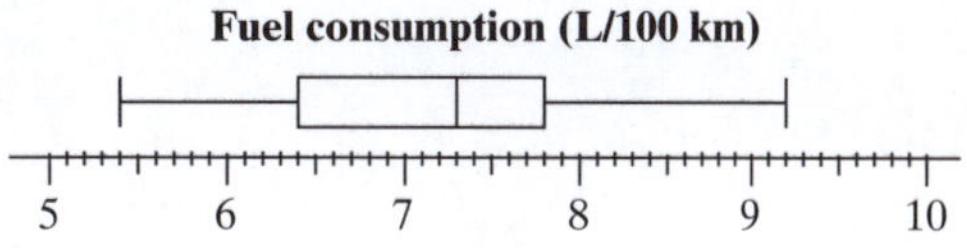

A 1.4 **B** 1.9 **C** 2.8 **D** 3.8 (1 mark)

2 Referring to the box plot in question 1, what percentage of the cars has a fuel consumption of less than 7.8 litres per 100 km?

A 25% **B** 50% **C** 75% **D** 85% (1 mark)

3 Referring to the box plot in question 1, a total of 84 cars were studied. Approximately how many of these had a fuel consumption that was less than 6.4 litres per 100 km?

A 12 **B** 21 **C** 30 **D** 42 (1 mark)

4 The heights and ages of 8 students were plotted. Although one of the youngest students, Mick is quite tall. Which point would represent Mick?

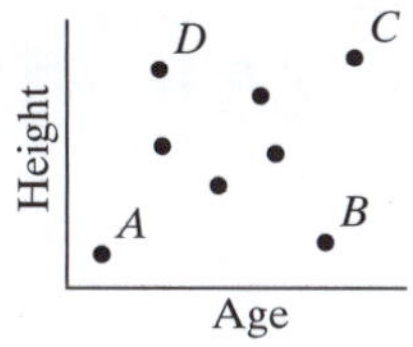

A A **B** B
C C **D** D (1 mark)

Part B Short Answer

5 A survey was done of the ages (in months) of cows when they had their first calf. The five-point summary is 23, 25, 26, 30, 33.

a Draw a box-and-whisker plot to display this information. *Hint 1* (2 marks)

b Comment on any conclusions about the data that can be seen from this plot. *Hint 2* (2 marks)

6 The marks that a group of students received in literacy and numeracy tests are shown on the scatterplot.

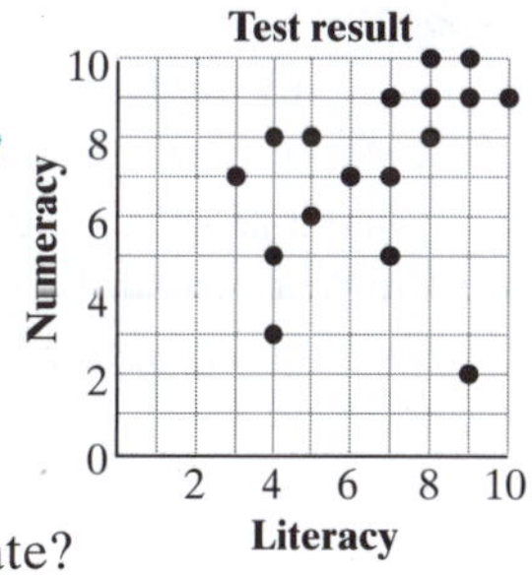

a Sally's numeracy mark was 6. What was her literacy mark? *Hint 3* (1 mark)

b Sally described the association between the scores as very strong positive. Do you agree? Justify your answer. (2 marks)

c Sally said everyone who did well in literacy also did well in numeracy. Give one example where this is not the case. (1 mark)

d Jim scored 7 in the literacy test but missed the numeracy test. What mark do you think the teacher should give Jim as an estimate? Justify your answer. *Hint 4* (2 marks)

7 These two dot plots show the marks received by two separate groups in a competition.

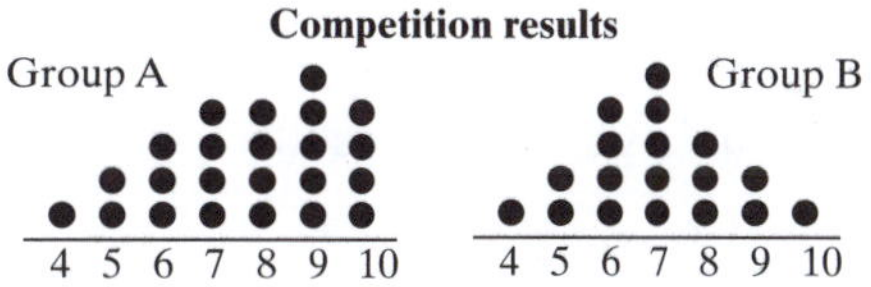

a On the same scale, draw box plots to match both dot plots. *Hint 5* (4 marks)

b Briefly comment on the merits of each type of plot. (2 marks)

Hint 1: You have been given the lower extreme, lower quartile, median, upper quartile and upper extreme.
Hint 2: Is it symmetrical or skewed?
Hint 3: There is only one dot when the literacy is 6.
Hint 4: Make a prediction based on the pattern in the graph.
Hint 5: The median, extremes and quartiles can easily be found because the scores are in order.

Your Feedback

$\frac{\square}{20} \times 100\% = \square\%$

PAGES 162–163
PAGE 176

BOX PLOTS AND SCATTER PLOTS

Data Representation and Interpretation

40 MINUTES

ADVANCED TEST

1 These are the ages of students at a camp.

8 15 11 9 16 15 13 12
10 7 14 13 15 16 18 12

a Determine the five-number summary. (2 marks)

b Draw a box plot for the data. (2 marks)

c What percentage of students are older than 10? (1 mark)

2 These two box plots show the daily maximum temperatures in a certain town over summer and winter.

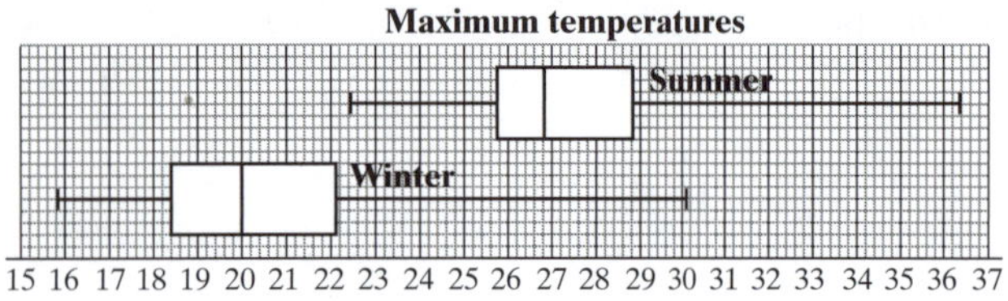

a What was the median summer temperature? (1 mark)

b What was the range in winter? (1 mark)

c Between what temperatures were the middle 50% of summer maximums? (1 mark)

d How much hotter was the hottest day in summer than the hottest day in winter? (1 mark)

e About how many days in winter was the temperature 20° or more? (1 mark)

f About how many days in summer had temperatures less than 25.7°? (1 mark)

g Comment on similarities and differences between the two box plots. (2 marks)

h The five-number summary, in °C, for spring for the town was 18.9, 23.1, 25.1, 27.2, 36.7.

i Comment on the spring temperatures using this information. (1 mark)

ii What benefits could be achieved by drawing a box plot? (1 mark)

3 A group of 12 students sat for two tests marked out of 10. The table shows the results.

Name	Test 1	Test 2	Name	Test 1	Test 2
Max	7	8	Ali	8	10
Jo	5	6	Jay	5	8
Jill	8	8	Ned	7	9
Sara	8	9	Rose	6	10
Will	7	7	Zane	9	9
Mary	6	6	Amy	6	7

a Show the scores on a scatter plot. (2 marks)

b Give one advantage of the scatter plot over the table. (1 mark)

c Give one advantage of the table over the scatter plot. (1 mark)

d Comment on any trends or patterns. (1 mark)

e Draw separate dot plots for test 1 and test 2. (2 marks)

f Comment on any similarities or differences between the two dot plots. (1 mark)

4 Some patients at a hospital were asked to rank their pain out of 10. These are the results:

8 7 5 7 7 9 6 8 4 7
10 7 9 3 8 4 6 8 6 7

a Draw up a table showing the frequency for each score. (2 marks)

b Draw a frequency histogram. (2 marks)

c Determine the five-number summary. (2 marks)

d Draw a box plot. (2 marks)

e Compare and contrast the histogram and box plot for this data. (2 marks)

5 These are the five-number summaries for two sets of related data.

[1, 5, 7, 12, 18]
[5, 7, 11, 15, 17]

a Draw a box plot for each data set using the same scale. (2 marks)

b How could you describe each box plot in terms of symmetry or skewness? (2 marks)

c What similarities and differences are there in the data? (2 marks)

d Alan said, 'There are exactly the same number of scores between the lower quartile and median in the first set as between the lower extreme and lower quartile in the second set.' Is he correct? Justify your answer. (1 mark)

Your Feedback

$\frac{\square}{40} \times 100\% = \square\%$

PAGES 163–165

PAGE 176

DATA INVOLVING TIME; EVALUATING REPORTS

Data Representation and Interpretation

STUDY NOTES

1 Univariate data uses just one variable but **bivariate data** uses two, for example, when considering changes in population over time. (Both time and population are changing).

2 When considering bivariate data one variable is **independent** and the other **dependent**. Time is usually an independent variable.

3 **Line graphs** are the most appropriate to show changes over time.

4 A **travel graph** shows the distance travelled, or the distance from a particular point, at a given time. Time is shown on the horizontal axis and distance on the vertical axis. A lot of **information** can be seen immediately from the graph without any need for calculations. Differences in speed over the journey are shown by the **different gradients** of the line segments. The higher the speed, the steeper the graph. A negative gradient shows that the journey is proceeding in the opposite direction. A horizontal line segment shows that the journey has stopped. For example, the speed in the journey shown by this travel graph is much greater in the first hour than in the third hour. The traveller was stopped between 10 am and 11 am and the return journey begins at noon.

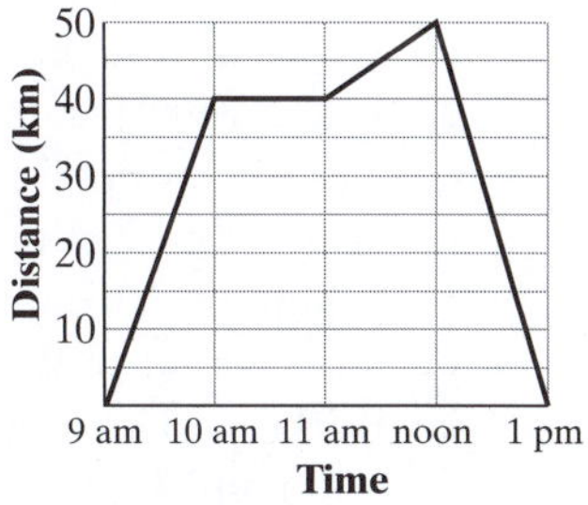

5 It is impossible to be in two **different places at the same time** so a graph should not show two different values at the same time.

6 A **constant** relationship is denoted by a **straight line** segment and a **variable** relationship is denoted by a **curved line** on a graph. A line or curve is **increasing** if it leans to the right at the top and **decreasing** if it leans to the left.

7 A **curve** that is increasing can be further **described** according to the **rate of change of the graph**. The steeper the graph the faster it is increasing. If the graph increases quickly at first and then slows, it is increasing at a decreasing rate. If it increases slowly and then more quickly it is increasing at an increasing rate. Similarly, a curve that is decreasing may be decreasing at an increasing rate or decreasing at a decreasing rate.

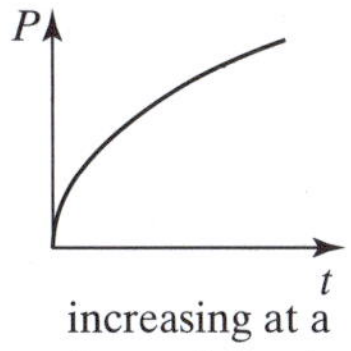

increasing at a decreasing rate

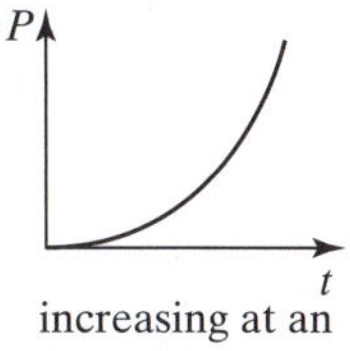

increasing at an increasing rate

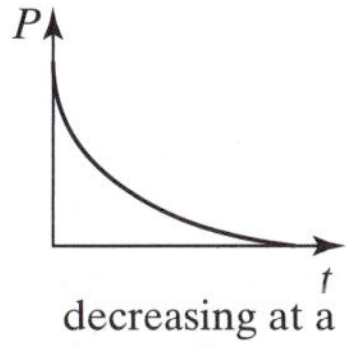

decreasing at a decreasing rate

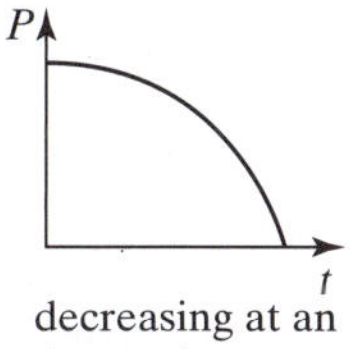

decreasing at an increasing rate

8 When **drawing a graph**, the axes must be clearly labelled and the scale on each axis must be equally spaced. The independent variable is placed on the horizontal axis and the dependent variable on the vertical axis. Use the information supplied to draw a clear well-sized graph.

9 A lot of **information** can be **gained from a graph**. You must be able to read and interpret the information supplied. Simple, clear explanations are required. You are not expected to write an essay of explanation.

10 **Misleading information** often arises and **incorrect conclusions** are often made from statistical displays and data. When reading about or viewing statistics in the media, for example, don't accept everything as stated. Look at any graphs and see if the conclusions drawn are appropriate. Question where the material came from (was it a reputable source, was the sample large enough and unbiased) and do your own evaluation of the data.

Checklist

Can you:

1 *Understand the terms univariate, bivariate, independent variable and dependent variable?* ☐
2 *Read information from a graph and interpret that information?* ☐
3 *Draw graphs?* ☐
4 *Critically evaluate reports in the media?* ☐

DATA INVOLVING TIME; EVALUATING REPORTS

Data Representation and Interpretation

SKILLS CHECK

1 The graph shows the distance Yasmin is from home during a hike she took recently.

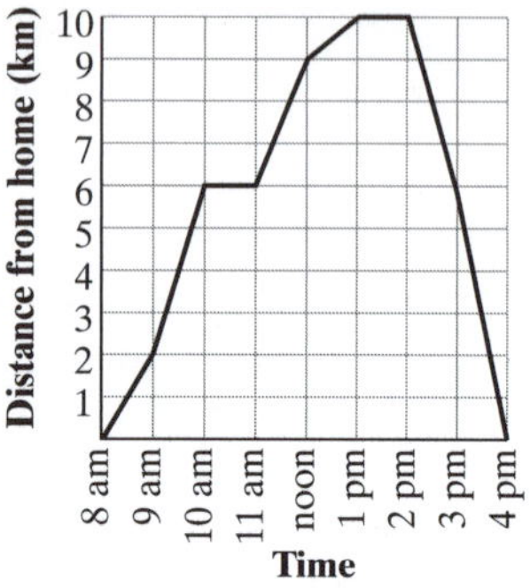

- **a** How far was Yasmin from home at 9 am?
- **b** What was her distance from home at 1 pm?
- **c** What distance was travelled between 11 am and noon?
- **d** When did Yasmin leave home?
- **e** At what time did she get back home?
- **f** When was Yasmin exactly 6 km from home?
- **g** Yasmin stopped twice during her journey. When did this happen?
- **h** How is the fact that Yasmin is stopped shown on the graph?
- **i** When did Yasmin start to travel home?
- **j** How is that shown on the graph?
- **k** Without calculating her speed, was Yasmin travelling faster between 9 and 10 o'clock or between noon and 1 o'clock?
- **l** How is it possible to tell when she is travelling fastest?
- **m** How far did Yasmin travel altogether?

2 Each graph below shows population over time. Which graph (A, B, C, D or E) shows the population:

- **a** constant?
- **b** increasing constantly?
- **c** decreasing constantly?
- **d** increasing at a constant rate, then at a faster constant rate?
- **e** increasing at a constant rate, then at a slower constant rate?

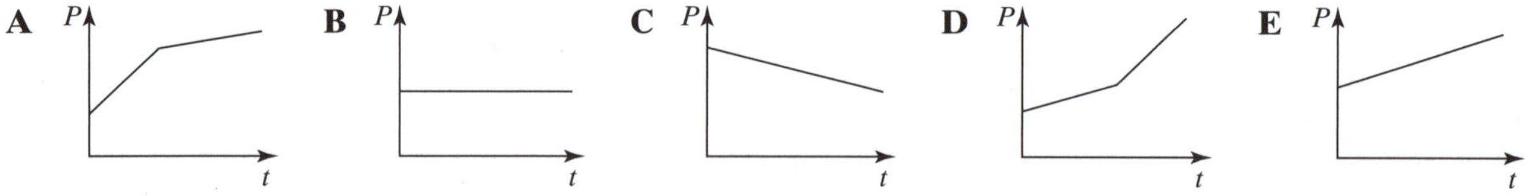

3 The step graph shows the prices charged for hiring skates at Hudson's Grass Ski Park.

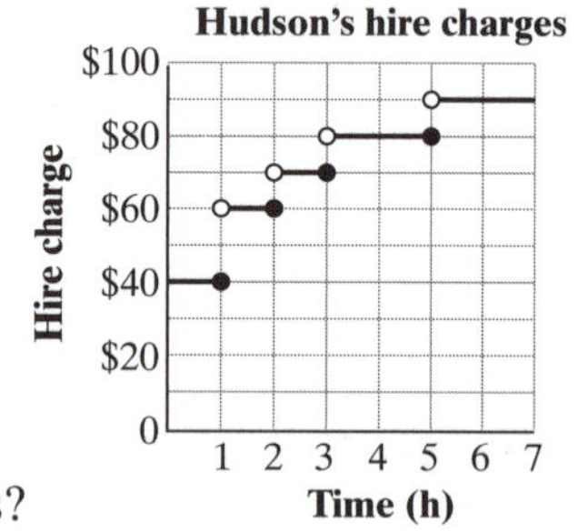

- **a** How much does it cost to hire skates for 4 hours?
- **b** How much does it cost to hire skates for 80 minutes?
- **c** How much does it cost to hire skates for 5 hours?
- **d** Penny paid $70 when she hired the skates. For how long did Penny have the skates?
- **e** Mervyn wanted the skates for 5 hours, but returned them 5 minutes late. How much extra did Mervyn pay because he was late returning the skates?

4 Each of these displays is misleading. Briefly explain why.

a

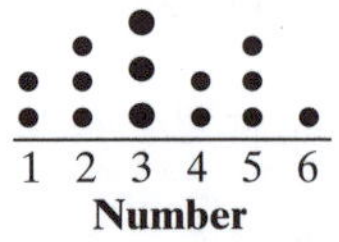

b

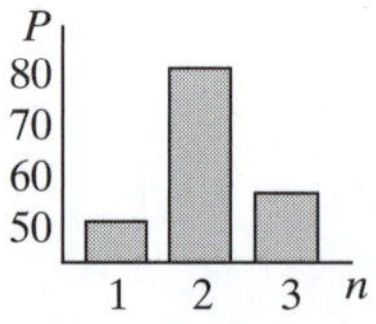

c

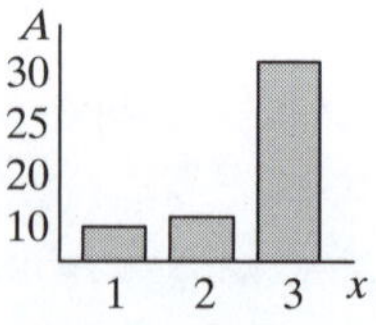

d

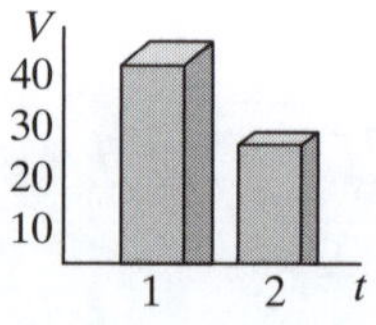

PAGES 165–166

Answers **1 a** 2 km **b** 10 km **c** 3 km **d** 8 am **e** 4 pm **f** between 10 am and 11 am, 3 pm **g** from 10 am to 11 am and 1 pm to 2 pm **h** horizontal line **i** 2 pm **j** negative gradient **k** between 9 am and 10 am **l** bigger gradient **m** 20 km **2 a** B **b** E **c** C **d** D **e** A **3 a** $80 **b** $60 **c** $80 **d** more than 2 hours but less than or equal to 3 hours **e** $10 **4** See worked solutions

DATA INVOLVING TIME; EVALUATING REPORTS

Data Representation and Interpretation

15 MINUTES

INTERMEDIATE TEST

Part A Multiple Choice

1 Which graph could <u>not</u> represent an actual journey?

A
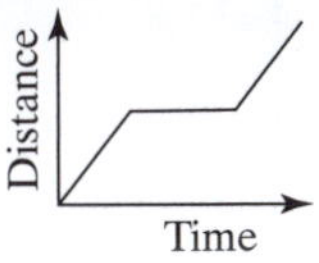

B
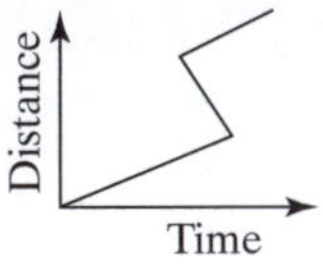

C
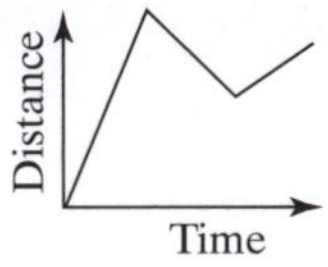

D
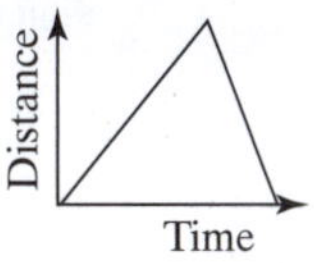

(1 mark)

2 This travel graph shows the distance from its depot of a train as time passes. Which section of the graph represents the period when the train travels at the fastest average speed? *Hint 1*

A OP B PQ
C QR D RS

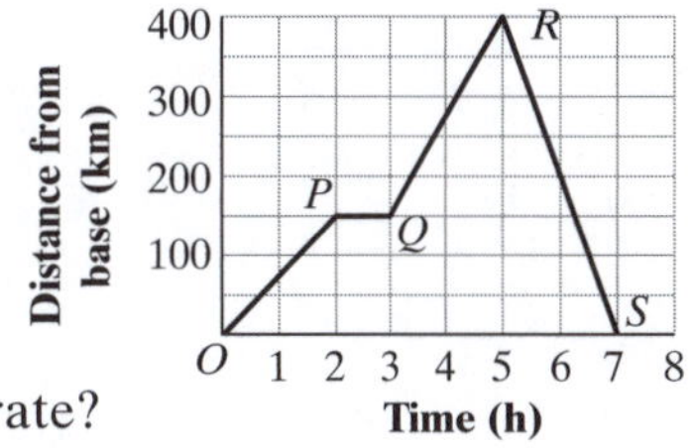

(1 mark)

3 Which graph shows the production decreasing at an increasing rate?

A
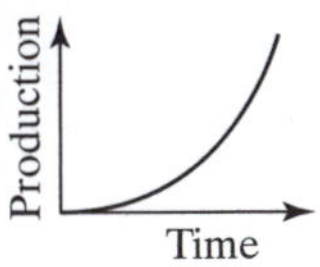

B
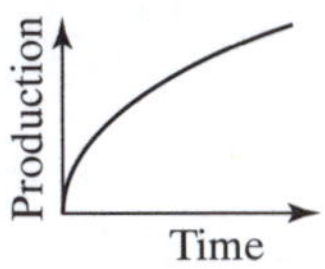

C
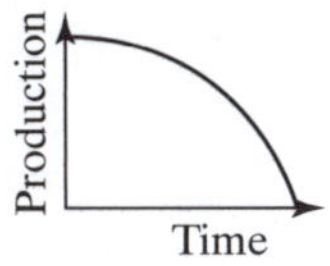

D
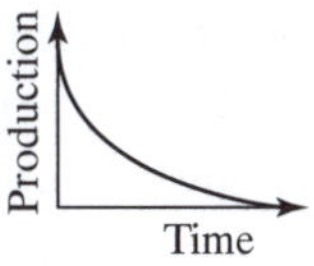

(1 mark)

4 Water is poured into this container at a constant rate. Which diagram shows the height of the water level as time passes? *Hint 2*

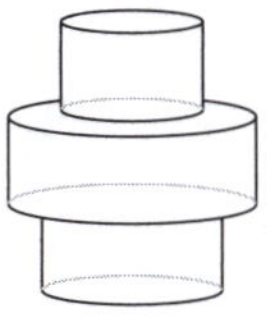

A
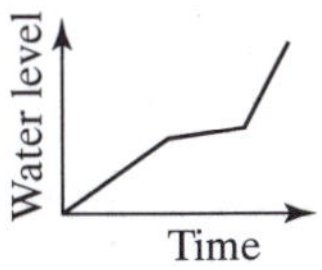

B
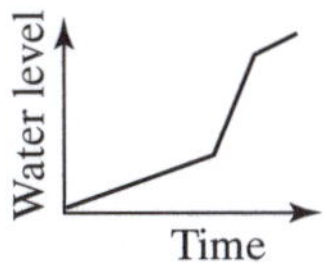

C
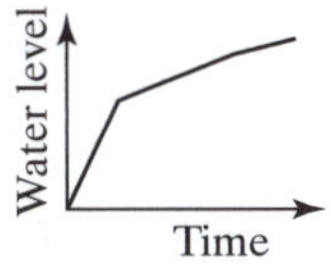

D
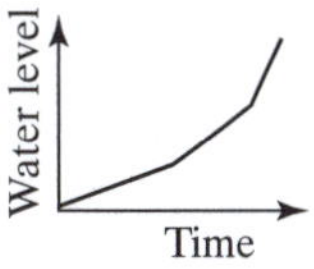

(1 mark)

Part B Short Answer

5 An automatic cleaning device pumps water through pipes to clean them. The graph shows the water level as the machine goes through its cycle. Explain what happens over the cycle. (3 marks)

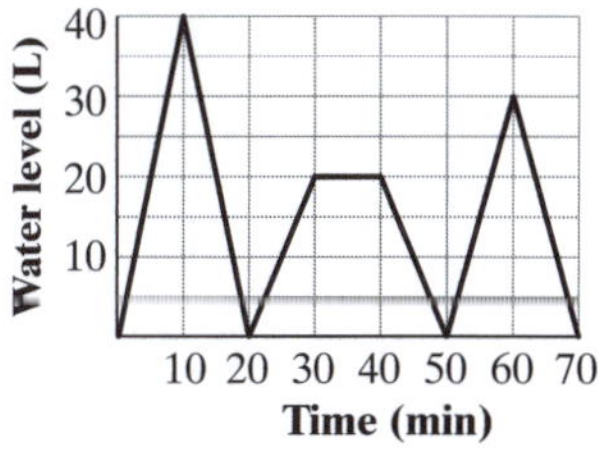

6 One night a television program showed a report on crime and then asked the viewers to respond to a survey by phoning a number to reply 'yes' or 'no' to the question 'Should police have more powers?'. The following night the program reported that 83% of respondents had said yes and 17% no, concluding that an overwhelming majority of the population believed that police should have more powers. Give three reasons why this conclusion might not be valid. *Hint 3* (3 marks)

7 The graph shows the value of three assets over the last 6 years.

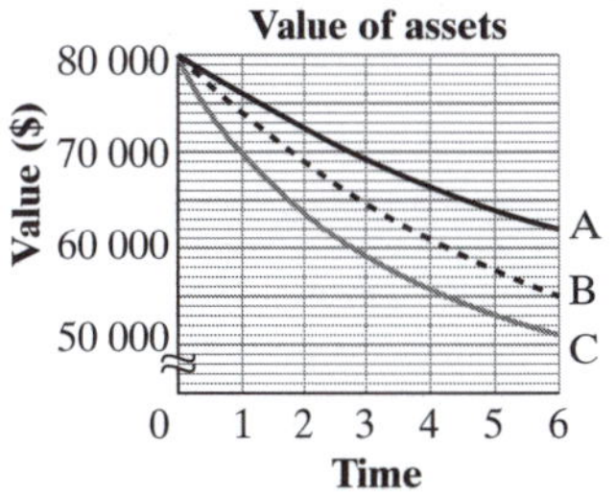

a What was the initial value of each of the assets? (1 mark)
b What was the value of asset C after 5 years? (1 mark)
c When was the value of asset B \$61 000? (1 mark)
d How much value had asset A lost after 1 year? (1 mark)
e Briefly explain why the graph could be misleading. (1 mark)

Hint 1: There is no need to find the different speeds.
Hint 2: When will the water level rise most quickly? When will it rise the slowest?
Hint 3: Consider the sample.

Your Feedback

$\frac{\square}{15} \times 100\% = \square\%$

PAGE 166
PAGE 176

DATA INVOLVING TIME; EVALUATING REPORTS

Data Representation and Interpretation

25 MINUTES

ADVANCED TEST

1 An article in a local newspaper discussed the huge increase in house prices in the area over the last year. This graph accompanied the article.

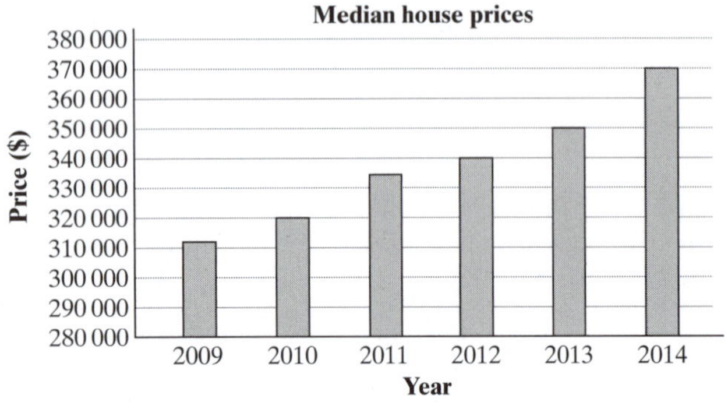

a Why do figures about house prices often use the median price rather than the mean? (1 mark)

b Why is the graph misleading? (1 mark)

c Redraw the graph as it should appear. (3 marks)

d Why might the newspaper prefer to use its original graph? (1 mark)

e Do you agree that there has been a huge increase in prices over the final year? Justify your answer. (2 marks)

2 This graph shows the population of Australia from 1900 until 2010.

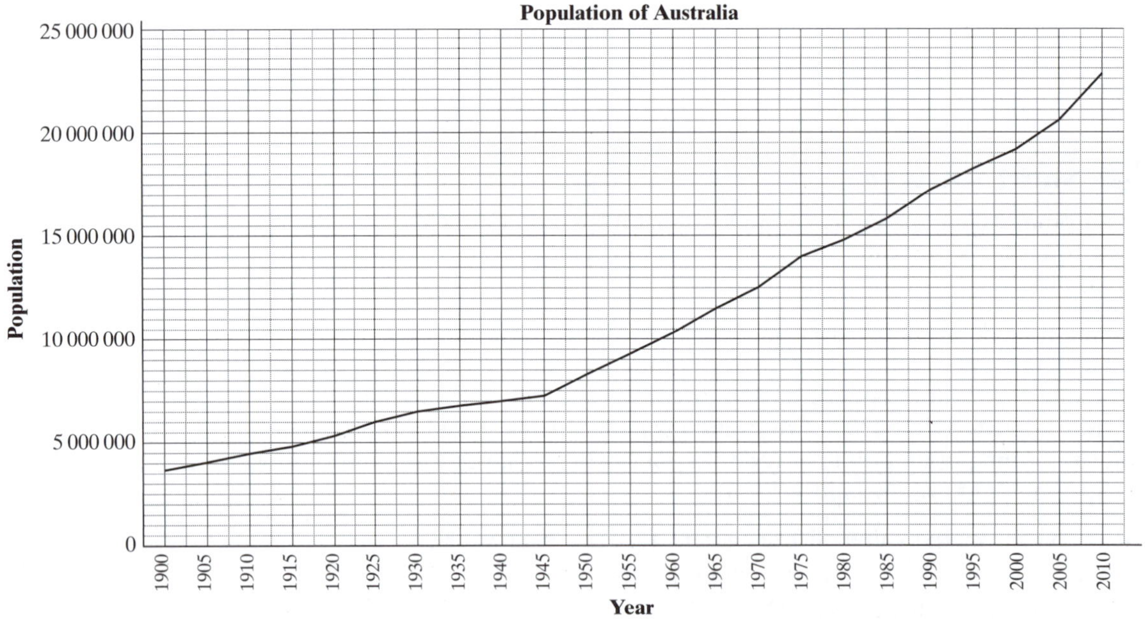

Based on Australian Bureau of Statistics data

a What is the independent variable? (1 mark)

b When did the population first exceed 15 million? (1 mark)

c In what fifteen-year period did the population increase the least? Give a reason why this might be the case. (2 marks)

d In 1960 the population of Queensland was about one and a half million. About what percentage of Australians lived in Queensland in 1960? (1 mark) e In 2010 the population of Queensland was about four and a half million. Briefly compare the growth of Queensland's population to that of Australia over those 50 years. (2 marks)

f Sulafah predicted that the population of Australia would reach 30 million in 2040. Is this a reasonable prediction? (1 mark)

3 This graph shows information about cotton production and exports in Australia as well as the gin-gate return (the amount per bale of cotton received by the grower).

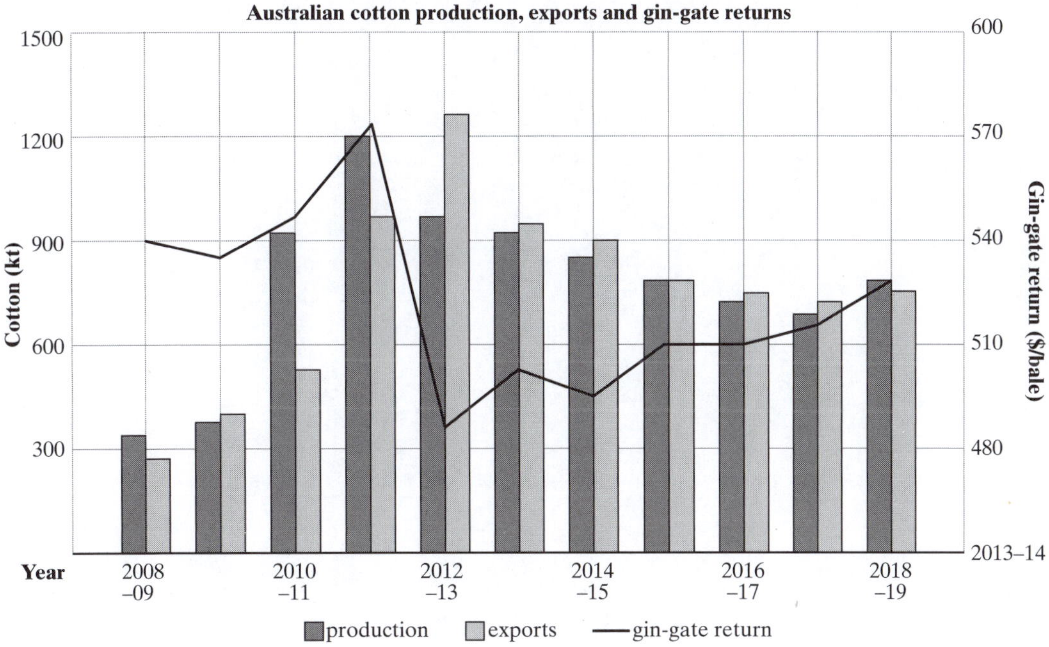

Source: ABARES 2014, *Agricultural Commodities: March quarter 2014*, CC BY 3.0.

a In what financial year were exports highest? (1 mark)

b About how many tonnes of cotton were produced in Australia in 2011–12? (1 mark)

c About what percentage of production were exports in 2010–11? (1 mark)

d In some years exports exceed production. How might this be possible? (1 mark)

e From 2014–15 onwards the figures are a prediction. Comment on expected trends. (1 mark)

f The gin-gate return uses the axis on the right-hand side. What was the approximate gin-gate return in 2008–09? (1 mark)

g By how much did the return decrease from 2011–12 to 2012–13? (1 mark)

h What do you notice about the scale on the right axis? Comment on the effects on the graph of choices made about this axis? (2 marks)

Your Feedback

$\frac{\square}{25} \times 100\% = \square\%$

PAGES 166–167
PAGE 176

TIPS FOR THE SAMPLE EXAM PAPERS

STUDY TIPS

Know what to expect

- **Find out** from your teacher exactly what topics will be assessed in the exam.
- Yearly exams normally test **all work covered** throughout the year.
- **Determine what format** is being used for the exam. Are there multiple-choice questions? Will calculators be allowed in any part of the examination?

Show all working

- **Read the instructions** on the exam paper—marks may be allocated for working.
- Even when **marks are not allocated**, working is important and it is rewarded, even though your answer may be incorrect.
- **Never use correction fluid**—just put a line through any incorrect working. Let your teacher see all your work.

Allocate your time

- Ensure you are **working through the paper efficiently** and not spending too much time on each question, only to find you run out of time at the end of the exam.

Understand mark allocations

- **Questions worth more marks** are often more complex and difficult. Working is crucial in these types of questions.

Re-read and check

- Once you have **completed a question**, rather than moving on to the next, re-read that question to make sure you have in fact answered the correct question. This will only take a second or two.
- If you **finish the exam** with time still remaining, check your answers—often mistakes are found and can be corrected.

Use quality diagrams

- Diagrams should be of **good quality, large and drawn with a lead pencil**. Use an eraser, not correction fluid, to delete mistakes.

Be ready

- Finally, you need to be **prepared for the exam**.
- Always **study mathematics actively**. Active study means using pen and paper to make notes, writing down difficult questions and their solutions, and recording rules to learn.

SAMPLE EXAM PAPERS

Level of difficulty—Average

50 MINUTES

PAPER 1

Part A Multiple Choice

(25 marks)

1 $7x^2 + 5x - x^2 = ?$

A $8x^2 + 5x$ **B** $6x^2 + 5x$ **C** $7 + 5x$ **D** $11x$ (1 mark)

2 The volume of a cylinder of radius 4 cm and height 7 cm is closest to:

A 352 cm³ **B** 226 cm³ **C** 176 cm³ **D** 117 cm³ (1 mark)

3 The value of θ, to the nearest whole degree is:

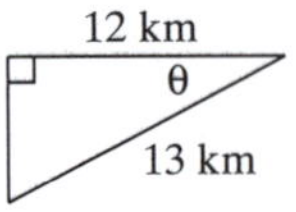

A 23° **B** 32°
C 43° **D** 67° (1 mark)

4 If $t = -2$, $4t^2 = ?$

A –16 **B** –64 **C** 16 **D** 64 (1 mark)

5 The solution of the inequality $\frac{x}{2} - 3 < 1$ is:

A $x < 8$ **B** $x < 2$ **C** $x < -2$ **D** $x < -8$ (1 mark)

6 AB is parallel to CD. CB bisects $\angle ACD$. $\angle DCE = 74°$. $\angle ACB = ?$

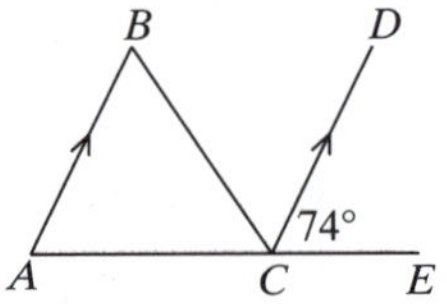

A 74° **B** 60°
C 53° **D** 37° (1 mark)

7 $\frac{x}{2} + \frac{x}{3} = ?$

A $\frac{x}{3}$ **B** $\frac{x}{5}$ **C** $\frac{2x}{5}$ **D** $\frac{5x}{6}$ (1 mark)

8 \$800 is invested for three years at 5% p.a. interest compounded annually. What is the compound interest?

A \$120 **B** \$126.10 **C** \$127.42 **D** \$400 (1 mark)

9 $a^2b^3 \div ab^2 = ?$

A ab **B** a^2b **C** ab^2 **D** $(ab)^2$ (1 mark)

10 The gradient of a line l is $\frac{2}{3}$. The gradient of any line perpendicular to l is:

A $\frac{2}{3}$ **B** $-\frac{2}{3}$ **C** $\frac{3}{2}$ **D** $-\frac{3}{2}$ (1 mark)

11 $(x - 3)(x - 7) = ?$

A $x^2 - 4x - 21$ **B** $x^2 - 4x + 21$ **C** $x^2 - 10x - 21$ **D** $x^2 - 10x + 21$ (1 mark)

12 $2(2a + 3) + 4(a - 2) = ?$

A $8a + 1$ **B** $8a + 4$ **C** $8a + 14$ **D** $8a - 2$ (1 mark)

13 $x = ?$

A 10 **B** 65
C 70 **D** 110

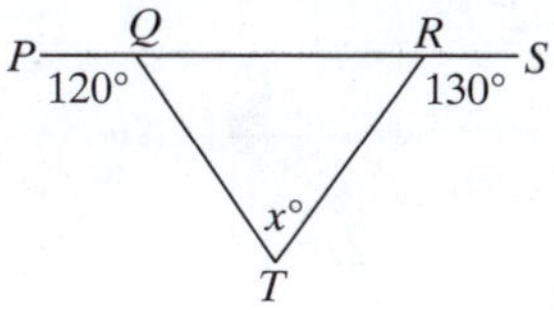

(1 mark)

14 If $x + 3y = 10$ and $x - 3y = 6$ then $x = ?$

A 2 **B** 3 **C** 5 **D** 8 (1 mark)

15 A cube has sides of length 5 cm. The surface area of the cube is:

A 100 cm^2 **B** 125 cm^2 **C** 150 cm^2 **D** 180 cm^2 (1 mark)

16 When factorised fully, $8x^2 - 6xy = ?$

A $x(8x - 6y)$ **B** $2(4x^2 - 3xy)$ **C** $2x(4x - 3y)$ **D** $2x^2y(4y - 3)$ (1 mark)

17 $PR = 28$ cm, $RS = 30$ cm and $RT = 24$ cm.
What is the length of QR?

A 22.4 cm **B** 26 cm
C 34 cm **D** 35 cm

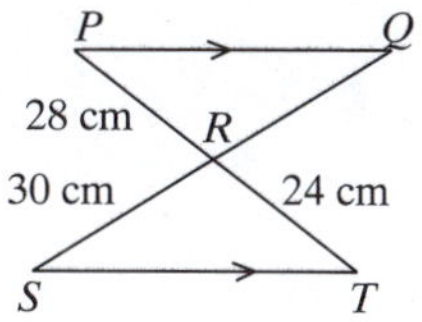

(1 mark)

18 There is a mistake in this solution to this equation.

$3(x + 2) - 5x = 18$ (Line 1)
$3x + 6 - 5x = 18$ (Line 2)
$-2x + 6 = 18$ (Line 3)
$-2x = 24$ (Line 4)
$x = -12$ (Line 5)

The mistake occurs between which lines?

A Line 1 and Line 2 **B** Line 2 and Line 3
C Line 3 and Line 4 **D** Line 4 and Line 5 (1 mark)

19 The equation $y = \frac{x}{4}$ is the equation of a:

A straight line **B** exponential curve
C parabola **D** none of these (1 mark)

20 Which does not show the circle $x^2 + y^2 = 25$?

A

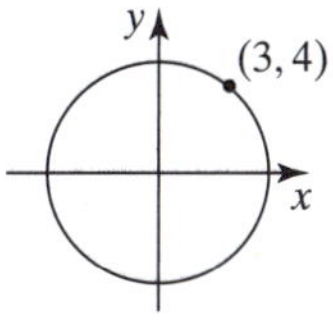

B

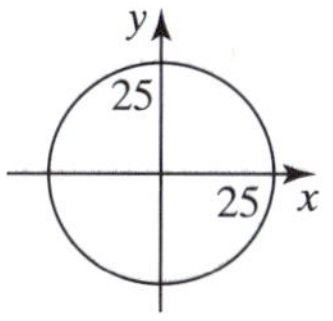

C

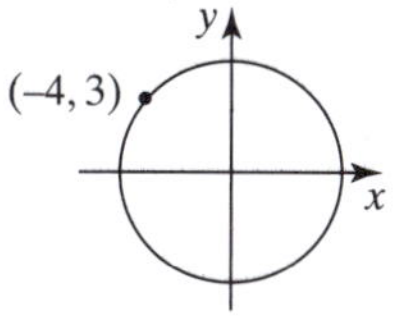

D

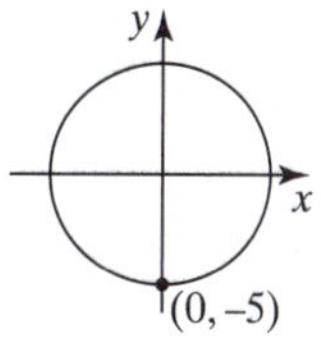

(1 mark)

21 A plane left Edgetown (E) on a bearing of 064° and flew 93 km to Fredericks (F). From Fredericks it flew 74 km on a bearing of 196° to Glensloy (G). Which diagram best fits this information?

A

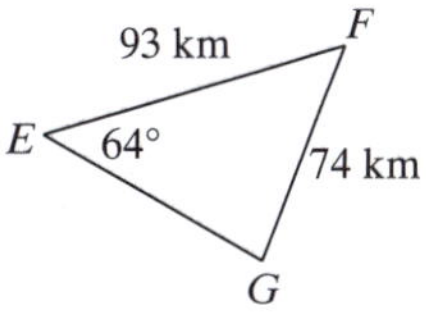

B

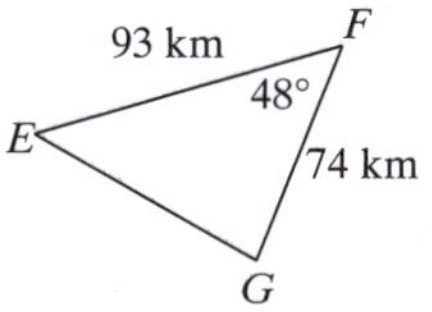

C

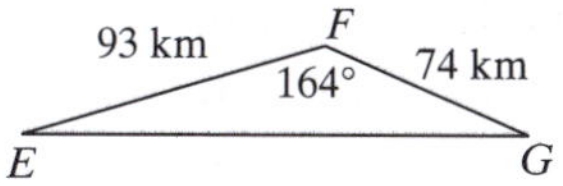

D

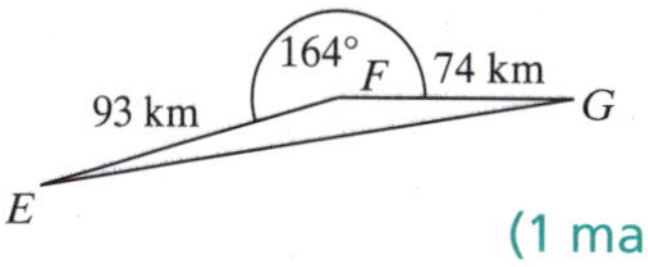

(1 mark)

22 Which diagram could show the graph of $y = 2^{-x}$?

A
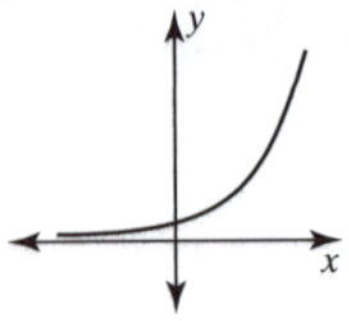

B
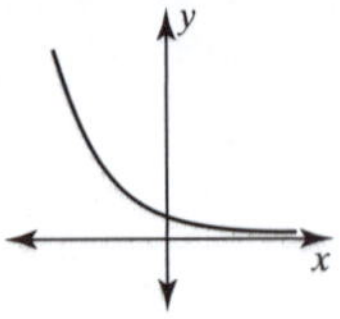

C
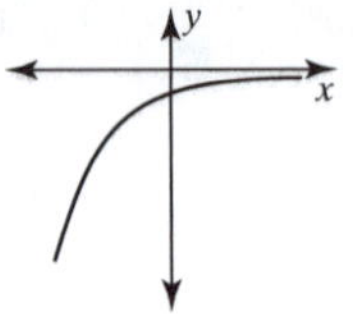

D
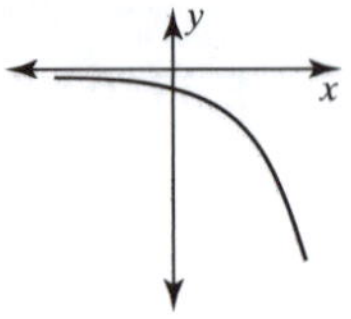

(1 mark)

23 If $\tan\theta = 0.369$, then the value of θ to the nearest minute is:

A $20°25'$ B $20°15'$ C $20°59'$ D $23°19'$ (1 mark)

24 The volume of this prism is:

A 90 m^3 B 150 m^3
C 180 m^3 D 390 m^3

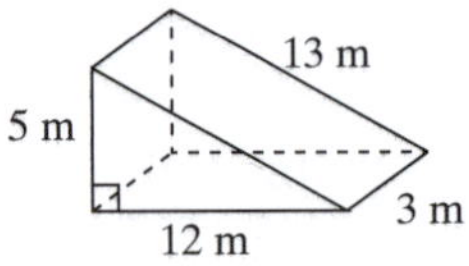

(1 mark)

25 $\dfrac{2x + 3y}{6x + 9y} = ?$

A 3 B $\dfrac{1}{6}$ C $\dfrac{2}{3}$ D $\dfrac{1}{3}$ (1 mark)

Your Score

$\dfrac{\quad}{25}$

PAGES 167–169

Part B Short Answer (25 marks)

26 The formula $C = \frac{5}{9}(F - 32)$ gives the temperature in °C (Celsius) when the temperature in °F (Fahrenheit) is known. What is the temperature in °C when it is 95 °F? (1 mark)

27 Solve $\dfrac{8x - 1}{5} \geq 3$ graphing the solution on a number line. (2 marks)

28 These are the marks Allan scored in his last 18 tests:
14, 11, 16, 13, 14, 12, 17, 16, 19, 14, 11, 17, 20, 17, 17, 19, 18, 20

a What is the median? (1 mark)
b What is the range? (1 mark)
c Draw a box plot to illustrate the data. (3 marks)

29 A box holds 1 red, 1 purple and 2 blue pens. Two pens are chosen at random from the box, one after the other without replacement.

a Draw a tree diagram to show the possible outcomes. (1 mark)
b What is the probability that both pens are blue? (1 mark)
c What is the probability that at least one of the pens is blue? (1 mark)

30 The graph shows the amount of petrol in a car's fuel tank during a day's journey.

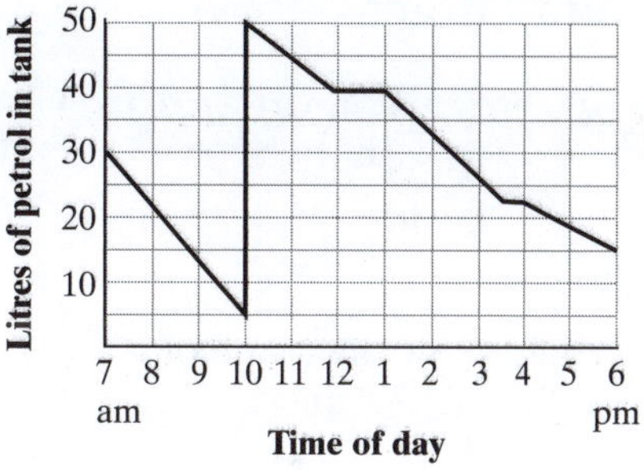

- **a** How much petrol was bought during the journey? (1 mark)
- **b** The whole trip covered a distance of 750 km. What is the car's average fuel consumption, in litres per 100 km? (2 marks)

31 Jennifer has a swimming pool with a rectangular surface of 25 m by 8 m. The pool is 1 m deep at one end and 2 m deep at the other end, as shown in the diagram.

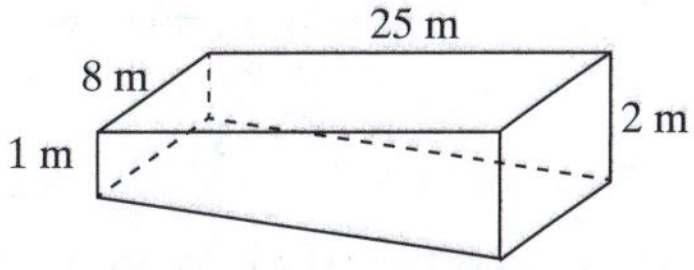

- **a** Find the volume of the pool. (1 mark)
- **b** How many litres of water will the pool hold when full? (1 mark)
- **c** 'The bottom of the pool is a rectangle 25 m long and 8 m wide,' Jennifer told her friend. Is she correct? Justify your answer. (2 marks)
- **d** The four walls and the bottom of the pool need to be painted. What is the area that needs painting? (2 marks)

32 *ABCD* is a rectangle. *E* is a point on *AB* and *F* is a point on *DC* such that $ED = BF$.

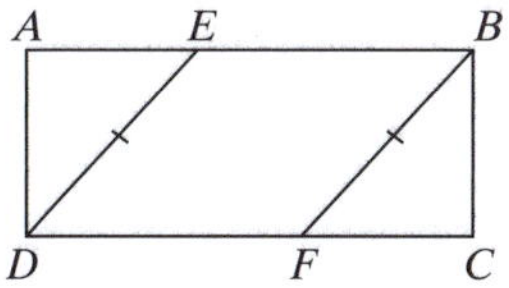

- **a** Prove that triangle *AED* is congruent to triangle *CFB*. (3 marks)
- **b** What type of quadrilateral is *EBFD*? Justify your answer. (2 marks)

Your Score

___ / 25

PAGES 169–170

Your Feedback

Part A + Part B = ___ / 50 = ___ %

PAGE 176

SAMPLE EXAM PAPERS

Level of difficulty—Above average

50 MINUTES

PAPER 2

Part A Multiple Choice **(25 marks)**

1 In ΔPQR, $\angle PQR = 70°$. $PQ = QR$. $\angle PRQ = ?$

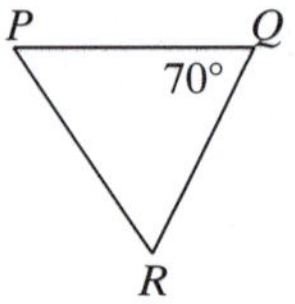

A 70° **B** 55°
C 50° **D** 40° (1 mark)

2 Which diagram could show the graph of $y = \dfrac{x}{2}$?

A
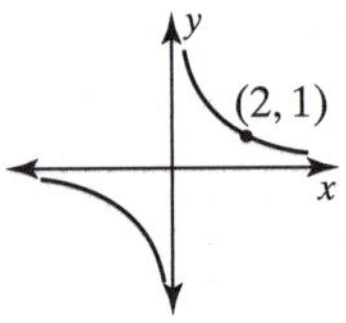

B
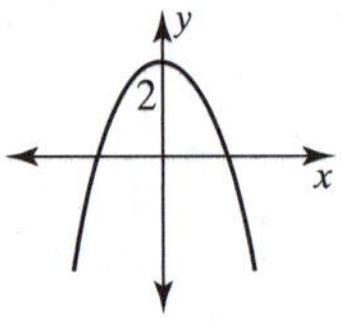

C
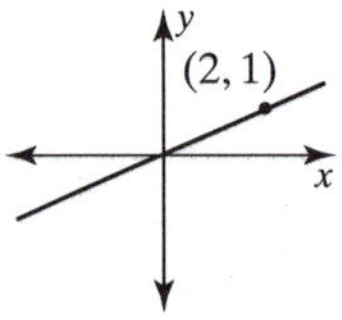

D
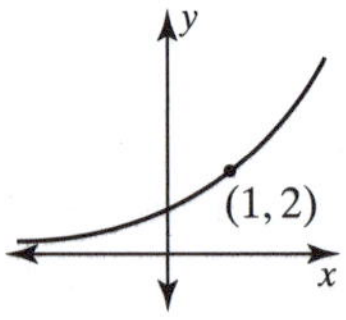

(1 mark)

3 Three coins are tossed together. What is the probability that exactly one shows a head?

A $\dfrac{1}{2}$ **B** $\dfrac{1}{3}$ **C** $\dfrac{1}{8}$ **D** $\dfrac{3}{8}$ (1 mark)

4 Which number line graph shows the solution to $-x > 2$?

A
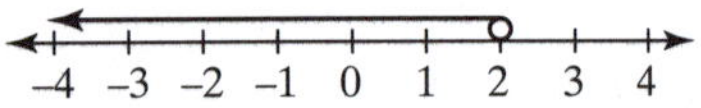

B
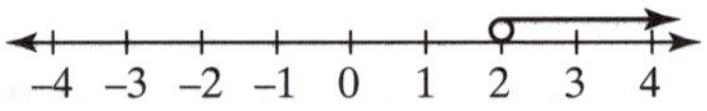

C
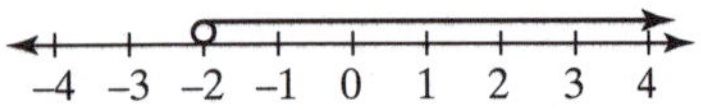

D
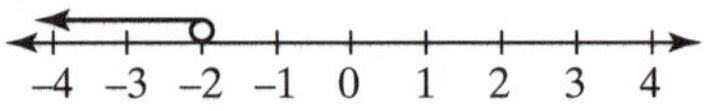

(1 mark)

5 $\dfrac{x}{2} + \dfrac{2x}{5} = ?$

A $\dfrac{3x}{7}$ **B** $\dfrac{3x}{10}$ **C** $\dfrac{9x}{7}$ **D** $\dfrac{9x}{10}$ (1 mark)

6 Which scatter plot shows a strong positive association between the variables?

A
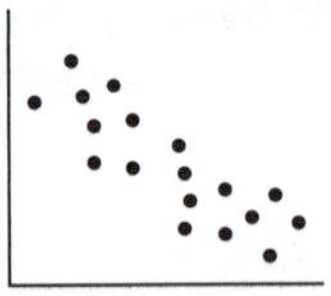

B
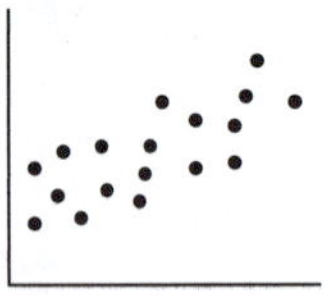

C
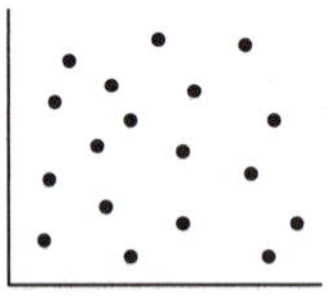

D
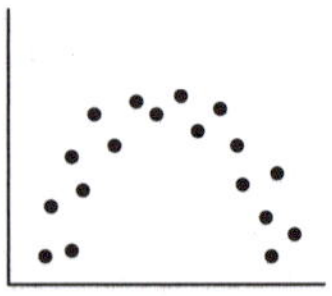

(1 mark)

7 The results of a quiz were illustrated with a box plot.

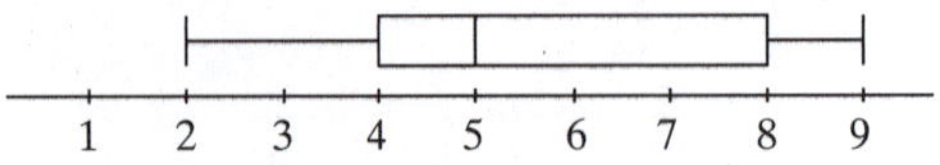

Which statement is <u>not</u> correct?

A The range is 7. **B** The median is 5.
C The interquartile range is 3. **D** The upper quartile is 8. (1 mark)

8 The surface area of a closed cylinder of radius 9 cm and height 50 cm is closest to:

A 2827 cm^2 **B** 3082 cm^2 **C** 3336 cm^2 **D** 5768 cm^2 (1 mark)

9 \$4000 is invested for 3 years. Interest of 2% per quarter is compounded quarterly. How much compound interest is earned?

A \$245 **B** \$507 **C** \$1039 **D** \$1073 (1 mark)

10 $\frac{2ab^2}{8a^2b^2} = ?$

A $\frac{1}{4a}$ **B** $\frac{a}{4}$ **C** $\frac{b}{4a}$ **D** $4a$ (1 mark)

11 $x = ?$

A 40 **B** 50
C 60 **D** 70

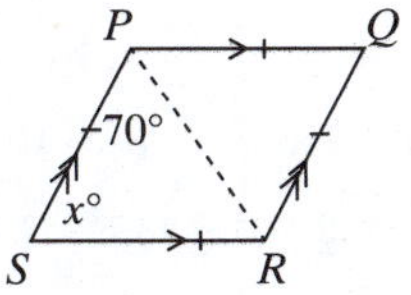

(1 mark)

12 Which is a correct factorisation of $x^2 + 5x - 6$?

A $(x-2)(x-3)$ **B** $(x+5)(x-1)$ **C** $(x+6)(x-1)$ **D** $(x+2)(x-3)$ (1 mark)

13 Amanda is standing on the roof of a building. The angle of depression of a car on the ground is 37°. Amanda knows that she is 48 m above the ground. Let x m be the distance between the car and the base of the building (on level ground). What is the value of x?

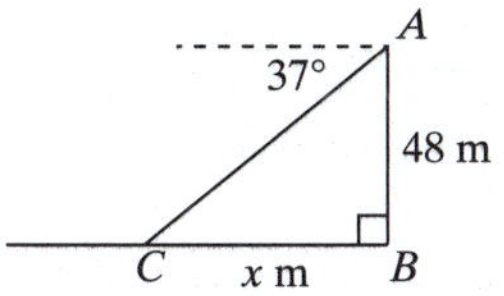

A 64 **B** 38
C 36 **D** 28 (1 mark)

14 Which graph shows the population increasing at a decreasing rate?

A
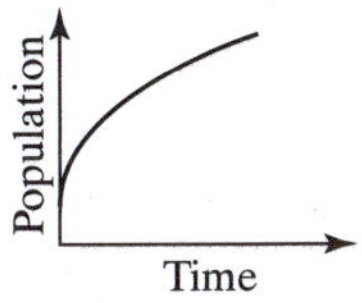

B
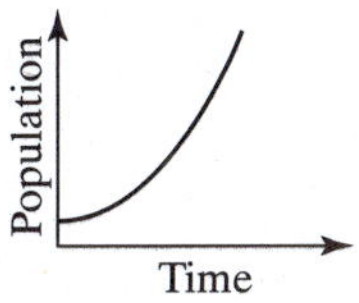

C
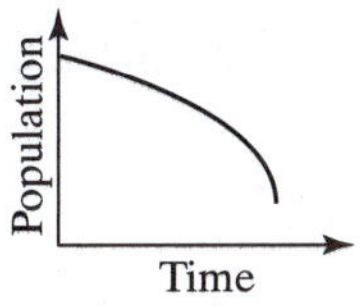

D
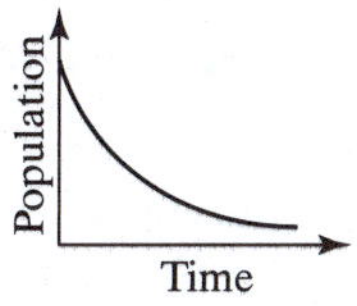

(1 mark)

15 If $A = \frac{1}{2}h(a + b)$, and $A = 108$ when $a = 7$ and $h = 12$, what is the value of b?

A 29 **B** 25 **C** 15 **D** 11 (1 mark)

16 $16m^{-6} \div 8m^2 = ?$

A $2m^{-8}$ **B** $2m^{-4}$ **C** $2m^{-3}$ **D** $2m^4$ (1 mark)

17 The solution to $3 - 2x \geq 5$ is:

A $x \leq -1$ **B** $x \geq -1$ **C** $x \leq 1$ **D** $x \geq 1$ (1 mark)

18 Which line is perpendicular to $y = \frac{1}{2}x + 5$?

A $y = \frac{1}{2}x + 7$ **B** $y = -\frac{1}{2}x + 3$ **C** $y = -2x + 1$ **D** $y = 2x - 4$ (1 mark)

19 Which rectangular prism has the greatest capacity?

A
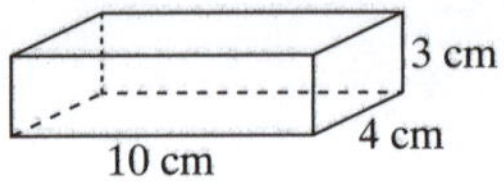

B
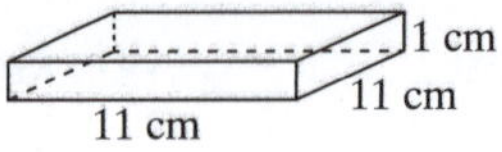

C
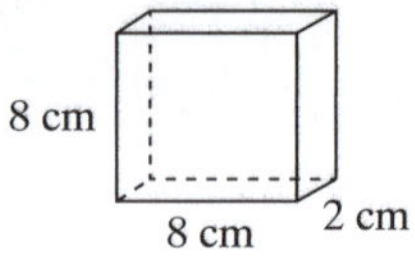

D
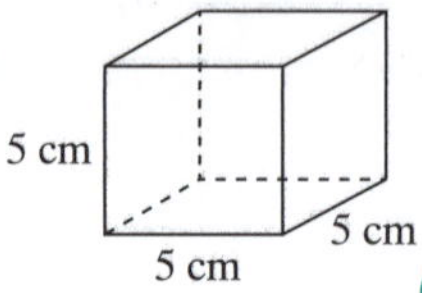

(1 mark)

20 For the scores 1, 4, 5, 6, 6, 7, 7, 7, 8, 9, 9, 10 the interquartile range is:

A 2.5 **B** 3 **C** 7 **D** 9 (1 mark)

21 The probability of choosing a red peg from one bag is $\frac{1}{4}$ and the probability of choosing a red peg from another bag is $\frac{1}{3}$. If a peg is chosen from each bag what is the probability that they are both red?

A $\frac{1}{12}$ **B** $\frac{1}{7}$ **C** $\frac{2}{7}$ **D** $\frac{7}{12}$ (1 mark)

22 In which diagram are the two triangles not necessarily congruent?

A
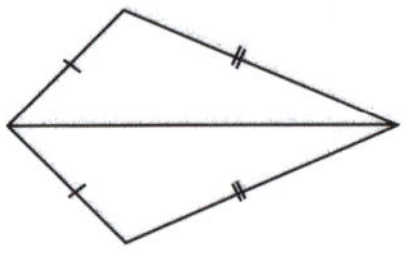

B
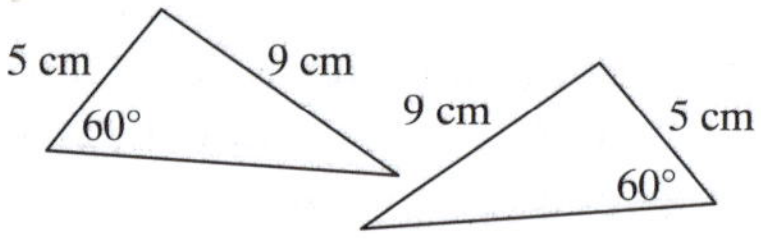

C
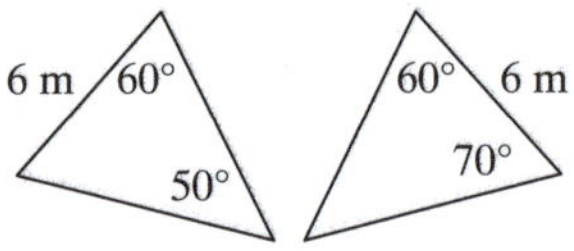

D
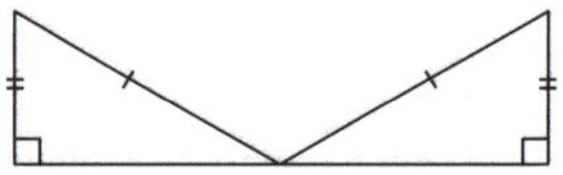

(1 mark)

23 $(2a - 3)(2a + 3) = ?$

A $4a^2 - 12a + 9$ **B** $4a^2 - 12a - 9$ **C** $4a^2 - 9$ **D** $4a^2 + 9$ (1 mark)

24 $QT = 10$ m. $TS = 24$ m and $RS = 25$ m.
What is the length of PT?

A 9 m **B** 9.6 m
C 12 m **D** 16 m

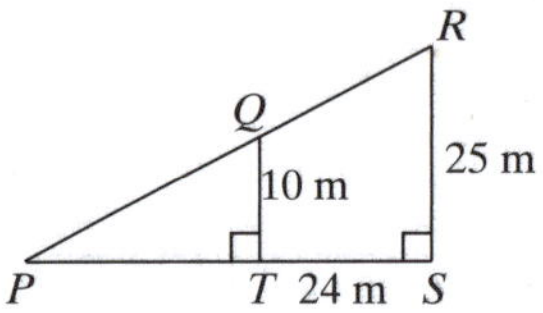

(1 mark)

25 $\frac{2a - 4}{a^2 - 4} = ?$

A $\frac{2}{a}$ **B** $\frac{2}{a + 2}$ **C** $\frac{1}{a + 1}$ **D** $\frac{1}{a - 1}$ (1 mark)

Your Score

/25

PAGES 170–171

Part B Short Answer (25 marks)

26 Factorise fully: $9m^2 - 6m$ (1 mark)

27 R is a point 750 m due south of Q. The bearing of Q from P is 052° and the bearing of R from P is 142°.

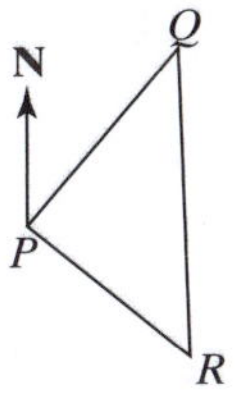

a Explain why $\angle QPR = 90°$. (1 mark)

b Explain why $\angle PQR = 52°$. (1 mark)

c Find the distance, to the nearest metre, from P to R. (2 marks)

28 Find the equation of the line that passes through the point (2, –4) and which is parallel to the line $y = 3x - 1$. (2 marks)

29 Solve $x^2 - 10x - 24 = 0$. (2 marks)

30 Solve simultaneously: $3x + 4y = 9$ and $5x - 2y = 41$. (2 marks)

31 P and Q are two towns 300 km apart. One afternoon Travis travels from P to Q and Saskia travels from Q to P. The travel graph of their journeys has been drawn.

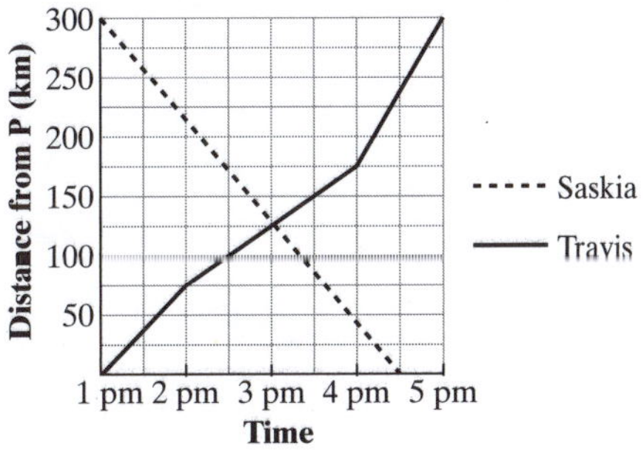

a At what time are they both the same distance from P and what is that distance? (2 marks)

b What is Travis's average speed for the whole journey? (1 mark)

c Who is travelling faster between 2.30 pm and 3.30 pm? Justify your answer. (2 marks)

32 A tractor was bought for \$48 000 five years ago. It has been depreciating at the rate of 15% per year. What is the value of the tractor now? (Give the answer to the nearest \$100.) (2 marks)

33 In the diagram, $\angle BAC = \angle CDB$ and $BE = CE$.

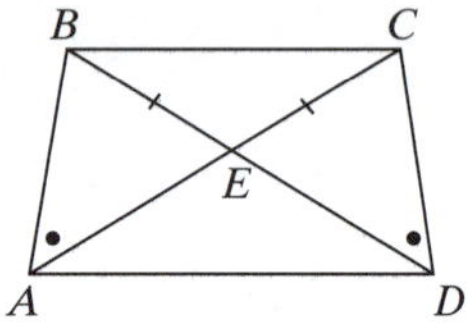

a Why does $\angle ACB = \angle DBC$? (1 mark)

b Prove that triangle BAC is congruent to triangle CDB. (2 marks)

c Prove that $\angle CAD = \angle BDA$. (2 marks)

d Is AD parallel to BC? Justify your answer. (2 marks)

Your Score

$\frac{\square}{25}$

PAGES 171–172

Your Feedback

$\frac{\square}{\text{Part A}} + \frac{\square}{\text{Part B}} = \frac{\square}{50} = \square\%$

PAGE 176

SAMPLE EXAM PAPERS
Level of difficulty—Difficult

PAPER 3

Part A Multiple Choice (25 marks)

1 These are the scores in a test.
17 14 18 15 14 12 24 18 16 25 12 19 21 13 15 14
What is the interquartile range?
A 4 **B** 4.5 **C** 5 **D** 5.5 (1 mark)

2 The simultaneous solution of the equations $3x - y + 11 = 0$ and $y = x + 3$ is:
A $x = 4, y = 1$ **B** $x = 1, y = 14$
C $x = -1, y = 2$ **D** $x = -4, y = -1$ (1 mark)

3 $-2x^2(3x - 5) - 3x(x - 5) = ?$
A $-6x^3 + 7x^2 + 15x$ **B** $-6x^3 + 7x^2 - 15x$
C $-6x^3 - 13x^2 + 15x$ **D** $-6x^3 - 13x^2 - 15x$ (1 mark)

4 The equation of this graph could be:
A $y = x^2$ **B** $y = \frac{1}{2}x^2$
C $y = 2x^2$ **D** $y = x^2 - 2$

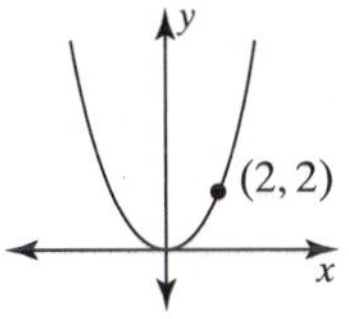

(1 mark)

5 This box plot could have been drawn from the data in which dot plot?

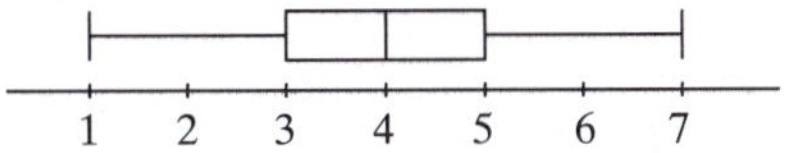

A

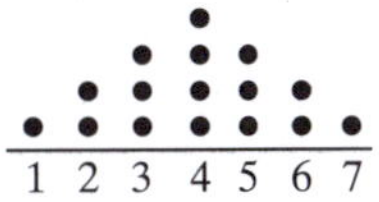

B

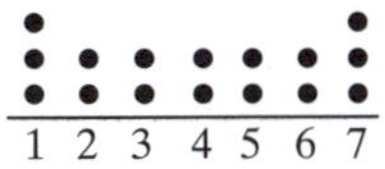

C

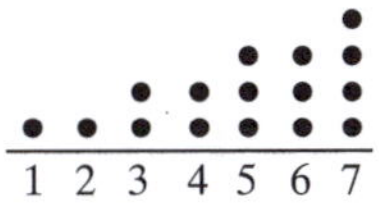

D

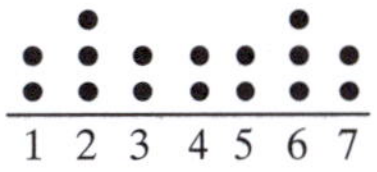

(1 mark)

6 $\frac{3t}{5} - \frac{2t}{3} = ?$
A $\frac{t}{2}$ **B** $\frac{t}{15}$ **C** $\frac{1}{2}$ **D** $-\frac{t}{15}$ (1 mark)

7 $\frac{a^2 - 25}{a^2 - 10a + 25} = ?$
A $\frac{1}{5a}$ **B** -1 **C** $\frac{a-5}{a+5}$ **D** $\frac{a+5}{a-5}$ (1 mark)

8 The solution to the inequality $2 - 3k \geq -10$ is:
A $k \leq 4$ **B** $k \geq 4$ **C** $k \leq -4$ **D** $k \geq -4$ (1 mark)

9 The equation of this line is:
A $y = -\frac{2}{3}x + 2$ **B** $y = -\frac{3}{2}x + 2$
C $y = \frac{2}{3}x + 2$ **D** $y = \frac{3}{2}x + 2$

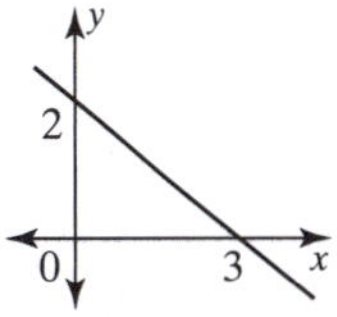

(1 mark)

10 $(3x^{-4})^{-2} \div x^{-5} = ?$

A $\dfrac{x^3}{9}$ **B** $\dfrac{x^{13}}{9}$ **C** $\dfrac{1}{9x}$ **D** $\dfrac{1}{9x^{11}}$ (1 mark)

11 In the diagram $\angle ACE = 30°$. $\angle ABD = x°$ and $\angle EDB = y°$.
$x° + y° = ?$

A 120° **B** 150°
C 180° **D** 210°

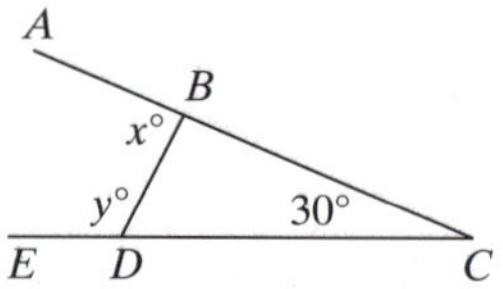

(1 mark)

12 A bag holds 5 red, 4 green and 3 blue balls. If three balls are taken at random from the bag, one after the other without replacement, what is the probability that they are all blue?

A $\dfrac{1}{27}$ **B** $\dfrac{1}{64}$ **C** $\dfrac{1}{220}$ **D** $\dfrac{1}{270}$ (1 mark)

13 $4x - 2y + \dfrac{2}{3}y - \dfrac{x}{2} = ?$

A $\dfrac{7x}{2} - \dfrac{8y}{3}$ **B** $\dfrac{7x}{2} - \dfrac{4y}{3}$ **C** $\dfrac{9x}{2} - \dfrac{8y}{3}$ **D** $\dfrac{9x}{2} - \dfrac{4y}{3}$ (1 mark)

14 Which statement is correct?

A $x = 3.2 \cos 54°$ **B** $x = 4.4 \sin 54°$
C $x = \dfrac{3.2}{\cos 54°}$ **D** $x = \dfrac{\sin 54°}{4.4}$

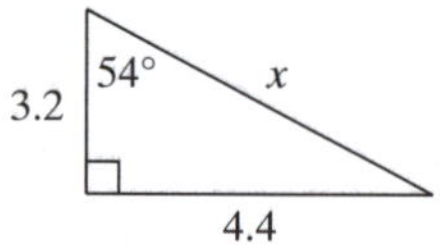

(1 mark)

15 Which is a correct rearrangement of the formula $R = \dfrac{1}{a} + \dfrac{1}{b}$?

A $b = \dfrac{a}{1 - Ra}$ **B** $b = \dfrac{1 - Ra}{a}$ **C** $b = \dfrac{Ra - 1}{a}$ **D** $b = \dfrac{a}{Ra - 1}$ (1 mark)

16 Riley borrowed \$9000 and repaid it with 18 equal monthly instalments of \$563.08. Interest was compounded quarterly on the loan. What was the annual rate of interest?

A 8% **B** 10% **C** 12% **D** 20% (1 mark)

17 One solution to the equation $\dfrac{x^2}{4} - 1 = 0$ is:

A $x = \dfrac{1}{2}$ **B** $x = -2$ **C** $x = 4$ **D** $x = 16$ (1 mark)

18 The graph shows the water level in a container over time as it is being filled. Which could be the container?

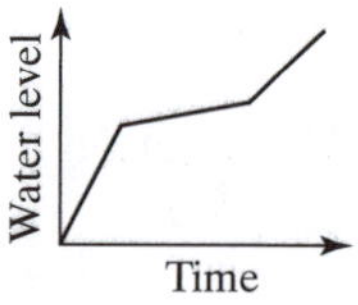

A
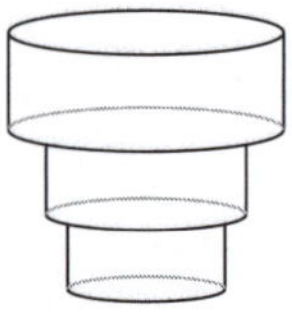

B
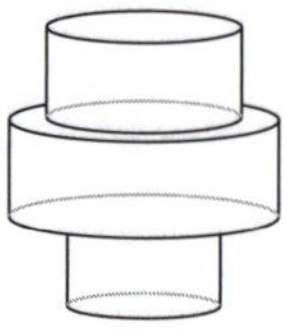

C
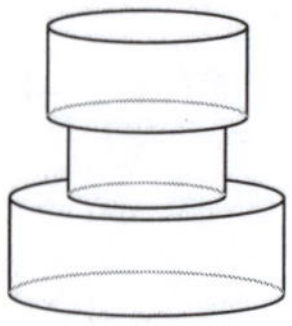

D
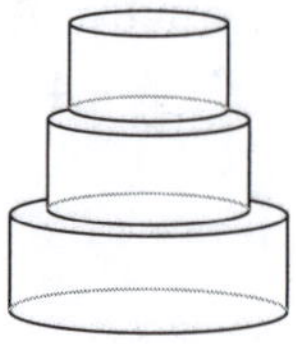

(1 mark)

19 Two similar figures have volumes in the ratio 1:8. The surface area of the larger figure is 72 m^2. What is the surface area of the smaller figure?

A 9 m^2 **B** 16 m^2 **C** 18 m^2 **D** 36 m^2 (1 mark)

20 The travel graph shows the distance of a truck from its depot. Between which times is the speed of the truck greatest?

A 8 am and 9 am **B** 9 am and 10 am
C 10 am and 11 am **D** 11 am and 12 noon

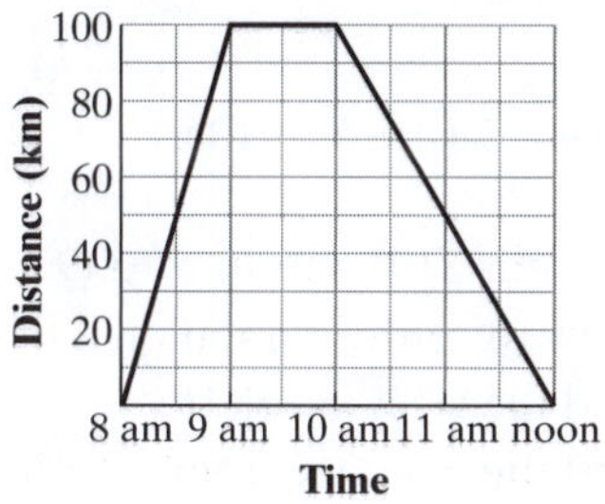

(1 mark)

21 What is the volume of this prism?

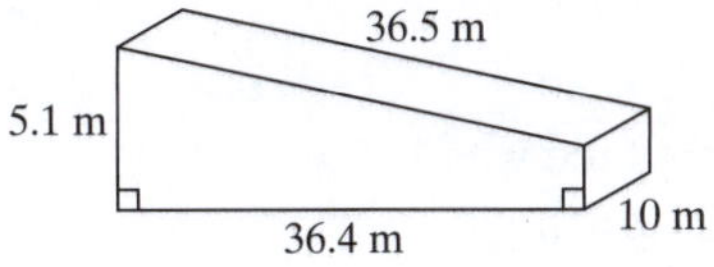

A 1365 m^3 **B** 1327 m^3 C 1859 m^3 **D** 1420 m^3 (1 mark)

22 The diagram shows a sector of a circle, centre O, radius 3 cm. The perimeter of the sector is closest to:

A 6.9 cm **B** 7.3 cm
C 7.9 cm **D** 8.6 cm

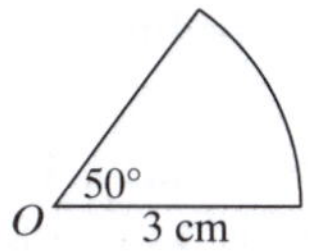

(1 mark)

23 $ABCEF$ is a regular pentagon. BC is produced to D. BE is perpendicular to ED. What is the size of $\angle BDE$?

A 36° **B** 54°
C 63° **D** 72°

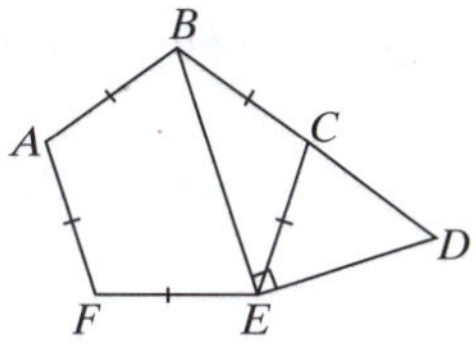

(1 mark)

24 C is the midpoint of AE. BD is perpendicular to both AB and ED. Which test could be used to prove that triangle ABC is congruent to triangle EDC?

A SSS **B** SAS
C RHS **D** AAS

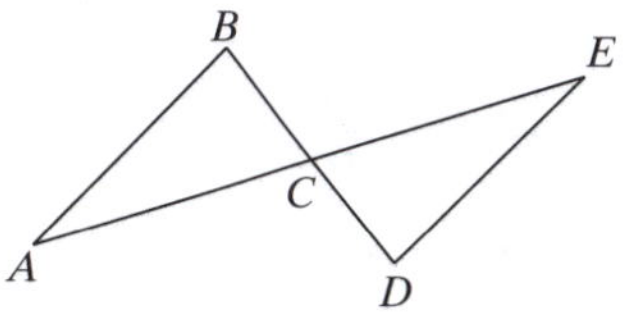

(1 mark)

25 A circle has equation $x^2 + y^2 = 25$. Where does the point (4, 3) lie?

A inside the circle **B** on the circle
C outside the circle **D** There is not enough information. (1 mark)

Your Score

/25

PAGES 172–174

Part B Short Answer (25 marks)

26 Factorise fully:

a $x^2 + xy - x - y$ (1 mark)

b $x^2 - x^4$ (1 mark)

27 Solve $(n - 3)(n + 5) = 105$. (2 marks)

28 A boat sails 21 km from P to Q on a bearing of 041°. From Q it sails 28 km on a bearing of 131° to R.

a Show this information on a diagram. (1 mark)

b Find the size of $\angle PQR$. (1 mark)

c Find the distance from P to R. (1 mark)

d Find the size of $\angle PRQ$ to the nearest degree. (1 mark)

e Find the bearing of P from R. (1 mark)

29 Oscar rolled a fair die three times. Each time the number on the uppermost face was 1. Before Oscar rolled the die a fourth time he commented that the chance of getting a 1 this time was less than $\frac{1}{6}$ because the die was very unlikely to roll the same number four times.

a Explain why Oscar is not correct. (1 mark)

b Find the probability of rolling a 1 four times in a row. (1 mark)

30 There were 46 children at a party. There were 8 more girls than boys. Let a be the number of girls and let b be the number of boys.

a Write a pair of equations to show this information. (1 mark)

b How many boys and how many girls were at the party? (2 marks)

31 $AD = 14.4$ m and $BD = 14.28$ m. BC is perpendicular to BD and BD is perpendicular to AD.
$\angle BAD = \angle CDB$

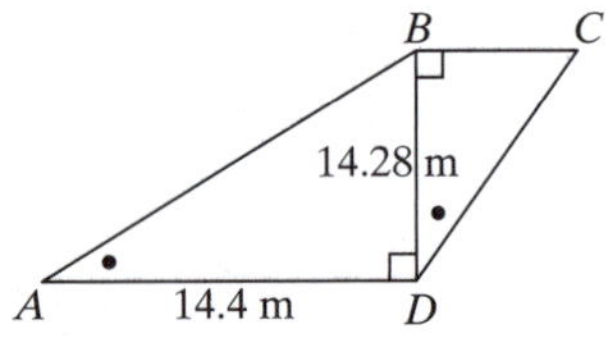

Find the length of CD. (3 marks)

32 The shaded face is made up of an isosceles triangle on a square.
It forms the cross-section of a prism.

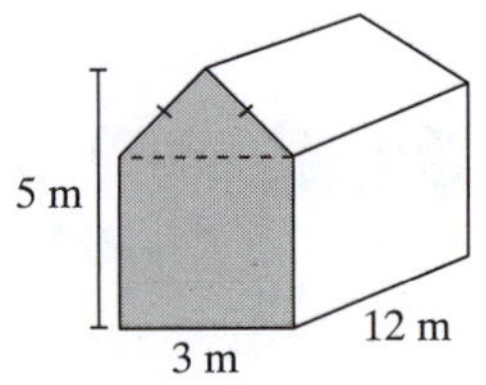

Find the surface area of the prism. (3 marks)

33 A, B and C are the points (5, –2), (3, 5) and (–4, –1) respectively.

a Find the equation, in general form, of the line l through C perpendicular to the line joining A and B. (2 marks)

b Where does line l cut the x-axis? (1 mark)

34 This graph was used by the Hungry Horse company to show how sales of MTGG products had increased.

a Briefly explain what is wrong with the graph. (1 mark)

b Because sales had doubled, each dimension on the 'before' diagram was doubled to produce the 'now' diagram. By what factor should each dimension be multiplied to produce an image that is twice as large? Give the answer correct to two decimal places. (1 mark)

Your Score

/ 25

PAGES 174–175

Your Feedback

Part A + Part B = / 50 = %

PAGE 176

COMPOUND INTEREST
SKILLS CHECK PAGE 2

1 a 10% of $800 = 0.1 × $800
= $80

b 7% of $6000 = 0.07 × $6000
= $420

c 3% of $200 000
= 0.03 × $200 000
= $6000

2 a 7 years = 7 × 12 months
= 84 months

b 5 years = 5 × 4 quarters
= 20 quarters

3 a 6% p.a.
= (6 ÷ 12)% per month
= 0.5% per month

b 6% p.a.
= (6 ÷ 4)% per quarter
= 1.5% per quarter

4 Interest = 4 × $360
= $1440

5 Annual interest = $1860 ÷ 6
= $310

6 a 8% = 8 ÷ 100
= 0.08

b 12% = 12 ÷ 100
= 0.12

c $2\frac{1}{2}$% = 2.5 ÷ 100
= 0.025

d 0.5% = 0.5 ÷ 100
= 0.005

7 a $P = \$5000, R = 0.07, N = 3$
$I = PRN$
= $5000 × 0.07 × 3
= $1050

b $P = \$240000, R = 0.08, N = 5$
$I = PRN$
= $240 000 × 0.08 × 5
= $96 000

c $P = \$12500, R = 0.008,$
$N = 24$
[The interest is charged each month so the number of periods must be the number of months. 2 years = 24 months]
$I = PRN$
= $12 500 × 0.008 × 24
= $2400

8 a Interest = 0.09 × $10 000
= $900

b Balance = $10 000 + $900
= $10 900

c Interest = 0.09 × $10 900
= $981

d Balance = $10 900 + $981
= $11 881

e Interest = 0.09 × $11 881
= $1069.29
Balance
= $11 881 + $1069.29
= $12 950.29

9 a $P = \$5000, R = 0.06, n = 4$
$A = P(1 + R)^n$
$= \$5000(1 + 0.06)^4$
$= \$5000(1.06)^4$
= $6312.3848
= $6312 (nearest dollar)

b $P = \$8750, R = 0.1, n = 7$
$A = P(1 + R)^n$
$= \$8750(1 + 0.1)^7$
= $17 051.27463...
= $17 051 (nearest dollar)

c $P = \$125000, R = 0.0075,$
$n = 144$
$A = P(1 + R)^n$
$= \$125000(1.0075)^{144}$
= $366 604.5967...
= $366 605
(nearest dollar)

10 $P = \$25000, R = 0.04, n = 10$
$A = P(1 + R)^n$
$= \$25000(1.04)^{10}$
= $37 006.107 12...
= $37 006 (nearest dollar)

11 a $P = \$8500, R = 0.05, n = 4$
$A = P(1 + R)^n$
$= \$8500(1 + 0.05)^4$
$= \$8500(1.05)^4$
= $10 331.803 125
= $10 331.80
(nearest cent)

b I = $10 331.80 – $8500
= $1831.80 (nearest cent)

12 $P = \$7000, R = 0.1, n = 3$
$A = P(1 - R)^n$
$= \$7000(1 - 0.1)^3$
$= \$7000(0.9)^3$
= $5103

COMPOUND INTEREST
INTERMEDIATE TEST PAGE 3

1 $P = \$7000, R = 0.1, n = 3$
$A = P(1 + R)^n$
$= \$7000(1 + 0.1)^3$
= $9317
∴ [B] ✓ (1 mark)

2 Try each option.
A: $P = \$12000, R = 0.08, N = 4$
$I = PRN$
= $12 000 × 0.08 × 4
= $3840
B: $P = \$8000, R = 0.12, N = 4$
$I = PRN$
= $8000 × 0.12 × 4
= $3840
C: $P = \$9000, R = 0.095, n = 4$
$A = P(1 + R)^n$
$= \$9000(1 + 0.095)^4$
= $12 939 (nearest dollar)
I = $12 939 – $9000
= $3939
D: $P = \$9500, R = 0.09, n = 4$
$A = P(1 + R)^n$
$= \$9500(1 + 0.09)^4$
= $13 410 (nearest dollar)
I = $13 410 – $9500
= $3910
$3939 is the greatest amount of interest.
∴ [C] ✓ (1 mark)

3 Try each option.
A: I = $1000 × 0.085 × 2
= $170
B: $A = \$1000(1.02)^8$
= $1171.66 (nearest cent)
I = $171.66 (nearest cent)
C: $A = \$1000(1.006)^{24}$
= $1154.39 (nearest cent)
I = $154.39 (nearest cent)
D: $A = \$1000(1.075)^2$
= $1155.63 (nearest cent)
I = $155.63 (nearest cent)
$171.66 is the largest amount.
∴ [B] ✓ (1 mark)

4 $P = \$50000$, $R = 1\%$,
$n = 300$ [25×12]
$A = P(1 + R)^n$
$= \$50000(1.01)^{300}$
$= \$989423.3131\ldots$
$= \$989423$ (nearest dollar)
$I = A - P$
$= \$989423 - \$50000$
$= \$939423$
$\therefore$ **[D]** ✓ (1 mark)

5 $A = \$20000$, $R = 0.0225$, $n = 12$
$A = P(1 + R)^n$
$\$20000 = P(1.0225)^{12}$
$= P \times 1.3060499\ldots$
$P = \$20000 \div 1.3060499\ldots$
$= \$15313.3495\ldots$
Of the options, \$15315 is the closest.
$\therefore$ **[A]**
[Or try each of the options to see which gives a value of A close to \$20000.] ✓ (1 mark)

6 **a** $P = \$5000$, $R = 0.06$, $n = 4$
$A = P(1 + R)^n$
$= \$5000(1 + 0.06)^4$
$= \$6312.3848$
$= \$6312.38$
(nearest cent) ✓
$I = \$6312.38 - \$5000$
$= \$1312.38$ ✓ (2 marks)

b $P = \$5000$,
$R = 0.06 \div 12 = 0.005$,
$n = 4 \times 12 = 48$
$A = P(1 + R)^n$
$= \$5000(1 + 0.005)^{48}$
$= \$6352.44580\ldots$
$= \$6352.45$
(nearest cent) ✓
$I = \$6352.45 - \$5000$
$= \$1352.45$ ✓
Difference
$= \$1352.45 - \$1312.38$
$= \$40.07$ ✓
[It wasn't necessary to actually find the interest because the difference between the accumulated amounts is also the difference in interest.]
(3 marks)

7 **a** $P = \$90000$, $R = 0.08$, $N = 3$
$I = PRN$
$= \$90000 \times 0.08 \times 3$
$= \$21600$ ✓
Total to be repaid
$= \$90000 + \$21600$
$= \$111600$ ✓
Each payment
$= \$111600 \div 3$
$= \$37200$ ✓ (3 marks)

b $I = 8\%$ of \$90000
$= 0.08 \times \$90000$
$= \$7200$ ✓
Amount owing
$= \$90000 + \$7200 - \$35000$
$= \$62200$ ✓ (2 marks)

c $I = 8\%$ of \$62200
$= 0.08 \times \$62200$
$= \$4976$ ✓
Amount owing
$= \$62200 + \$4976 - \$35000$
$= \$32176$ ✓ (2 marks)

d $I = 8\%$ of \$32176
$= 0.08 \times \$32176$
$= \$2574.08$ ✓
Amount owing
$= \$32176 + \$2574.08$
$= \$34750.08$
Mike must pay \$34750.08 ✓ (2 marks)

e Total Mike repays
$= 2 \times \$35000 + \$34750.08$
$= \$104750.08$ ✓
Lily repays \$111600
Difference
$= \$111600 - \$104750.08$
$= \$6849.92$
Lily pays \$6849.92 more than Mike. ✓ (2 marks)

8 **a** $A = P(1 - R)^n$
$= \$32000(1 - 0.1)^4$
$= \$20995.20$ ✓ (1 mark)

b $A = \$32000(1 - 0.1)^n$
$= \$32000(0.9)^n$
We require $A = \$10000$
$\$10000 = \$32000(0.9)^n$ ✓
If $n = 10$, $A = \$11158$ (nearest dollar)
If $n = 12$, $A = \$9038$ (nearest dollar)
If $n = 11$, $A = \$10042$ (nearest dollar)
After 11 years the car would be worth approximately \$10000. ✓
[You can begin with any guess for the value of n. If it is too high, choose a lower value. If it is too low choose a higher value and so on until you have found the closest value.] (2 marks)

9 $P = \$30000$, $R = 9\%$, $N = 5$
a $I = PRN$
$= \$30000 \times 0.09 \times 5$
$= \$13500$ ✓
Total to be repaid
$= \$30000 + \$13500$
$= \$43500$ ✓
5 years = 60 months
Each instalment
$= \$43500 \div 60$
$= \$725$ ✓ (3 marks)

b $P = \$30000$, $R = 0.7\%$,
$n = 60$
$A = P(1 + R)^n$
$= \$30000(1.007)^{60}$
$= \$45592.088\ldots$
$= \$45592$ ✓
(nearest dollar)
Interest $= \$15592$
(nearest dollar)
Neil would make a profit. ✓
The interest he would receive is greater than the interest he has to pay. ✓
(3 marks)
(Total: 30 marks)

COMPOUND INTEREST
ADVANCED TEST PAGE 4

1 $A = P(1 + R)^n$, $P = \$50000$
a $R = 0.09 \div 12 = 0.0075$
$n = 7 \times 12 = 84$ ✓
$A = \$50000(1 + 0.0075)^{84}$
$= \$93660.10$
(nearest cent) ✓ (2 marks)

b $R = 0.06 \div 4 = 0.015$
$n = 12 \times 4 = 48$ ✓
$A = \$50000(1 + 0.015)^{48}$
$= \$102173.91$
(nearest cent) ✓ (2 marks)

c $R = 0.075 \div 2 = 0.0375$
$n = 25 \times 2 = 50$ ✓
$A = \$50000(1 + 0.0375)^{50}$
$= \$315046.95$
(nearest cent) ✓ (2 marks)

d $R = 0.0005$, $n = 418$
$A = \$50000(1 + 0.0005)^{418}$
$= \$61619.03$
(nearest cent) ✓ (1 mark)

2 **a** $P = \$20000$,
$R = 0.08 \div 4 = 0.02$
$n = 7 \times 4 = 28$
$A = P(1 + R)^n$
$= \$20000(1 + 0.02)^{28}$
$= \$34820.48$
(nearest cent) ✓
$I = \$34820.48 - \$20000$
$= \$14820.48$ ✓ (2 marks)

b $P = \$350000$, $R = 0.07 \div 12$,
$n = 30 \times 12 = 360$
$A = P(1 + R)^n$
$= \$350000(1 + \frac{0.07}{12})^{360}$
$= \$2840774.12$
(nearest cent) ✓
$I = \$2840774.12 - \$350000$
$= \$2490774.12$ ✓ (2 marks)

3 **a** $I = \$6620$, $P = \$20000$,
$n = 3$
$A = \$20000 + \$6620$
$= \$26620$
$A = P(1 + R)^n$
$\$26620 = \$20000(1 + R)^3$
$(1 + R)^3 = \frac{\$26620}{\$20000}$
$= 1.331$ ✓
$1 + R = \sqrt[3]{1.331}$
$= 1.1$
$R = 1.1 - 1$
$= 0.1$
So the rate of interest is 10% p.a. ✓ (2 marks)

b $P = \$26620$, $R = 0.1$, $n = 3$
$A = P(1 + R)^n$
$= \$26620(1 + 0.1)^3$
$= \$35431.22$
$I = \$35431.22 - \$26620$
$= \$8811.22$ ✓ (1 mark)

4 $R = 0.06 \div 12 = 0.005$
$A = P(1 + R)^n$
The amount will have doubled when $A = 2P$.
$2P = P(1 + 0.005)^n$
So $(1.005)^n = 2$ ✓
Now $1.005^{100} < 2$
but $1.005^{200} > 2$
So $100 < n < 200$
By trial and error,
$1.005^{139} = 2.000242\ldots$
So $n = 139$
139 months = 11 years and 7 months ✓ (2 marks)

5 **a** $P = \$35000$
$R = 0.09 \div 4 = 0.0225$
$n = 8 \times 4 = 32$
$A = P(1 + R)^n$
$= \$35000(1 + 0.0225)^{32}$
$= \$71333.61$
(nearest cent) ✓
$I = \$71333.61 - \$35000$
$= \$36333.61$ ✓ (2 marks)

b $I = \$36333.61$, $P = \$35000$,
$n = 8$
$I = PRn$
$\$36333.61 = \$35000 \times R \times 8$ ✓
$r = \frac{\$36333.61}{\$35000 \times 8}$
$= 0.12976\ldots$
$= 0.13$ (2 d.p.)
The simple interest rate is 13% p.a. ✓ (2 marks)

6 $A = \$8000$, $R = 0.04 \div 12$,
$n = 6 \times 12 = 72$
$A = P(1 + R)^n$
$\$8000 = P(1 + \frac{0.04}{12})^{72}$ ✓
$P = \$8000 \div (1 + \frac{0.04}{12})^{72}$
$= \$6295.54$
(nearest cent) ✓ (2 marks)

7 **a** Interest = 6% of \$20 000
= \$1200 ✓ (1 mark)

b Repayment
= \$20 000 + \$1200
− \$13 717.80
= \$7482.20 ✓ (1 mark)

c Interest in second year
= 6% of \$13 717.80
= \$823.07 (nearest cent) ✓
Amount owing at beginning of third year
= \$13 717.80 + \$823.07
− \$7482.20
= \$7058.67 ✓ (2 marks)

d Interest in third year
= 6% of \$7058.67
= \$423.52 (nearest cent) ✓
Amount owing at the end of third year
= \$7058.67 + \$423.52
− \$7482.20
= −\$0.01
So there is nothing owing at the end of the third year, so the loan has been repaid. ✓ (2 marks)

e The amount on which the interest is being charged is not the same each year. ✓ (1 mark)
[It is actually reducing each time, and hence the interest is reducible interest.]

f Total interest
= \$1200 + \$823.07
+ \$423.52
= \$2446.59
$I = PRn$
$\$2446.59 = \$20000 \times r \times 3$
$R = \frac{\$2446.59}{\$20000 \times 3}$
$= 0.040776\ldots$
So the simple interest rate is about 4% p.a. ✓ (1 mark)

8 **a** Total depreciation
= \$12 800 − \$512
= \$12 288
Annual depreciation
= \$12 288 ÷ 16
= \$768 ✓ (1 mark)

b The value would be negative and that makes no sense. ✓ (1 mark)

c Percentage
$= \frac{768}{12800} \times 100\%$
$= 6\%$ ✓ (1 mark)

d **i** $A = P(1 - R)^n$
$= \$12800 \times (1 - 0.06)^{16}$
$= \$4756$
(nearest dollar) ✓ (1 mark)

ii No, under that method the machine would always have some value, even though it might not be very much. ✓ (1 mark)

(Total: 35 marks)

PRODUCTS AND QUOTIENTS
SKILLS CHECK PAGE 6

1 **a** $2 \times 3x = 6x$
b $x + x = 2x$
c $x \times x = x^2$
d $a \times b = ab$
e $2x \times y = 2xy$
f $3a \times 2b = 6ab$

g $5x \times 4x = 20x^2$

h $4 \times 2x^2 = 8x^2$

i $q \times 7p = 7pq$ [or $7qp$]

j $2xy \times 2z = 4xyz$

k $2a \times 3b \times 4c = 24abc$

l $3k \times 2k \times 4t = 24k^2t$

2 **a** $8a \div 2 = 4a$

b $8a \div a = 8$

c $8a \div 2a = 4$

d $20xy \div 4 = 5xy$

e $20xy \div x = 20y$

f $20xy \div 4y = 5x$

g $20xy \div xy = 20$

h $20xy \div 4xy = 5$

3 **a** $\frac{4p}{2} = 2p$ $\left[\frac{{}^2\cancel{4}p}{{}^1\cancel{2}}\right]$

b $\frac{3ab}{3} = ab$ $\left[\frac{{}^1\cancel{3}ab}{{}^1\cancel{3}}\right]$

c $\frac{3ab}{a} = 3b$ $\left[\frac{3\,{}^1\cancel{a}b}{{}^1\cancel{a}}\right]$

d $\frac{3ab}{3b} = a$ $\left[\frac{{}^1\cancel{3}a\cancel{b}^1}{{}^1\cancel{3}\cancel{b}^1}\right]$

e $\frac{4x}{8y} = \frac{x}{2y}$
[dividing both numerator and denominator by 4]

f $\frac{2t^2}{5t} = \frac{2t}{5}$
[dividing both numerator and denominator by t]

g $\frac{6pqr}{9pq} = \frac{2r}{3}$
[dividing both numerator and denominator by $3pq$]

h $\frac{6x}{12x} = \frac{1}{2}$
[dividing both numerator and denominator by $6x$]

4 **a** If $x = 4$, $x + 3 = 4 + 3$
$= 7$

b If $x = 4$, $3x = 3 \times 4$
$= 12$

c If $x = 4$, $4x - 1 = 4 \times 4 - 1$
$= 15$

d If $x = 4$, $24 \div x = 24 \div 4$
$= 6$

e If $x = 4$, $8 - 2x = 8 - 2 \times 4$
$= 8 - 8$
$= 0$

f If $x = 4$, $x^2 = 4^2$
$= 16$

g If $x = 4$, $2x^2 = 2 \times 4^2$
$= 2 \times 16$
$= 32$

h If $x = 4$, $\frac{3x}{2} = \frac{3 \times 4}{2}$
$= 6$

5 **a** $x^4 \times x^5 = x^{4+5}$
$= x^9$

b $t^6 \times t = t^7$ [t^{6+1}]

c $5p \times p^3 = 5p^4$ [$5p^{1+3}$]

d $2a^2 \times 3a^3 = 6a^5$
[$2 \times 3 = 6$, $a^2 \times a^3 = a^{2+3}$]

e $-8m^3 \times 2m^7 = -16m^{10}$
[$-8 \times 2 = -16$, $m^3 \times m^7 = m^{3+7}$]

f $ab^2 \times 3b = 3ab^3$

g $3x^2 \times 3x^2 = 9x^4$

h $x^3y^2 \times x^4y^6 = x^7y^8$
[$x^3 \times x^4 = x^{3+4}$, $y^2 \times y^6 = y^{2+6}$]

i $6x^2y^3 \times 2x^3y^5 = 12x^5y^8$
[$6 \times 2 = 12$, $x^2 \times x^3 = x^{2+3}$,
$y^3 \times y^5 = y^{3+5}$]

j $4p^3q^7 \times 4pq^2 = 16p^4q^9$

k $3p^2qr^3 \times 5pqr^4 = 15p^3q^2r^7$
[$p^2 \times p = p^{2+1}$, $q \times q = q^2$]

l $4a^2 \times 3a^7 \times 2a^3 = 24a^{12}$
[$4 \times 3 \times 2 = 24$,
$a^2 \times a^7 \times a^3 = a^{2+7+3}$]

6 **a** $x^{12} \div x^2 = x^{12-2}$
$= x^{10}$

b $12m^{12} \div 2m^2 = 6m^{10}$
[$12 \div 2 = 6$, $m^{12} \div m^2 = m^{12-2}$]

c $16x^6 \div 4x = 4x^5$
[$x^6 \div x = x^{6-1} = x^5$]

d $9m^4n^3 \div 3m^2n^2 = 3m^2n$

e $\frac{7n^6}{14n^3} = \frac{n^3}{2}$
[dividing both numerator and denominator by $7n^3$]

f $\frac{x^4y^5}{x^3y^7} = \frac{x}{y^2}$
[There is one more factor of x in the numerator and 2 more factors of y in the denominator.]

g $\frac{9a^2b^7}{6a^2b^3} = \frac{3b^4}{2}$
[dividing by $3a^2b^3$]

h $\frac{8a^3b^6c^2}{4abc^2} = 2a^2b^5$
[The numerator is divisible by the denominator.]

7 **a** $(x^4)^2 = x^{4\times2}$
$= x^8$

b $(3t^6)^3 = 27t^{18}$
[$3^3 = 27$, $(t^6)^3 = t^{6\times3}$]

c $5(p^4)^9 = 5p^{36}$
[5 is not in the bracket; it does not get raised to the ninth power]

d $(2m^2n^4)^4 = 16m^8n^{16}$

PRODUCTS AND QUOTIENTS
INTERMEDIATE TEST PAGE 7

1 $2a^2 \times 6a^6 = 12a^8$
[$2 \times 6 = 12$, $a^2 \times a^6 = a^{2+6}$]
∴ [C] ✓ (1 mark)

2 $5n^8 \div 5n^4 = n^4$
[$5 \div 5 = 1$, $n^8 \div n^4 = n^{8-4}$]
∴ [B] ✓ (1 mark)

3 $(2x^4)^3 = 8x^{12}$
[$2^3 = 8$, $(x^4)^3 = x^{4\times3}$]
∴ [D] ✓ (1 mark)

4 $a^2b^4c^3 \times a^2bc^5 = a^4b^5c^8$
[$a^2 \times a^2 = a^{2+2}$, $b^4 \times b = b^{4+1}$,
$c^3 \times c^5 = c^{3+5}$]
∴ [C] ✓ (1 mark)

5 $4a^7 + 2a^3 \times a^4$
$= 4a^7 + 2a^7$
$= 6a^7$
∴ [B] ✓ (1 mark)

6 $\frac{15e^6}{25e^7} = \frac{3}{5e}$
[Dividing both numerator and denominator by $5e^6$.]
∴ [A] ✓ (1 mark)

7 $(x^3y^2)^4 = x^{12}y^8$
$[(x^3)^4 = x^{3\times4},\ (y^2)^4 = y^{2\times4}]$
$\therefore$ [B] ✓ (1 mark)

8 $9p^3 + 3p^3 \times 5p = 9p^3 + 15p^4$
[Multiply before you add; unlike terms cannot be added together.]
$\therefore$ [D] ✓ (1 mark)

9 $5x^2y^2(3x^3 - 4y^4)$
$= 15x^5y^2 - 20x^2y^6$
$\therefore$ [A] ✓ (1 mark)

10 $\dfrac{6a^7b^4c^9}{(abc)^5} = \dfrac{6a^7b^4c^9}{a^5b^5c^5}$
$= \dfrac{6a^2c^4}{b}$
$\therefore$ [C] ✓ (1 mark)

11 $3x^2 - 2x \times 4x = 3x^2 - 8x^2$
$= -5x^2$ ✓ (1 mark)

12 $8x^4 \div 2x^6 = 4x^{-2}$ ✓
$= \dfrac{4}{x^2}$ ✓ (2 marks)

13 $2a^8 \times 6a^4 \div 4a^5$
$= 12a^{12} \div 4a^5$ ✓
$= 3a^7$ ✓ (2 marks)

14 $(4k^3)^2 \div (24k^7 \div 3k)$
$= 16k^6 \div 8k^6$ ✓✓
$= 2$ ✓ (3 marks)

15 $4x^2(3x^5 + 7) - 3x(4x - 9)$
$= 12x^7 + 28x^2 - 12x^2 + 27x$ ✓
$= 12x^7 + 16x^2 + 27x$ ✓
(2 marks)

16 $(8x^6 \div 4x^2)^5 = (2x^4)^5$ ✓
$= 32x^{20}$ ✓ (2 marks)

17 $\dfrac{15x^{15} \div 3x^3}{10x^{10} \div 2x^2} = \dfrac{5x^{12}}{5x^8}$ ✓✓
$= x^4$ ✓ (3 marks)
(Total: 25 marks)

PRODUCTS AND QUOTIENTS ADVANCED TEST PAGE 8

1 a $9x^{-2} \times 5x^4 = 45x^2$ ✓(1 mark)
b $3a^7 \times 4a^{-4} = 12a^3$ ✓(1 mark)
c $7m^{-5} \times 2m^{-6}$
$= 14m^{-11}$ ✓ (1 mark)
d $3d^5 \times 8d^{-11}$
$= 24d^{-6}$ ✓ (1 mark)
e $24k^3 \div 6k^8 = 4k^{-5}$ ✓(1 mark)
f $40p^5 \div 5p^{-3} = 8p^8$ ✓(1 mark)
g $36m^{-3} \div 9m^7$
$= 4m^{-10}$ ✓ (1 mark)
h $70u^{-8} \div 10u^{-7}$
$= 7u^{-1}$ ✓ (1 mark)
i $18y^{-4} \div 3y^{-9}$
$= 6y^5$ ✓ (1 mark)

2 a $(5p^4)^3 = 125p^{12}$ ✓ (1 mark)
b $2(x^2y^3)^4 = 2x^8y^{12}$ ✓ (1 mark)
c $(11a^0)^2 = 121$ ✓ (1 mark)
d $(3x^{-2})^2 = 9x^{-4}$ ✓ (1 mark)
e $5(a^5)^{-2} = 5a^{-10}$ ✓ (1 mark)
f $\left(\dfrac{1}{2}e^{-3}\right)^{-4} = 16e^{12}$ ✓
(1 mark)

3 a $\dfrac{12x^3y^2 \times 4x^5y}{16x^7y^3 \div 2xy^2}$
$= \dfrac{48x^8y^3}{8x^6y}$ ✓
$= 6x^2y^2$ ✓ (2 marks)
b $\dfrac{32a^5b^3c \div 4abc^2}{6a^2bc^4 \times 4abc^2}$
$= \dfrac{8a^4b^2c^{-1}}{24a^3b^2c^6}$ ✓
$= \dfrac{a}{3c^7}$ ✓ (2 marks)
c $(9x^3y^2)^2 \times 5(xy^3)^3$
$= 81x^6y^4 \times 5x^3y^9$ ✓
$= 405x^9y^{13}$ ✓ (2 marks)
d $15p^6q^4 \div 3pq \div p^8q^9 \times 2p^3q^3$
$= 5p^5q^3 \div p^8q^9 \times 2p^3q^3$ ✓
$= 5p^{-3}q^{-6} \times 2p^3q^3$
$= 10q^{-3}$ ✓ (2 marks)
[Or $\dfrac{10}{q^3}$]

4 a $4a^2(a^2 + 3a + 1) - 5a^2$
$= 4a^4 + 12a^3 + 4a^2 - 5a^2$ ✓
$= 4a^4 + 12a^3 - a^2$ ✓
(2 marks)
b $7x^2y^2(x^2 - y) + 2xy^3(3x - 2y)$
$= 7x^4y^2 - 7x^2y^3 + 6x^2y^3$
$- 4xy^4$ ✓
$= 7x^4y^2 - x^2y^3 - 4xy^4$ ✓
(2 marks)
c $3p(5q - 2r) - 4q(2p + 3r)$
$- r(8p - 7q)$
$= 15pq - 6pr - 8pq$
$- 12qr - 8pr + 7qr$ ✓
$= 7pq - 14pr - 5qr$ ✓
(2 marks)
d $(2k)^2(2k - 3m) - k(8k^2 - 12km)$
$= 4k^2(2k - 3m)$
$- k(8k^2 - 12km)$
$= 8k^3 - 12k^2m - 8k^3$
$+ 12k^2m$ ✓
$= 0$ ✓ (2 marks)

5 a $3a^{-1} = \dfrac{3}{a}$ ✓ (1 mark)
b $(4x)^{-1} = \dfrac{1}{4x}$ ✓ (1 mark)
c $\left(\dfrac{x}{2}\right)^{-1} = \dfrac{2}{x}$ ✓ (1 mark)
d $5\left(\dfrac{1}{x}\right)^{-1} = 5x$ ✓ (1 mark)
e $\dfrac{1}{2}m^{-1} = \dfrac{1}{2m}$ ✓ (1 mark)

6 a $x(x^2 + ax + a^2) - ax(x + a)$
$= x^3 + ax^2 + a^2x - ax^2 - a^2x$
$= x^3$ ✓ (1 mark)
b When $x = 5$,
$x(x^2 + ax + a^2) - ax(x + a)$
$= x^3$
$= 5^3$
$= 125$ ✓ (1 mark)
c The expression can be simplified to x^3 which is independent of a. So, it doesn't matter what the value of a is, the expression will always be the cube of the value of x. ✓ (1 mark)
d No, Michael is not correct. Although his statement is true for linear equations, it is not for expressions involving higher powers of the unknown. For example, in this question an expression of $5x^2$ will give the same value as x^3 when $x = 5$, but the two expressions are not the same. ✓ (1 mark)
(Total: 40 marks)

FACTORISING SKILLS CHECK PAGE 10

1 Factors of 24:
1, 2, 3, 4, 6, 8, 12, 24

2 $90 = 9 \times 10$
$= 3 \times 3 \times 2 \times 5$

3 $3a^2b^3c$
$= 3 \times a \times a \times b \times b \times b \times c$

4 a Factors of 12: 1, 2, 3, 4, 6, 12
Factors of 18: 1, 2, 3, 6, 9, 18
Highest common factor = 6

b Factors of 36:
1, 2, 3, 4, 6, 9, 12, 18, 36
Factors of 48:
1, 2, 3, 4, 6, 8, 12, 16, 24, 48
Highest common factor
= 12

c Factors of 15:
1, 3, 5, 15
Factors of 30:
1, 2, 3, 5, 6, 10, 15, 30
Highest common factor
= 15

5 a $4x^2 = 2 \times 2 \times x \times x$
$6x = 2 \times 3 \times x$
Greatest common factor
$= 2 \times x$
$= 2x$

b $3ab^2c = 3 \times a \times b \times b \times c$
$9a^2bc^3 = 3 \times 3 \times a \times a \times b \times c \times c \times c$
Greatest common factor
$= 3 \times a \times b \times c$
$= 3abc$

c Highest common factor of 8 and 12 is 4.
Greatest common factor of x^2 and x^3 is x^2.
Greatest common factor of y^3 and y^2 is y^2.
Greatest common factor of z^5 and z^3 is z^3.
$\therefore$ the greatest common factor of $8x^2y^3z^5$ and $12x^3y^2z^3$ is $4x^2y^2z^3$.

6 a $5(x + 2) = 5 \times x + 5 \times 2$
$= 5x + 10$

b $6(2x - 3) = 6 \times 2x - 6 \times 3$
$= 12x - 18$

c $4(3x + y) = 4 \times 3x + 4 \times y$
$= 12x + 4y$

7 a $7x + 14 = 7 \times x + 7 \times 2$
$= 7(x + 2)$

b $10x - 15 = 5 \times 2x - 5 \times 3$
$= 5(2x - 3)$

c $6x + 3y = 3 \times 2x + 3 \times y$
$= 3(2x + y)$

d $2p + 2q = 2(p + q)$

e $5t - 5 = 5(t - 1)$

f $12m - 15n = 3(4m - 5n)$

8 a $x(x + 4) = x \times x + x \times 4$
$= x^2 + 4x$

b $a(a - 2) = a \times a - a \times 2$
$= a^2 - 2a$

c $p(p + q) = p \times p + p \times q$
$= p^2 + pq$

9 a $x^2 + 5x = x \times x + x \times 5$
$= x(x + 5)$

b $a^2 - 3a = a \times a - a \times 3$
$= a(a - 3)$

c $t^2 + tu = t \times t + t \times u$
$= t(t + u)$

d $2x^2 + 3x = x(2x + 3)$

e $5p^2 - 2pq = p(5p - 2q)$

f $x^3 + x^2 + x = x(x^2 + x + 1)$

10 a $-2(3x + 5)$
$= -2 \times 3x + (-2) \times 5$
$= -6x - 10$

b $-k(2k - 3m)$
$= -k \times 2k - (-k) \times 3m$
$= -2k^2 + 3km$

c $-(2t - 7)$
$= -1 \times 2t - (-1) \times 7$
$= -2t + 7$

11 a $-8x - 6 = -2 \times 4x + (-2) \times 3$
$= -2(4x + 3)$

b $-4k + 12 = -4(k - 3)$

c $-3x^2 + xy = -x \times 3x - (-x) \times y$
$= -x(3x - y)$

12 a $-5a - 4 = -1 \times 5a + (-1) \times 4$
$= -(5a + 4)$

b $-9t + 8 = -(9t - 8)$

c $-2a - 3b + 4c$
$= -(2a + 3b - 4c)$

13 a $2x + 8 = 2(x + \mathbf{4})$

b $6y - 12z = \mathbf{6}(y - 2z)$

c $a^2b + abc = \mathbf{ab}(a + c)$

d $6x^2y + 9xy^2 = 3xy(\mathbf{2x + 3y})$

e $3p^2 + 3p = 3p(p + \mathbf{1})$

f $12k + 8m - 20n$
$= 4(\mathbf{3k + 2m - 5n})$

FACTORISING
INTERMEDIATE TEST PAGE 11

1 $8 - 2x = 2(4 - x)$
Of the options only $4 - x$ is a factor.
$\therefore$ [C] ✓ (1 mark)

2 $6x^2 - 18xy = 6x(x - 3y)$
[The other answers are correct factorisations of $6x^2 - 18xy$ but only in D is the expression fully factorised.]
$\therefore$ [D] ✓ (1 mark)

3 Try each option.
$3(3x^2y^3 - 2xy^4) = 9x^2y^3 - 6xy^4$
$3xy(3xy^2 - 2y^3) = 9x^2y^3 - 6xy^4$
$xy^2(9xy - 6y^2) = 9x^2y^3 - 6xy^4$
$3xy^3(3x - 2) = 9x^2y^3 - 6xy^3$
$3xy^3(3x - 2)$ is not a correct factorisation of $9x^2y^3 - 6xy^4$.
$\therefore$ [D] ✓ (1 mark)

4 $2x^2$ and $8x$ have factors of both 2 and x
$\therefore$ [D] ✓ (1 mark)

5 Try each option.
A: $4x^2 + 8x = 4(x^2 + 2x)$
$= 4x(x + 2)$
B: $15p - 18q = 3(5p - 6q)$
C: $6a - 24 = 2(3a - 12)$
$= 6(a - 4)$
D: $16t^2 - 8t = 4t(4t - 2)$
$= 8t(2t - 1)$
Only $15p - 18q$ is fully factorised.
$\therefore$ [B] ✓ (1 mark)

6 $16t^2u - 12tu^2 = 4tu(4t - 3u)$
[not $4tu(4t - 3u^2)$]
$\therefore$ [C] ✓ (1 mark)

7 $8b - 3a = -(-8b + 3a)$
$= -(3a - 8b)$
$\therefore$ [A] ✓ (1 mark)

8 $\dfrac{5x + 10}{5} = \dfrac{5(x + 2)}{5}$
$= x + 2$
$\therefore$ [B] ✓ (1 mark)

9 $(x + 2)(x + 3) + (x + 2)(x - 3)$
$= (x + 2)(x + 3 + x - 3)$
$= (x + 2)(2x)$
$= 2x(x + 2)$
$\therefore$ [A] ✓ (1 mark)

10 $6a^6 + 3a = 3a(2a^5 + 1)$
[The largest common factor is $3a$, $6a^6 \div 3a = 2a^5$, $3a \div 3a = 1$]
✓✓ (2 marks)

11 $5xy(3x - 2y) = 15x^2y - 10xy^2$ ✓
The given factorisation is incorrect because y^3 is a common factor not just y. It would be correct if the factor included y^3 not y.
$15x^2y^3 - 10xy^4$
$= 5xy^3(3x - 2y)$ ✓
[$15x^2y^3 - 10xy^4 = 5xy(3xy^2 - 2y^3)$ is a correct factorisation but it is not fully factorised.] (2 marks)

12 $3a + 9ab - 12a^2$
$= 3a(1 + 3b - 4a)$ ✓✓
[Both 3 and a are common factors. $3a$ divides into $3a$ once so the first term in the brackets must be 1. Both $3(a + 3ab - 4a^2)$ and $a(3 + 9b - 12a)$ are correct factorisations but the expression has not been fully factorised. (One mark for either of those answers.)]
(2 marks)

13 a $8x^2 - 12x = 4x(2x - 3)$ ✓✓
(2 marks)

b $3x^3 + 15x^8 = 3x^3(1 + 5x^5)$ ✓✓
(2 marks)

c $ab^2 + a^2b = ab(b + a)$ ✓✓
(2 marks)

d $9xy^2 - 6xyz = 3xy(3y - 2z)$
✓✓ (2 marks)

e $12x^2y^4z^3 - 21xy^3z^2$
$= 3xy^3z^2(4xyz - 7)$ ✓✓
(2 marks)

f $10ab^2 + 20a^2b - 5ab - 25a^2b^2$
$= 5ab(2b + 4a - 1 - 5ab)$ ✓✓
(2 marks)

14 $\dfrac{8x^3 + 24x^2}{2x^2 + 6x} = \dfrac{8x^2(x + 3)}{2x(x + 3)}$ ✓✓
$= 4x$ ✓ (3 marks)
(Total: 30 marks)

FACTORISING ADVANCED TEST PAGE 12

1 a $3p^2 - p$
$= p(3p - 1)$ ✓ (1 mark)

b $x^4 + xy^2$
$= x(x^3 + y^2)$ ✓ (1 mark)

c $9a^2b - 5b^2c$
$= b(9a^2 - 5bc)$ ✓(1 mark)

d $xe^x + e^x$
$= e^x(x + 1)$ ✓ (1 mark)

e $2000 + 2000(1.08) + 2000(1.08)^2$
$= 2000(1 + 1.08 + 1.08^2)$
✓ (1 mark)

f $5\sin\theta - 10\cos\theta$
$= 5(\sin\theta - 2\cos\theta)$ ✓
(1 mark)

2 Jarod's factorisation is not incorrect but it is not fully factorised. There is still a common factor of xy^2.
$9x^3y^4z^2 + 12x^2y^3z^2$
$= 3x^2y^3z^2(3xy + 4)$ ✓ (1 mark)

3 a $x^3 - x^2y^2$
$= x^2(x - y^2)$ ✓ (1 mark)

b $8a^2b^3c - 12ab^4c^3$
$= 4ab^3c(2a - 3bc^2)$ ✓
(1 mark)

c $24x^9 - 36x^{12}$
$= 12x^9(2 - 3x^3)$ ✓ (1 mark)

d $12x^3y^2 - 18xy^3$
$= 6xy^2(2x^2 - 3y)$ ✓(1 mark)

e $-4a^2 - 18ab$
$= -2a(2a + 9b)$ ✓ (1 mark)

f $2\pi r^2 + 2\pi rh$
$= 2\pi r(r + h)$ ✓ (1 mark)

g $\dfrac{1}{2}a^2 - \dfrac{ab}{2}$
$= \dfrac{a}{2}(a - b)$ ✓ (1 mark)

h $16a^2b^2 - 8ab - 2b$
$= 2b(8a^2b - 4a - 1)$ ✓
(1 mark)

i $15pq^2 - 3pq + 5p^2q$
$= pq(15q - 3 + 5p)$ ✓
(1 mark)

j $15a^4b^9cd^3 + 30ab^5cd^7$
$= 15ab^5cd^3(a^3b^4 + 2d^4)$ ✓
(1 mark)

k $35x^7y^4z^3 - 49x^5y^5z^6 + 28x^4y^6z^4$
$= 7x^4y^4z^3(5x^3 - 7xyz^3 + 4y^2z)$
✓ (1 mark)

4 a A further factor of 2 can be taken out.
$40a^2b - 24ab^2$
$= 8ab(5a - 3b)$ ✓ (1 mark)

b A further factor of n^2 can be taken out.
$9m^2n^3 + 12mn^4$
$= 3mn^3(3m + 4n)$ ✓
(1 mark)

5 a There is only a single factor of y in the expression. Only z is cubed.
$4x^2yz^3 + 12xyz^3$
$= 4xyz^3(x + 3)$ ✓(1 mark)

b A negative factor has been taken out so the sign in the bracket must be +.
$25p^2q^2r^3 - 15pqr^2$
$= -5pqr^2(5pqr + 3)$ ✓
(1 mark)

6 a $x(x + 4) + 7(x + 4)$
$= (x + 4)(x + 7)$ ✓
(1 mark)

b $2x(3x - 5) - 3(3x - 5)$
$= (3x - 5)(2x - 3)$ ✓
(1 mark)

c $(a + b)^2 + 6(a + b)$
$= (a + b)(a + b + 6)$ ✓
(1 mark)

d $t(t - 1) + t - 1$
$= (t - 1)(t + 1)$ ✓
(1 mark)

7 a $(x^3 - 7x^2) + (5x - 35)$
$= x^2(x - 7) + 5(x - 7)$
$= (x - 7)(x^2 + 5)$ ✓
(1 mark)

b $(pq - 2pr) - (3q^2 - 6qr)$
$= p(q - 2r) - 3q(q - 2r)$
$= (q - 2r)(p - 3q)$ ✓
(1 mark)

8 a $\dfrac{4x^2 - 8x}{2y} = \dfrac{4x(x - 2)}{2y}$ ✓
$= \dfrac{2x(x - 2)}{y}$ ✓
(2 marks)

b $\dfrac{7x - 14y}{7x + 21y} = \dfrac{7(x - 2y)}{7(x + 3y)}$ ✓
$= \dfrac{x - 2y}{x + 3y}$ ✓
(2 marks)

c $\dfrac{3ab - 9b^2}{6a^2b} = \dfrac{3b(a - 3b)}{6a^2b}$ ✓
$= \dfrac{a - 3b}{2a^2}$ ✓
(2 marks)

d $\dfrac{10mn-5n}{2m-1}=\dfrac{5n(2m-1)}{2m-1}$ ✓

$=5n$ ✓ (2 marks)

e $\dfrac{3t-4}{6t^2-8t}=\dfrac{3t-4}{2t(3t-4)}$ ✓

$=\dfrac{1}{2t}$ ✓ (2 marks)

f $\dfrac{12-2q}{3q-18}=\dfrac{2(6-q)}{3(q-6)}$ ✓

$=-\dfrac{2}{3}$ ✓ (2 marks)

(Total: 40 marks)

ALGEBRAIC FRACTIONS
SKILLS CHECK PAGE 14

1 **a** $\dfrac{1}{3}+\dfrac{1}{3}=\dfrac{1+1}{3}$

$=\dfrac{2}{3}$

b $\dfrac{a}{3}+\dfrac{a}{3}=\dfrac{a+a}{3}$

$=\dfrac{2a}{3}$

c $\dfrac{a}{3}+\dfrac{b}{3}=\dfrac{a+b}{3}$

d $\dfrac{2}{7}+\dfrac{3}{7}=\dfrac{2+3}{7}$

$=\dfrac{5}{7}$

e $\dfrac{2x}{7}+\dfrac{3x}{7}=\dfrac{2x+3x}{7}$

$=\dfrac{5x}{7}$

f $\dfrac{2x}{7}+\dfrac{3y}{7}=\dfrac{2x+3y}{7}$

2 **a** $\dfrac{4}{5}-\dfrac{1}{5}=\dfrac{4-1}{5}$

$=\dfrac{3}{5}$

b $\dfrac{4t}{5}-\dfrac{t}{5}=\dfrac{4t-t}{5}$

$=\dfrac{3t}{5}$

c $\dfrac{4t}{5}-\dfrac{u}{5}=\dfrac{4t-u}{5}$

d $\dfrac{8}{9}-\dfrac{7}{9}=\dfrac{8-7}{9}$

$=\dfrac{1}{9}$

e $\dfrac{8x}{9}-\dfrac{7x}{9}=\dfrac{8x-7x}{9}$

$=\dfrac{x}{9}$

f $\dfrac{8}{9a}-\dfrac{7}{9a}=\dfrac{8-7}{9a}$

$=\dfrac{1}{9a}$

3 **a** $\dfrac{4}{6}=\dfrac{2}{3}$

[divide numerator and denominator by 2]

b $\dfrac{4x}{6}=\dfrac{2x}{3}$ [divide by 2]

c $\dfrac{4a}{6b}=\dfrac{2a}{3b}$ [divide by 2]

d $\dfrac{12e}{20}=\dfrac{3e}{5}$ [divide by 4]

e $\dfrac{3u}{5u}=\dfrac{3}{5}$ [divide by u]

f $\dfrac{8m}{2}=\dfrac{4m}{1}$ [divide by 2]

$=4m$

g $\dfrac{15}{20a}=\dfrac{3}{4a}$ [divide by 5]

h $\dfrac{7n^2}{3n}=\dfrac{7n}{3}$ [divide by n]

i $\dfrac{8e}{6e}=\dfrac{4}{3}=1\dfrac{1}{3}$

[divide numerator and denominator by $2e$]

j $\dfrac{3}{6k}=\dfrac{1}{2k}$

[divide numerator and denominator by 3]

k $\dfrac{4a}{8a}=\dfrac{1}{2}$

[divide numerator and denominator by $2a$]

l $\dfrac{9e}{3e}=\dfrac{3}{1}=3$

[divide numerator and denominator by $3e$]

4 **a** $\dfrac{2}{3}=\dfrac{4}{6}$

[The denominator has been multiplied by 2, so the numerator must be multiplied by 2.]

b $\dfrac{2x}{3}=\dfrac{4x}{6}$ [multiply by 2]

c $\dfrac{2ab}{3}=\dfrac{4ab}{6}$ [multiply by 2]

d $\dfrac{5a}{8}=\dfrac{15a}{24}$ [multiply by 3]

e $\dfrac{x}{5}=\dfrac{3x}{15}$ [multiply by 3]

f $\dfrac{3}{a}=\dfrac{6}{2a}$ [multiply by 2]

5 **a** $\dfrac{2}{3}\times\dfrac{4}{5}=\dfrac{2\times 4}{3\times 5}$

$=\dfrac{8}{15}$

b $\dfrac{2a}{3}\times\dfrac{4b}{5}=\dfrac{2a\times 4b}{3\times 5}$

$=\dfrac{8ab}{15}$

c $\dfrac{2x}{3}\times\dfrac{4x}{5}=\dfrac{2x\times 4x}{3\times 5}$

$=\dfrac{8x^2}{15}$

d $\dfrac{a}{4}\times\dfrac{b}{5}=\dfrac{a\times b}{4\times 5}$

$=\dfrac{ab}{20}$

e $\dfrac{x}{2}\times\dfrac{x}{3}=\dfrac{x\times x}{2\times 3}$

$=\dfrac{x^2}{6}$

f $\dfrac{3a}{4}\times\dfrac{2bc}{9}=\dfrac{6abc}{36}$

$=\dfrac{abc}{6}$

6 **a** $\dfrac{3}{5}\div\dfrac{2}{3}=\dfrac{3}{5}\times\dfrac{3}{2}$

$=\dfrac{9}{10}$

b $\dfrac{3a}{5}\div\dfrac{2}{3}=\dfrac{3a}{5}\times\dfrac{3}{2}$

$=\dfrac{9a}{10}$

c $\dfrac{3a}{5}\div\dfrac{2a}{3}=\dfrac{3a}{5}\times\dfrac{3}{2a}$

$=\dfrac{9}{10}$

[The a's cancel.]

ALGEBRAIC FRACTIONS
INTERMEDIATE TEST PAGE 15

1 $\frac{x}{3}+\frac{x}{4}=\frac{4x}{12}+\frac{3x}{12}$

$=\frac{7x}{12}$

$\therefore$ **[C]** ✓ (1 mark)

2 $\frac{8ab}{12a}=\frac{2b}{3}$

[Dividing both numerator and denominator by $4a$. All the options are equivalent fractions but only D is in simplest form.]

$\therefore$ **[D]** ✓ (1 mark)

3 $\frac{7a}{8}-\frac{3a}{5}=\frac{35a}{40}-\frac{24a}{40}$

$=\frac{11a}{40}$

$\therefore$ **[C]** ✓ (1 mark)

4 Try each option.

$\frac{12x}{20y}=\frac{3x}{5y}$

[dividing both numerator and denominator by 4]

$\frac{27x}{45y}=\frac{3x}{5y}$

[dividing both numerator and denominator by 9]

$\frac{36x}{60y}=\frac{3x}{5y}$

[dividing both numerator and denominator by 12]

All are equivalent.

$\therefore$ **[D]** ✓ (1 mark)

5 $\frac{2x}{3}\times\frac{3}{4}=\frac{6x}{12}$

$=\frac{x}{2}$

[or cancel before multiplying]

$\therefore$ **[A]** ✓ (1 mark)

6 $\frac{2a}{3}\div\frac{1}{9}=\frac{2a}{3}\times\frac{9}{1}$

$=6a$

$\therefore$ **[D]** ✓ (1 mark)

7 $\frac{m}{6}+\frac{n}{5}=\frac{5m}{30}+\frac{6n}{30}$ ✓

$=\frac{5m+6n}{30}$ ✓ (2 marks)

8 $\frac{3a}{5}-\frac{a}{2}=\frac{6a}{10}-\frac{5a}{10}$ ✓

$=\frac{a}{10}$ ✓ (2 marks)

9 $\frac{2x}{3}+\frac{5x}{6}=\frac{4x}{6}+\frac{5x}{6}$ ✓

$=\frac{9x}{6}$

$=\frac{3x}{2}$ ✓

[$1\frac{1}{2}x$ is correct but we usually leave the answer as an improper fraction] (2 marks)

10 $\frac{3a}{4b}\div\frac{2a}{5c}=\frac{3a}{4b}\times\frac{5c}{2a}$ ✓

$=\frac{15ac}{8ab}$

$=\frac{15c}{8b}$ ✓ (2 marks)

11 $\frac{5pq}{2}\times\frac{4}{10p}=\frac{20pq}{20p}$ ✓

$=\frac{q}{1}$

$=q$ ✓

[or cancel before multiplying] (2 marks)

12 $\frac{7e}{12}-\frac{5e}{9}=\frac{21e}{36}-\frac{20e}{36}$ ✓

$=\frac{e}{36}$ ✓

[The lowest common denominator is 36.] (2 marks)

13 $\frac{7a}{20}+\frac{2a}{15}=\frac{21a}{60}+\frac{8a}{60}$ ✓

$=\frac{29a}{60}$ ✓ (2 marks)

(Total: 20 marks)

ALGEBRAIC FRACTIONS
ADVANCED TEST PAGE 16

1 **a** $\frac{12x^6}{3x^3}=4x^3$ ✓ (1 mark)

b $\frac{2ab}{6a^2b^3}=\frac{1}{3ab^2}$ ✓ (1 mark)

c $\frac{8p^3q^4}{4p^5q^2}=\frac{2q^2}{p^2}$ ✓ (1 mark)

d $\frac{16ab^5c^6}{24a^5bc^7}=\frac{2b^4}{3a^4c}$ ✓ (1 mark)

2 **a** $\frac{2x}{5}+\frac{3x}{4}=\frac{8x}{20}+\frac{15x}{20}$

$=\frac{23x}{20}$ ✓ (1 mark)

b $\frac{3a}{5}+\frac{2b}{7}=\frac{21a}{35}+\frac{10b}{35}$

$=\frac{21a+10b}{35}$ ✓ (1 mark)

c $\frac{4x^2}{3}+\frac{x}{6}=\frac{8x^2}{6}+\frac{x}{6}$

$=\frac{8x^2+x}{6}$ ✓ (1 mark)

d $\frac{5}{2m}+\frac{2}{3m}=\frac{15}{6m}+\frac{4}{6m}$

$=\frac{19}{6m}$ ✓ (1 mark)

e $\frac{3}{a}+\frac{7}{b}=\frac{3b}{ab}+\frac{7a}{ab}$

$=\frac{7a+3b}{ab}$ ✓ (1 mark)

f $\frac{5a}{2b}+\frac{3a}{b}=\frac{5a}{2b}+\frac{6a}{2b}$

$=\frac{11a}{2b}$ ✓ (1 mark)

3 **a** $\frac{8a}{5}-\frac{3a}{2}=\frac{16a}{10}-\frac{15a}{10}$

$=\frac{a}{10}$ ✓ (1 mark)

b $\frac{7n}{2}-\frac{n}{3}=\frac{21n}{6}-\frac{2n}{6}$

$=\frac{19n}{6}$ ✓ (1 mark)

c $\frac{5a^2b^3}{9}-\frac{2ab^4}{3}$

$=\frac{5a^2b^3}{9}-\frac{6ab^4}{9}$

$=\frac{5a^2b^3-6ab^4}{9}$ ✓ (1 mark)

d $\frac{4}{a}-\frac{1}{4a}=\frac{16}{4a}-\frac{1}{4a}$

$=\frac{15}{4a}$ ✓ (1 mark)

e $\frac{6pq}{r}-\frac{5pr}{q}$

$=\frac{6pq^2}{qr}-\frac{5pr^2}{qr}$

$=\frac{6pq^2-5pr^2}{qr}$ ✓ (1 mark)

f $\frac{4a}{3b} - \frac{2a}{5b} = \frac{20a}{15b} - \frac{6a}{15b}$
$= \frac{14a}{15b}$ ✓ (1 mark)

4 a $\frac{3mn}{4} \times \frac{2}{5m} = \frac{6mn}{20m}$
$= \frac{3n}{10}$ ✓(1 mark)

b $\frac{6x^2}{5} \times \frac{4x}{9} = \frac{24x^3}{45}$
$= \frac{8x^3}{15}$ ✓ (1 mark)

c $\frac{8a^2b^2}{9} \times \frac{1}{12ab^2} = \frac{8a}{108}$
$= \frac{2a}{27}$ ✓ (1 mark)

d $\frac{2a}{5b} \times \frac{3b^2}{8} = \frac{6ab^2}{40b}$
$= \frac{3ab}{20}$ ✓ (1 mark)

e $\frac{4x^2y^3}{7} \times \frac{3x}{10y^2} = \frac{12x^3y^3}{70y^2}$
$= \frac{6x^3y}{35}$ ✓ (1 mark)

f $\frac{5p^2q}{3r} \times \frac{9qr}{10p^3} = \frac{45p^2q^2r}{30p^3r}$
$= \frac{3q^2}{2p}$ ✓ (1 mark)

5 a $\frac{3}{a} \div \frac{6}{a} = \frac{3}{a} \times \frac{a}{6}$
$= \frac{1}{2}$ ✓ (1 mark)

b $\frac{x^2}{5} \div \frac{x}{4} = \frac{x^2}{5} \times \frac{4}{x}$
$= \frac{4x}{5}$ ✓ (1 mark)

c $\frac{2x}{3y} \div \frac{x}{y} = \frac{2x}{3y} \times \frac{y}{x}$
$= \frac{2}{3}$ ✓ (1 mark)

d $\frac{8a^2}{5b^2} \div \frac{2a}{3b^2} = \frac{8a^2}{5b^2} \times \frac{3b^2}{2a}$
$= \frac{12a}{5}$ ✓(1 mark)

e $\frac{6a^2b^3}{7c^4} \div \frac{3a^2bc}{7}$
$= \frac{6a^2b^3}{7c^4} \times \frac{7}{3a^2bc}$
$= \frac{2b^2}{c^5}$ ✓ (1 mark)

f $\frac{p^2q}{pq^3} \div \frac{4a^2p^3}{8a^5}$
$= \frac{p^2q}{pq^3} \times \frac{8a^5}{4a^2p^3}$
$= \frac{2a^3}{p^2q^2}$ ✓ (1 mark)

6 a $\frac{a^2b^3}{5} \times \frac{6}{ab} \times \frac{2a}{3b} = \frac{12a^3b^3}{15ab^2}$ ✓
$= \frac{4a^2b}{5}$ ✓
(2 marks)

b $\frac{4p^2}{5q^2} \times \frac{6q^3}{p} - \frac{pq}{8}$
$= \frac{24pq}{5} - \frac{pq}{8}$ ✓
$= \frac{192pq}{40} - \frac{5pq}{40}$
$= \frac{187pq}{40}$ ✓ (2 marks)

c $\frac{x}{3} + \frac{5x^2}{2} \times \frac{8}{x}$
$= \frac{x}{3} + 20x$ ✓
$= \frac{x}{3} + \frac{60x}{3}$
$= \frac{61x}{3}$ ✓ (2 marks)

d $\frac{a}{5} \div \frac{b}{5} + \frac{b}{a}$
$= \frac{a}{5} \times \frac{5}{b} + \frac{b}{a}$ ✓
$= \frac{a}{b} + \frac{b}{a}$
$= \frac{a^2 + b^2}{ab}$ ✓ (2 marks)

e $\frac{9a^2}{3b} \div \frac{2a^2}{6b} + \frac{3x^3}{14y^2} \div \frac{9x^3}{7y^2}$
$= \frac{9a^2}{3b} \times \frac{6b}{2a^2} + \frac{3x^3}{14y^2} \times \frac{7y^2}{9x^3}$ ✓
$= 9 + \frac{1}{6}$
$= 9\frac{1}{6}$ ✓ (2 marks)

f $\frac{5x - 4y}{2} - \frac{3x + 2y}{5}$
$= \frac{5(5x - 4y) - 2(3x + 2y)}{10}$
$= \frac{25x - 20y - 6x - 4y}{10}$ ✓
$= \frac{19x - 24y}{10}$ ✓ (2 marks)

(Total: 40 marks)

BINOMIAL PRODUCTS
SKILLS CHECK PAGE 18

1 a $-5a + 3b + 2a - 6b + 3$
$= -3a - 3b + 3$

b $x^2 + 7x - 3x + 4$
$= x^2 + 4x + 4$

c $2a^2 - 6a + 8a - 5$
$= 2a^2 + 2a - 5$

2 a $2(3x - 5) = 6x - 10$
$[2 \times 3x - 2 \times 5]$

b $x(x + 6) = x^2 + 6x$
$[x \times x + x \times 6]$

c $2p(3p - 7q) = 6p^2 - 14pq$

3 a $5(3x + 2) + 4(x - 2)$
$= 15x + 10 + 4x - 8$
$= 19x + 2$

b $2(4a + 3) + 3(5a + 1)$
$= 8a + 6 + 15a + 3$
$= 23a + 9$

c $6(4p - q) + 5(2p + 3q)$
$= 24p - 6q + 10p + 15q$
$= 34p + 9q$

d $3(2e - 7) - 2(3e - 2)$
$= 6e - 21 - 6e + 4$
$= -17$

e $-4(5a + 2b) - 3(6a + 5)$
$= -20a - 8b - 18a - 15$
$= -38a - 8b - 15$

f $-2(8x - y) - (3x + y)$
$= -16x + 2y - 3x - y$
$= -19x + y$

4 a $x(x + 3) + 4(x + 3)$
$= x^2 + 3x + 4x + 12$
$= x^2 + 7x + 12$

b $a(a + 5) - 2(a + 5)$
$= a^2 + 5a - 2a - 10$
$= a^2 + 3a - 10$

c $m(m - 4) + 7(m - 4)$
$= m^2 - 4m + 7m - 28$
$= m^2 + 3m - 28$

d $n(n-2)-5(n-2)$
$= n^2-2n-5n+10$
$= n^2-7n+10$

5

	x	7
x	A_1	A_3
5	A_2	A_4

a A_1: $x \times x = x^2$
b A_2: $5 \times x = 5x$
c A_3: $x \times 7 = 7x$
d A_4: $5 \times 7 = 35$

6 $(x+7)(x+5)$
$= x^2+5x+7x+35$
$= x^2+12x+35$

7 **a** $(x+2)(x+5)$
$= x^2+5x+2x+\mathbf{10}$
b $(a+4)(a-3)$
$= a^2-3a+\mathbf{4a}-12$
c $(t-7)(t+1)$
$= t^2+\mathbf{t}-7t-7$
d $(e-8)(e-5)$
$= e^2-5e-8e+\mathbf{40}$
e $(2p+3)(p+4)$
$= \mathbf{2p^2}+8p+3p+12$
f $(3k-1)(2k+7)$
$= \mathbf{6k^2}+21k-2k-\mathbf{7}$

8 **a** $(x+4)(x+2)$
$= x^2+2x+4x+8$
$= x^2+6x+8$
b $(a+3)(a-2)$
$= a^2-2a+3a-6$
$= a^2+a-6$
c $(2a+b)(3c-4d)$
$= 6ac-8ad+3bc-4bd$
d $(p-6)(p+4)$
$= p^2+4p-6p-24$
$= p^2-2p-24$
e $(3m-8)(m-5)$
$= 3m^2-15m-8m+40$
$= 3m^2-23m+40$
f $(5x-3)(2x+5)$
$= 10x^2+25x-6x-15$
$= 10x^2+19x-15$
g $(x-1)(x+1)$
$= x^2+x-x-1$
$= x^2-1$
h $(x+7)(x+7)$
$= x^2+7x+7x+49$
$= x^2+14x+49$
i $(3y+2)(3y-2)$
$= 9y^2-6y+6y-4$
$= 9y^2-4$

9 **a** $(x-6)(x+6)$
$= x^2+6x-6x-36$
$= x^2-36$
b $(a+4)(a-4)$
$= a^2-4a+4a-16$
$= a^2-16$
c $(2n+9)(2n-9)$
$= 4n^2-18n+18n-81$
$= 4n^2-81$
d $(5x-1)(5x+1)$
$= 25x^2+5x-5x-1$
$= 25x^2-1$

10 **a** $(x+9)^2$
$= (x+9)(x+9)$
$= x^2+9x+9x+81$
$= x^2+18x+81$
b $(y-6)^2$
$= (y-6)(y-6)$
$= y^2-6y-6y+36$
$= y^2-12y+36$
c $(4x-11)^2$
$= (4x-11)(4x-11)$
$= 16x^2-44x-44x+121$
$= 16x^2-88x+121$
d $(3p+4q)^2$
$= (3p+4q)(3p+4q)$
$= 9p^2+12pq+12pq+16q^2$
$= 9p^2+24pq+16q^2$

BINOMIAL PRODUCTS
INTERMEDIATE TEST PAGE 19

1 $(x-5)(y+2)$
$= xy+2x-5y-10$
$\therefore$ [C] ✓ (1 mark)

2 $(x+4)(x-5)$
$= x^2-5x+4x-20$
$= x^2-x-20$
$\therefore$ [A] ✓ (1 mark)

3 $(n-4)(n-2)$
$= n^2-2n-4n+8$
$= n^2-6n+8$
$\therefore$ [D] ✓ (1 mark)

4 $(3a-5)(2a-1)$
$= 6a^2-3a-10a+5$
$= 6a^2-13a+5$
$\therefore$ [A] ✓ (1 mark)

5 $(2m-5n)(2m+5n)$
$= 4m^2+10mn-10mn-25n^2$
$= 4m^2-25n^2$
$\therefore$ [B] ✓ (1 mark)

6 $(4x-3)^2$
$= (4x-3)(4x-3)$
$= 16x^2-12x-12x+9$
$= 16x^2-24x+9$
$\therefore$ [D] ✓ (1 mark)

7 $(3a+2)(b-3)$
$= 3ab-9a+2b-6$ ✓
(1 mark)

8 $(4x^2-3)(2x+1)$
$= 8x^3+4x^2-6x-3$ ✓
(1 mark)

9 $(6m-5p)(2m-3p)$
$= 12m^2-18mp-10mp+15p^2$ ✓
$= 12m^2-28mp+15p^2$ ✓
(2 marks)

10 **a** $(x-7)(x+3)$
$= x^2+3x-7x-21$
$= x^2-4x-21$ ✓ (1 mark)
b $(x+7)(x-3)$
$= x^2-3x+7x-21$
$= x^2+4x-21$ ✓ (1 mark)
c $(x-7)(x+3)-(x+7)(x-3)$
$= x^2-4x-21-(x^2+4x-21)$
$= x^2-4x-21-x^2-4x+21$
$= -8x$ ✓ (1 mark)

11 $(3x+2)(x+5)+(2x+5)(x+3)$
$= 3x^2+15x+2x+10$
$+ 2x^2+6x+5x+15$ ✓
$= 5x^2+28x+25$ ✓ (2 marks)

12 $(4x+5)(2x-3)-(3x-1)(x+7)$
$= 8x^2-12x+10x-15$
$-(3x^2+21x-x-7)$ ✓
$= 8x^2-2x-15-3x^2-20x+7$ ✓
$= 5x^2-22x-8$ ✓ (3 marks)

13 **a** If $n=2$,
$(2n-3)(n+5)+(3n-1)(n-1)$
$= (2\times2-3)(2+5)$
$+(3\times2-1)(2-1)$
$= 1\times7+5\times1$
$= 12$ ✓ (1 mark)

b $(2n-3)(n+5)+(3n-1)(n-1)$
$= 2n^2 + 10n - 3n - 15$
$+ 3n^2 - 3n - n + 1$ ✓
$= 5n^2 + 3n - 14$ ✓
(2 marks)

c If $n = 2$,
$5n^2 + 3n - 14$
$= 5 \times 2^2 + 3 \times 2 - 14$
$= 20 + 6 - 14$
$= 12$ ✓ (1 mark)

d The answer to part a must be the same as the answer to part **c** because the expressions are equal so after substituting the answers must also be equal. ✓ (1 mark)

14 **a** **i** $7^2 = 49$ and $6 \times 8 = 48$
ii $2^2 = 4$ and $1 \times 3 = 3$
iii $11^2 = 121$ and $10 \times 12 = 120$
iv $43^2 = 1849$ and $42 \times 44 = 1848$ ✓
(1 mark)

b With three consecutive whole numbers, the square of the middle number is equal to one more than the product of the first and last number. ✓ (1 mark)

c $x, x + 1, x + 2$
[The second number is one more than the first, so it is $x + 1$. The third number is one more than the second number so it is $x + 1 + 1$ or $x + 2$.] ✓ (1 mark)

d $x(x + 2) = x^2 + 2x$ ✓
(1 mark)

e $(x + 1)^2 = (x + 1)(x + 1)$
$= x^2 + x + x + 1$
$= x^2 + 2x + 1$ ✓
(1 mark)

f $(x + 1)^2$ is one more than $x(x + 2)$ ✓
It will not matter which three consecutive numbers are chosen, the square of the middle one will always be one more than the product of the first and last number. ✓ (2 marks)
(Total: 30 marks)

BINOMIAL PRODUCTS
ADVANCED TEST PAGE 20

1 **a** $(x + 7)(x + 8)$
$= x^2 + 8x + 7x + 56$
$= x^2 + 15x + 56$ ✓ (1 mark)

b $(x - y)(x + 11)$
$= x^2 + 11x - xy - 11y$ ✓
(1 mark)

c $(x + 4)(x - 12)$
$= x^2 - 12x + 4x - 48$
$= x^2 - 8x - 48$ ✓ (1 mark)

d $(a - 12)(b - 20)$
$= ab - 20a - 12b + 240$ ✓
(1 mark)

e $(2e + 3f)(5e - f)$
$= 10e^2 - 2ef + 15ef - 3f^2$
$= 10e^2 + 13ef - 3f^2$ ✓
(1 mark)

f $(9x - 5y)(5x - 9y)$
$= 45x^2 - 81xy - 25xy + 45y^2$
$= 45x^2 - 106xy + 45y^2$ ✓
(1 mark)

g $(x^2 - 1)(x - 4)$
$= x^3 - 4x^2 - x + 4$ ✓
(1 mark)

h $(3pq + 5r)(9p - 2qr)$
$= 27p^2q - 6pq^2r + 45pr$
$- 10qr^2$ ✓ (1 mark)

i $2(7m - 2n)(m + n)$
$= 2(7m^2 + 7mn - 2mn - 2n^2)$
$= 2(7m^2 + 5mn - 2n^2)$
$= 14m^2 + 10mn - 4n^2$ ✓
(1 mark)

j $(5 - a^3)(a + 1)$
$= 5a + 5 - a^4 - a^3$ ✓
(1 mark)

k $3(2m^2 - 5n)(4n^2 - 3m)$
$= 3(8m^2n^2 - 6m^3 - 20n^3$
$+ 15mn)$
$= 24m^2n^2 - 18m^3 - 60n^3$
$+ 45mn$ ✓ (1 mark)

l $(4a^2b^3c + 5abc)(2ab^2c^2 + 3b^2c)$
$= 8a^3b^5c^3 + 12a^2b^5c^2$
$+ 10a^2b^3c^3 + 15ab^3c^2$ ✓
(1 mark)

2 **a** $x(x + 4)(x - 3)$
$= x(x^2 - 3x + 4x - 12)$
$= x(x^2 + x - 12)$ ✓
$= x^3 + x^2 - 12x$ ✓ (2 marks)

b $3a^2(2a - 5)(a - 5)$
$= 3a^2(2a^2 - 10a - 5a + 25)$
$= 3a^2(2a^2 - 15a + 25)$ ✓
$= 6a^4 - 45a^3 + 75a^2$ ✓
(2 marks)

3 **a** $(x + 6)(x - 3) + (x + 5)(x - 2)$
$= x^2 + 6x - 3x - 18$
$+ x^2 - 2x + 5x - 10$ ✓
$= 2x^2 + 6x - 28$ ✓
(2 marks)

b $(a - 3)(a - 4) - (a - 5)(a - 6)$
$= a^2 - 4a - 3a + 12$
$- (a^2 - 6a - 5a + 30)$ ✓
$= a^2 - 7a + 12 - a^2 + 11a - 30$
$= 4a - 18$ ✓ (2 marks)

c $(3m + 2n)(m - 2n)$
$+ (4m - n)(5n - m)$
$= 3m^2 - 6mn + 2mn - 4n^2$
$+ 20mn - 4m^2 - 5n^2$
$+ mn$ ✓
$= -m^2 + 17mn - 9n^2$ ✓
(2 marks)

d $(5p - 2qr)(3p + 2qr)$
$- 4qr(1 - qr)$
$= 15p^2 + 10pqr - 6pqr$
$- 4q^2r^2 - 4qr + 4q^2r^2$ ✓
$= 15p^2 + 4pqr - 4qr$ ✓
(2 marks)

4 **a** $(x + 3)(x + y + 5)$
$= x(x + y + 5)$
$+ 3(x + y + 5)$ ✓
$= x^2 + xy + 5x + 3x + 3y + 15$
$= x^2 + xy + 8x + 3y + 15$ ✓
(2 marks)

b $(2x - 1)(x^2 + x - 2)$
$= 2x(x^2 + x - 2) - (x^2 + x - 2)$
$= 2x^3 + 2x^2 - 4x - x^2 - x + 2$ ✓
$= 2x^3 + x^2 - 5x + 2$ ✓
(2 marks)

5 **a** $2(m + 8)(m - 5)$
$= 2(m^2 + 3m - 40)$
$= 2m^2 + 6m - 80$ ✓
(1 mark)

b $(m+8)(m-7)+(m+8)(m-3)$
$= m^2 + m - 56 + m^2 + 5m$
$- 24$ ✓
$= 2m^2 + 6m - 80$ ✓
(2 marks)

c The expression in part **b** has a common factor of $m + 8$.
$(m+8)(m-7)+(m+8)(m-3)$
$= (m + 8)(m - 7 + m - 3)$
$= (m + 8)(2m - 10)$
$= 2(m + 8)(m - 5)$ ✓
(1 mark)

6 a $x^2(x-3)(x-2)$
$-7x(x+5)(x-1)$
$+2(x+3)(x+4)$
$= x^2(x^2-5x+6)$
$-7x(x^2+4x-5)$
$+2(x^2+7x+12)$ ✓
$= x^4-5x^3+6x^2-7x^3$
$-28x^2+35x+2x^2$
$+14x+24$ ✓
$= x^4-12x^3-20x^2$
$+49x+24$ ✓
(3 marks)

b $bc^4(4a^2b^2c-3)\times ab^2(5a^2c^2+2b)$
$= (4a^2b^3c^5-3bc^4)(5a^3b^2c^2$
$+2ab^3)$ ✓
$= 20a^5b^5c^7+8a^3b^6c^5$
$-15a^3b^3c^6-6ab^4c^4$ ✓
(2 marks)

c $(x+1)(x+2)(x+3)$
$= (x+1)(x^2+5x+6)$ ✓
$= x(x^2+5x+6)$
$+1(x^2+5x+6)$
$= x^3+5x^2+6x+x^2+5x$
$+6$ ✓
$= x^3+6x^2+11x+6$ ✓
(3 marks)
(Total: 40 marks)

FACTORISATION OF QUADRATIC EXPRESSIONS
SKILLS CHECK PAGE 22

1 a $9x-3y=3(3x-y)$
b $x^2-6x=x(x-6)$
c $10xy-5y=5y(2x-1)$
[Both 5 and y are common factors. To factorise fully we need to take out both factors.]
d $12x^2-8x=4x(3x-2)$
e $x^2y+xy^2=xy(x+y)$
f $8ab^2-4abc=4ab(2b-c)$
g $15x^7-3x^2=3x^2(5x^5-1)$
h $16p^3q^2-2p^4q+4pq^2$
$=2pq(8p^2q-p^3+2q)$

2 a $2+5=7,\ \ 2\times5=10$
The numbers are 2 and 5.
b $2+3=5,\ \ 2\times3=6$
The numbers are 2 and 3.
c $5+7=12,\ \ 5\times7=35$
The numbers are 5 and 7.
d $3+8=11,\ \ 3\times8=24$
The numbers are 3 and 8.

3 a $-3+5=2,\ \ -3\times5=-15$
The numbers are –3 and 5.
b $-4+7=3,\ \ -4\times7=-28$
The numbers are –4 and 7.
c $-1+5=4,\ \ -1\times5=-5$
The numbers are –1 and 5.
d $-4+4=0,\ \ -4\times4=-16$
The numbers are –4 and 4.

4 a $-11+5=-6,\ \ -11\times5=-55$
The numbers are –11 and 5.
b $-10+3=-7,\ \ -10\times3=-30$
The numbers are –10 and 3.
c $-6+3=-3,\ \ -6\times3=-18$
The numbers are –6 and 3.
d $-12+5=-7,\ \ -12\times5=-60$
The numbers are –12 and 5.

5 a $-4+(-5)=-9,\ \ -4\times-5=20$
The numbers are –4 and –5.
b $-2+(-3)=-5,\ \ -2\times-3=6$
The numbers are –2 and –3.
c $-3+(-12)=-15,\ \ -3\times-12=36$
The numbers are –3 and –12.
d $-6+(-6)=-12,\ \ -6\times-6=36$
The numbers are –6 and –6.

6 a $[3+7=10,\ \ 3\times7=21]$
$x^2+10x+21=(x+3)(x+7)$
[Or $(x+7)(x+3)$]
b $[3+4=7,\ \ 3\times4=12]$
$x^2+7x+12=(x+3)(x+4)$
[Or $(x+4)(x+3)$]
c $[4+8=12,\ \ 4\times8=32]$
$x^2+12x+32$
$=(x+4)(x+8)$
d $[-4+6=2,\ \ -4\times6=-24]$
$a^2+2a-24=(a-4)(a+6)$
[Or $(a+6)(a-4)$]
e $[-2+5=3,\ \ -2\times5=-10]$
$p^2+3p-10=(p-2)(p+5)$
[Or $(p+5)(p-2)$]
f $[-2+9=7,\ \ -2\times9=-18]$
$x^2+7x-18=(x-2)(x+9)$
g $[-6+2=-4,\ \ -6\times2=-12]$
$x^2-4x-12=(x-6)(x+2)$
[Or $(x+2)(x-6)$]
h $[-9+3=-6,\ \ -9\times3=-27]$
$a^2-6a-27=(a-9)(a+3)$
i $[-7+2=-5,\ \ -7\times2=-14$
$m^2-5m-14=(m-7)(m+2)$
j $[-1+(-7)=-8,\ \ -1\times-7=7]$
$n^2-8n+7=(n-1)(n-7)$
[Or $(n-7)(n-1)$]
k $[-4+(-9)=-13,\ -4\times-9=36]$
$x^2-13x+36=(x-4)(x-9)$
l $[-5+(-5)=-10,\ -5\times-5=25]$
$b^2-10b+25=(b-5)(b-5)$
[Or $(b-5)^2$]

7 a $x^2+8x+15=(x+3)(x+5)$
[Look for two numbers that add to 8 and multiply to 15. The numbers are 3 and 5.]
b $a^2-14a+48=(a-6)(a-8)$
c $p^2+2p-35=(p+7)(p-5)$
d $m^2-3m-28=(m-7)(m+4)$
e $n^2+n-6=(n+3)(n-2)$
f $k^2-5k-6=(k-6)(k+1)$

8 a $x^2+18x+81$
$=(x+9)(x+9)$
$=(x+9)^2$
b $y^2+12y+36$
$=(y+6)(y+6)$
$=(y+6)^2$
c $b^2-14b+49$
$=(b-7)(b-7)$
$=(b-7)^2$

9 a $x^2-49=(x+7)(x-7)$
[Look for two numbers that add to 0 and multiply to –49. They are 7 and –7.]
b $a^2-1=(a+1)(a-1)$
c $36-t^2=(6+t)(6-t)$
d $x^2-y^2=(x+y)(x-y)$

FACTORISATION OF QUADRATIC EXPRESSIONS
INTERMEDIATE TEST PAGE 23

1 $a^2-2a-63=(a-9)(a+7)$
[Look for two numbers that add to –2 and multiply to –63. The numbers are –9 and 7.]
∴ [C] ✓ (1 mark)

2 $x^2-5x+6=(x-3)(x-2)$
[What two numbers multiply to 6 and add to –5?]
∴ [A] ✓ (1 mark)

3 $x^2+17x-60=(x+20)(x-3)$
∴ [D] ✓ (1 mark)

4 $(x+5)(x-4)$
$= x^2 - 4x + 5x - 20$
$= x^2 + x - 20$
$\therefore$ **[B]** ✓ (1 mark)

5 $x^2 - 18x + 81$
$= (x-9)(x-9)$
$= (x-9)^2$
$\therefore$ **[B]** ✓ (1 mark)

6 Try each option.
$x^2 - 4x + 3 = (x-1)(x-3)$
$x^2 + 4x - 12 = (x+6)(x-2)$
$x^2 + 9x + 18 = (x+3)(x+6)$
$x^2 - 3x + 2 = (x-2)(x-1)$
The only one with a factor of $x + 3$ is $x^2 + 9x + 18$.
$\therefore$ **[C]** ✓ (1 mark)

7 $m^2 - 1 = m^2 - 1^2$
$= (m+1)(m-1)$
$\therefore$ **[C]** ✓ (1 mark)

8 Try each option.
$x^2 - 13x + 22 = (x-2)(x-11)$
$x^2 + 2x - 8 = (x-2)(x+4)$
$x^2 + 18x - 40 = (x-2)(x+20)$
$x^2 - 6x - 16 = (x-8)(x+2)$
Only $x^2 - 6x - 16$ does not have a factor of $x - 2$.
$\therefore$ **[D]** ✓ (1 mark)

9 $x^2 + 10x - 24 = (x+12)(x-2)$
Of the options only $x - 2$ is a factor
$\therefore$ **[A]** ✓ (1 mark)

10 $\dfrac{g^2 + 7g + 6}{g + 6} = \dfrac{(g+6)(g+1)}{g+6}$
$= g + 1$
$\therefore$ **[D]** ✓ (1 mark)

11 $x^2 + x - 90 = (x+10)(x-9)$ ✓ (1 mark)

12 $21 - 10x + x^2 = (7-x)(3-x)$ ✓
[Or $(x-7)(x-3)$] (1 mark)

13 **a** $x^2 + x - 6 = (x+3)(x-2)$ ✓
Both Megan and Xavier had the correct answer. ✓ (2 marks)

b Joe had the signs the wrong way around.
$(x+2)(x-3)$
$= x^2 - 3x + 2x - 6$
$= x^2 - x - 6$ ✓
Ali had two negative signs instead of one positive and one negative.
$(x-2)(x-3)$
$= x^2 - 3x - 2x + 6$
$= x^2 - 5x + 6$ ✓ (2 marks)

14 **a** $x^2 + 2x - 48$
$= (x+8)(x-6)$ ✓(1 mark)

b $a^2 + 15a + 56$
$= (a+7)(a+8)$ ✓(1 mark)

c $h^2 - 5h - 24$
$= (h-8)(h+3)$ ✓(1 mark)

d $k^2 - 6k + 72$
$= (k-12)(k+6)$ ✓ (1 mark)

e $x^2 - 17x + 72$
$= (x-8)(x-9)$ ✓ (1 mark)

f $q^2 + q - 42$
$= (q+7)(q-6)$ ✓(1 mark)

g $x^2 + 8x + 16$
$= (x+4)(x+4)$
$= (x+4)^2$ ✓ (1 mark)

h $p^2 - 64 = (p+8)(p-8)$ ✓ (1 mark)

i $x^2 - 2x + 1 = (x-1)(x-1)$
$= (x-1)^2$ ✓ (1 mark)

(Total: 25 marks)

FACTORISATION OF QUADRATIC EXPRESSIONS
ADVANCED TEST PAGE 24

1 **a** $x^2 + 9x + 20$
$= (x+4)(x+5)$ ✓(1 mark)

b $a^2 - 12a + 32$
$= (a-4)(a-8)$ ✓(1 mark)

c $x^2 - 2x - 24$
$= (x-6)(x+4)$ ✓(1 mark)

d $m^2 + 5m - 6$
$= (m-1)(m+6)$ ✓(1 mark)

e $t^2 + 17t - 60$
$= (t+20)(t-3)$ ✓(1 mark)

f $h^2 - 32h - 144$
$= (h-36)(h+4)$ ✓(1 mark)

2 **a** $3n^2 - 9n - 120$
$= 3(n^2 - 3n - 40)$
$= 3(n-8)(n+5)$ ✓ (1 mark)

b $5p^2 - 55p + 140$
$= 5(p^2 - 11p + 28)$
$= 5(p-4)(p-7)$ ✓(1 mark)

3 **a** $\dfrac{x^2 + 8x + 9}{x - 1}$
$= \dfrac{(x+9)(x-1)}{x-1}$ ✓
$= x + 9$ ✓ (2 marks)

b $\dfrac{a^2 + 10a + 24}{a^2 + 4a - 12}$
$= \dfrac{(a+6)(a+4)}{(a+6)(a-2)}$ ✓
$= \dfrac{a+4}{a-2}$ ✓ (2 marks)

c $\dfrac{n-2}{n^2 - 11n + 18}$
$= \dfrac{n-2}{(n-9)(n-2)}$ ✓
$= \dfrac{1}{n-9}$ ✓ (2 marks)

4 **a** $x^2 + xy + 5x + 5y$
$= x(x+y) + 5(x+y)$ ✓
$= (x+y)(x+5)$ ✓ (2 marks)

b $a^2 - 7a + ab - 7b$
$= a(a-7) + b(a-7)$ ✓
$= (a-7)(a+b)$ ✓ (2 marks)

c $pq + 3p - 8q - 24$
$= p(q+3) - 8(q+3)$ ✓
$= (q+3)(p-8)$ ✓ (2 marks)

d $5x^3 - 5xy - x^2y + y^2$
$= 5x(x^2 - y) - y(x^2 - y)$ ✓
$= (x^2 - y)(5x - y)$ ✓ (2 marks)

5 **a** $2m^2 + 13m + 15$
$= 2m^2 + 10m + 3m + 15$
$= 2m(m+5) + 3(m+5)$ ✓
$= (m+5)(2m+3)$ ✓ (2 marks)

b $3a^2 - ab - 2b^2$
$= 3a^2 - 3ab + 2ab - 2b^2$
$= 3a(a-b) + 2b(a-b)$ ✓
$= (a-b)(3a+2b)$ ✓ (2 marks)

c $21p^2 + 29p - 10$
$= 21p^2 + 35p - 6p - 10$
$= 7p(3p+5) - 2(3p+5)$ ✓
$= (3p+5)(7p-2)$ ✓ (2 marks)

6 **a** $\frac{m^2+8m+15}{m+5} \times \frac{m-3}{m^2-5m+6}$

$= \frac{(m+5)(m+3)}{m+5} \times \frac{m-3}{(m-3)(m-2)}$ ✓

$= \frac{m+3}{m-2}$ ✓ (2 marks)

b $\frac{x^2+17x+60}{x^2-16x+60} \times \frac{x^2+2x-48}{x^2+20x+96}$

$= \frac{(x+5)(x+12)}{(x-6)(x-10)} \times \frac{(x+8)(x-6)}{(x+8)(x+12)}$ ✓

$= \frac{x+5}{x-10}$ ✓ (2 marks)

c $\frac{n^2+3n-4}{n^2-10n+21} \div \frac{n^2-5n-36}{n^2-6n-7}$

$= \frac{(n+4)(n-1)}{(n-7)(n-3)} \div \frac{(n-9)(n+4)}{(n-7)(n+1)}$ ✓

$= \frac{(n+4)(n-1)}{(n-7)(n-3)} \times \frac{(n-7)(n+1)}{(n-9)(n+4)}$

$= \frac{(n-1)(n+1)}{(n-3)(n-9)}$ ✓(2 marks)

7 **a** $\frac{1}{x^2+6x+8} + \frac{1}{x^2+7x+12}$

$= \frac{1}{(x+2)(x+4)} + \frac{1}{(x+3)(x+4)}$ ✓

$= \frac{x+3}{(x+2)(x+3)(x+4)}$

$+ \frac{x+2}{(x+2)(x+3)(x+4)}$

$= \frac{2x+5}{(x+2)(x+3)(x+4)}$ ✓

(2 marks)

b $\frac{3}{x^2+5x+4} - \frac{2}{x^2+4x+3}$

$= \frac{3}{(x+4)(x+1)} - \frac{2}{(x+3)(x+1)}$

$= \frac{3(x+3)-2(x+4)}{(x+3)(x+4)(x+1)}$ ✓

$= \frac{3x+9-2x-8}{(x+3)(x+4)(x+1)}$

$= \frac{x+1}{(x+3)(x+4)(x+1)}$

$= \frac{1}{(x+3)(x+4)}$ ✓ (2 marks)

c $\frac{x^2+7x+10}{x^2+4x-5} + \frac{x^2-9x+18}{x^2-4x+3}$

$= \frac{(x+2)(x+5)}{(x+5)(x-1)} + \frac{(x-3)(x-6)}{(x-3)(x-1)}$

$= \frac{x+2}{x-1} + \frac{x-6}{x-1}$ ✓

$= \frac{2x-4}{x-1}$

$= \frac{2(x-2)}{x-1}$ ✓ (2 marks)

(Total: 40 marks)

SPECIAL BINOMIAL PRODUCTS AND FACTORISATIONS

SKILLS CHECK PAGE 26

1 **a** $(x+3)(x+8)$
$= x^2+8x+3x+24$
$= x^2+11x+24$

b $(a-5)(a+9)$
$= a^2+9a-5a-45$
$= a^2+4a-45$

c $(2a+3)(5b+4)$
$= 10ab+8a+15b+12$

d $(6x-1)(x-8)$
$= 6x^2-48x-x+8$
$= 6x^2-49x+8$

e $(5n+3)(2n-3)$
$= 10n^2-15n+6n-9$
$= 10n^2-9n-9$

f $(x+y)(a+b)$
$= ax+bx+ay+by$

2 **a** $(x+3)^2$
$= x^2+2\times x\times 3+3^2$
$= x^2+6x+\mathbf{9}$

b $(x-6)^2$
$= x^2-2\times x\times 6+6^2$
$= x^2-\mathbf{12x}+36$

c $(3x+2)^2$
$= (3x)^2+2\times 3x\times 2+2^2$
$= 9x^2+\mathbf{12x}+4$

d $(2x-5)^2 = \mathbf{4x^2}-20x+25$
$[(2x)^2]$

3 **a** $(x+4)^2 = x^2+2\times x\times 4+4^2$
$= x^2+8x+16$

b $(x-9)^2 = x^2-18x+81$
$[x^2-2\times x\times 9+9^2]$

c $(3a-10)^2$
$= (3a)^2-2\times 3a\times 10+10^2$
$= 9a^2-60a+100$

d $(2p+3q)^2$
$= 4p^2+12pq+9q^2$
$[(2p)^2+2\times 2p\times 3q+(3q)^2]$

e $(7-x)^2$
$= 7^2-2\times 7\times x+x^2$
$= 49-14x+x^2$

f $(a+b)^2$
$= a^2+2\times a\times b+b^2$
$= a^2+2ab+b^2$

4 **a** $(p+3)(p-3) = p^2-3^2$
$= p^2-\mathbf{9}$

b $(2x-1)(2x+1) = (2x)^2-1^2$
$= \mathbf{4x^2}-1$

c $(3x+2y)(3x-2y)$
$= (3x)^2-(2y)^2$
$= 9x^2-\mathbf{4y^2}$

d $(x^2+2)(x^2-2) = (x^2)^2-2^2$
$= \mathbf{x^4}-4$

5 **a** $(x-9)(x+9) = x^2-9^2$
$= x^2-81$

b $(3x-1)(3x+1) = (3x)^2-1^2$
$= 9x^2-1$

c $(2n+7)(2n-7) = (2n)^2-7^2$
$= 4n^2-49$

d $(8-x)(8+x) = 64-x^2$

e $(7x+2y)(7x-2y)$
$= (7x)^2-(2y)^2$
$= 49x^2-4y^2$

f $(ab+2)(ab-2) = (ab)^2-2^2$
$= a^2b^2-4$

6 **a** $x^2+13x+42 = (x+6)(x+7)$
$[6+7=13, 6\times 7=42]$

b $x^2+3x-40 = (x+8)(x-5)$
[Or $(x-5)(x+8)$]

c $x^2-10x+16$
$= (x-2)(x-8)$

d $x^2-3x-54 = (x+6)(x-9)$

7 **a** $x^2+8x+16 = (x+\mathbf{4})^2$

b $a^2-10a+25 = (a-\mathbf{5})^2$

c $n^2+2n+1 = (n+\mathbf{1})^2$

d $x^2-16x+64 = (x-\mathbf{8})^2$

8 **a** $x^2+20x+100 = (x+10)^2$

b $y^2+4y+4 = (y+2)^2$

c $49+14n+n^2 = (7+n)^2$

d $a^2-2a+1 = (a-1)^2$

e $9-6m+m^2 = (3-m)^2$

f $x^2-2xy+y^2 = (x-y)^2$

9 **a** $x^2-49 = (x+7)(x-\mathbf{7})$

b $a^2-9 = (a+3)(\mathbf{a-3})$

c $25-m^2 = (\mathbf{5+m})(5-m)$

d $x^2-y^2 = (x+\mathbf{y})(x-\mathbf{y})$

10 **a** $x^2-81 = x^2-9^2$
$= (x+9)(x-9)$
[Or $(x-9)(x+9)$]

b $121-y^2 = 11^2-y^2$
$= (11+y)(11-y)$

c $m^2 - 900 = m^2 - 30^2$
$= (m + 30)(m - 30)$

d $p^2 - q^2 = (p + q)(p - q)$

11 **a** $x^2 + xy + 2x + 2y$
$= x(x + y) + 2(x + y)$
$= (x + y)(x + 2)$

b $ab + 3a - 5b - 15$
$= a(b + 3) - 5(b + 3)$
$= (b + 3)(a - 5)$

c $a^2 - ab - ac + bc$
$= a(a - b) - c(a - b)$
$= (a - b)(a - c)$

d $2mn - 2m + 3n - 3$
$= 2m(n - 1) + 3(n - 1)$
$= (n - 1)(2m + 3)$

e $xz - yz - x + y$
$= z(x - y) - 1(x - y)$
$= (x - y)(z - 1)$
[It is not necessary to write the 1 in the 2nd line but it can't be left out in the last line.]

f $6kn - 3k - 2n + 1$
$= 3k(2n - 1) - 1(2n - 1)$
$= (2n - 1)(3k - 1)$

SPECIAL BINOMIAL PRODUCTS AND FACTORISATIONS
INTERMEDIATE TEST PAGE 27

1 $(2x - 5)^2$
$= (2x)^2 - 2 \times 2x \times 5 + 5^2$
$= 4x^2 - 20x + 25$
∴ [C] ✓ (1 mark)

2 $(3x - y)(3x + y)$
$= (3x)^2 - y^2$
$= 9x^2 - y^2$
∴ [C] ✓ (1 mark)

3 $xy - 4x - 6y + 24$
$= x(y - 4) - 6(y - 4)$
$= (y - 4)(x - 6)$
∴ [A] ✓ (1 mark)

4 $x^2 + 34x + 289 = (x + 17)^2$
∴ [B] ✓ (1 mark)

5 $6k(k + 2) - (k + 2)$
$= 6k(k + 2) - 1(k + 2)$
$= (k + 2)(6k - 1)$
∴ [D] ✓ (1 mark)

6 $\dfrac{x^2 - 64}{8x - 64} = \dfrac{(x + 8)(x - 8)}{8(x - 8)}$
$= \dfrac{x + 8}{8}$
[After cancelling the common factor $(x - 8)$, the fraction $\dfrac{x + 8}{8}$ cannot be simplified. 8 cannot be cancelled because it is not a common factor.]
∴ [C] ✓ (1 mark)

7 Try each option.
$x^2 - 49 = (x + 7)(x - 7)$
not $(x - 7)^2$
$x^2 - 10x - 24 = (x - 12)(x + 2)$
not $(x - 4)(x - 6)$
$x^2 - 24x + 144 = (x - 12)^2$
not $(x + 12)^2$
$x^2 - 625 = (x + 25)(x - 25)$
∴ [D] ✓ (1 mark)

8 Try each option.
$x^2 + 40x + 400 = (x + 20)^2$
$x^2 - 38x + 361 = (x - 19)^2$
$x^2 + 42x + 441 = (x + 21)^2$
$x^2 - 44x + 464$ is not a perfect square.
∴ [D] ✓ (1 mark)

9 $(3x + 5)^2 - (3x - 5)^2$
$= (3x + 5 + 3x - 5)$
$\times (3x + 5 - (3x - 5))$
$= 6x \times 10$
$= 60x$
∴ [D] ✓ (1 mark)

10 $(3x + 8)^2 + (x - 8)(x + 8)$
$= 9x^2 + 48x + 64 + x^2 - 64$ ✓
$= 10x^2 + 48x$ ✓ (2 marks)

11 $a^2 + ab - a - b$
$= a(a + b) - 1(a + b)$ ✓
$= (a + b)(a - 1)$ ✓ (2 marks)

12 $a^2b[(2a + b)^2 - (2a - b)^2]$
$= a^2b[(4a^2 + 4ab + b^2)$
$- (4a^2 - 4ab + b^2)]$
$= a^2b[4a^2 + 4ab + b^2 - 4a^2$
$+ 4ab - b^2]$ ✓
$= a^2b[8ab]$
$= 8a^3b^2$ ✓ (2 marks)

13 $(2a - 1)^2$
$= (2a)^2 - 2 \times 2a \times 1 + 1^2$
$= 4a^2 - 4a + 1$ ✓ (1 mark)

14 **a** $7p^2 - 63$
$= 7(p^2 - 9)$ ✓
$= 7(p + 3)(p - 3)$ ✓
(2 marks)

b $6x^2 - 48x + 96$
$= 6(x^2 - 8x + 16)$ ✓
$= 6(x - 4)^2$ ✓ (2 marks)

15 **a** **i** When $x = 2$,
$2x(x + 3) - (x + 2)(x - 2)$
$- (x + 3)^2$
$= 2 \times 2 \times (2 + 3)$
$- (2 + 2) \times (2 - 2)$
$- (2 + 3)^2$
$= 20 - 0 - 25$
$= -5$ ✓ (1 mark)

ii When $x = 4$,
$2x(x + 3) - (x + 2)(x - 2)$
$- (x + 3)^2$
$= 2 \times 4 \times (4 + 3)$
$- (4 + 2) \times (4 - 2)$
$- (4 + 3)^2$
$= 56 - 12 - 49$
$= -5$ ✓ (1 mark)

iii When $x = -1$,
$2x(x + 3) - (x + 2)(x - 2)$
$- (x + 3)^2$
$= 2(-1)(-1 + 3)$
$- (-1 + 2)(-1 - 2)$
$- (-1 + 3)^2$
$= -4 + 3 - 4$
$= -5$ ✓ (1 mark)

b $2x(x + 3) - (x + 2)(x - 2)$
$- (x + 3)^2$
$= 2x^2 + 6x - (x^2 - 4)$
$- (x^2 + 6x + 9)$
$= 2x^2 + 6x - x^2 + 4 - x^2$
$- 6x - 9$ ✓
$= -5$
∴ $2x(x + 3) - (x + 2)(x - 2)$
$- (x + 3)^2$
will always equal -5. ✓
(2 marks)
(Total: 25 marks)

SPECIAL BINOMIAL PRODUCTS AND FACTORISATIONS
ADVANCED TEST PAGE 28

1 **a** $(4p + 5q)^2$
$= (4p)^2 + 2 \times 4p \times 5q + (5q)^2$
$= 16p^2 + 40pq + 25q^2$ ✓
(1 mark)

b $(12-p)(12+p)$
$= 12^2 - p^2$
$= 144 - p^2$ ✓ (1 mark)

c $(5x-3y)(5x+3y)$
$= (5x)^2 - (3y)^2$
$= 25x^2 - 9y^2$ ✓ (1 mark)

d $2(x-4)^2$
$= 2(x^2 - 8x + 16)$
$= 2x^2 - 16x + 32$ ✓ (1 mark)

e $3x(x+3)(x-3)$
$= 3x(x^2 - 9)$
$= 3x^3 - 27x$ ✓ (1 mark)

f $(4x^2y^3 - 7z)(4x^2y^3 + 7z)$
$= (4x^2y^3)^2 - (7z)^2$
$= 16x^4y^6 - 49z^2$ ✓ (1 mark)

2 **a** $81 - 4d^2 = (9-2d)(9+2d)$ ✓ (1 mark)

b $36 - 12c + c^2 = (6-c)^2$ ✓ (1 mark)

c $4n^2 - 12n + 9$
$= (2n-3)^2$ ✓ (1 mark)

d $x^4 - 25y^2$
$= (x^2 - 5y)(x^2 + 5y)$ ✓ (1 mark)

e $x^4 + 4x^2 + 4$
$= (x^2+2)^2$ ✓ (1 mark)

f $p^2 - p + \frac{1}{4} = (p - \frac{1}{2})^2$ ✓ (1 mark)

3 **a** $2x^2 - 18$
$= 2(x^2 - 9)$
$= 2(x+3)(x-3)$ ✓ (1 mark)

b $3x^2 + 30x + 75$
$= 3(x^2 + 10x + 25)$
$= 3(x+5)^2$ ✓ (1 mark)

c $x^4 - 1$
$= (x^2+1)(x^2-1)$
$= (x^2+1)(x+1)(x-1)$ ✓ (1 mark)

d $p - p^3$
$= p(1-p^2)$
$= p(1+p)(1-p)$ ✓ (1 mark)

e $2x^3 - 12x^2 + 18x$
$= 2x(x^2 - 6x + 9)$
$= 2x(x-3)^2$ ✓ (1 mark)

f $16a^4 - 72a^2 + 81$
$= (4a^2 - 9)^2$
$= [(2a+3)(2a-3)]^2$ ✓ (1 mark)

4 **a** $(5x-2)(7-x) + (3x-2)^2$
$= 35x - 5x^2 - 14 + 2x$
$+ 9x^2 - 12x + 4$ ✓
$= 4x^2 + 25x - 10$ ✓ (2 marks)

b $(x+y)(x-y) - (x-y)^2$
$= x^2 - y^2 - (x^2 - 2xy + y^2)$ ✓
$= x^2 - y^2 - x^2 + 2xy - y^2$
$= 2xy - 2y^2$ ✓ (2 marks)

c $9(m+5)(m-2)$
$- (3m-10)(3m+10)$
$= 9(m^2 + 3m - 10)$
$- (9m^2 - 100)$ ✓
$= 9m^2 + 27m - 90 - 9m^2$
$+ 100$
$= 27m + 10$ ✓ (2 marks)

d $(5a-2b)^2 - (5a+2b)^2$
$= 25a^2 - 20ab + 4b^2$
$- (25a^2 + 20ab + 4b^2)$ ✓
$= 25a^2 - 20ab + 4b^2 - 25a^2$
$- 20ab - 4b^2$
$= -40ab$ ✓ (2 marks)

5 **a** $\frac{9p^2+9}{9p^2+18p+9}$
$= \frac{9(p^2+1)}{9(p^2+2p+1)}$ ✓
$= \frac{p^2+1}{(p+1)^2}$ ✓ (2 marks)

b $\frac{n^2-100}{n^2-12n+20}$
$= \frac{(n+10)(n-10)}{(n-10)(n-2)}$ ✓
$= \frac{n+10}{n-2}$ ✓ (2 marks)

c $\frac{x^3-4x}{x^3-4x^2+4x}$
$= \frac{x(x^2-4)}{x(x^2-4x+4)}$
$= \frac{(x+2)(x-2)}{(x-2)^2}$ ✓
$= \frac{x+2}{x-2}$ ✓ (2 marks)

d $\frac{x^2-2x+1}{1-x^2}$
$= \frac{(x-1)^2}{(1+x)(1-x)}$ ✓
$= \frac{-(x-1)}{1+x}$
$= \frac{1-x}{1+x}$ ✓ (2 marks)

6 **a** $\frac{1}{x^2+20x+100} - \frac{1}{x^2-100}$
$= \frac{1}{(x+10)^2} - \frac{1}{(x+10)(x-10)}$ ✓
$= \frac{(x-10)-(x+10)}{(x+10)^2(x-10)}$
$= \frac{-20}{(x+10)^2(x-10)}$ ✓ (2 marks)

b $\frac{1}{a^3-1} \div \frac{1}{a^6-1}$
$= \frac{1}{a^3-1} \times \frac{a^6-1}{1}$ ✓
$= \frac{(a^3-1)(a^3+1)}{a^3-1}$
$= a^3 + 1$ ✓ (2 marks)

7 **a** $1636 = 1637 - 1$
and $1638 = 1637 + 1$
So 1636×1638
$= (1637-1)(1637+1)$
$= 1637^2 - 1$
$= 2\,679\,769 - 1$
$= 2\,679\,768$ ✓ (1 mark)

b 897×903
$= (900-3)(900+3)$
$= 900^2 - 3^2$
$= 810\,000 - 9$
$= 809\,991$ ✓ (1 mark)

(Total: 40 marks)

SUBSTITUTION, EQUATIONS AND FORMULAE
SKILLS CHECK PAGE 30

1 If $x = 4$,

a $x + 3 = 4 + 3$
$= 7$

b $x - 6 = 4 - 6$
$= -2$

c $9 - x = 9 - 4$
$= 5$

d $5x = 5 \times 4$
$= 20$

e $2x + 1 = 2 \times 4 + 1$
$= 9$

f $3x - 2 = 3 \times 4 - 2$
$= 10$

g $x^2 = 4^2$
$= 16$

h $2x^2 = 2 \times 4^2$
$= 2 \times 16$
$= 32$

2 If $a = 2, b = 5, c = 6,$

a $a + b = 2 + 5$
$= 7$

b $c - b = 6 - 5$
$= 1$

c $b + c - a = 5 + 6 - 2$
$= 9$

d $3a = 3 \times 2$
$= 6$

e $bc = 5 \times 6$
$= 30$

f $ab + c = 2 \times 5 + 6$
$= 16$

g $3ac = 3 \times 2 \times 6$
$= 36$

h $a^2 + b^2 = 2^2 + 5^2$
$= 4 + 25$
$= 29$

3 a $x + 13 = 25$
$x = 25 - 13$
$x = 12$

b $y - 9 = 37$
$y = 37 + 9$
$y = 46$

c $5a = 325$
$a = 325 \div 5$
$a = 65$

d $\frac{h}{6} = 12$
$h = 12 \times 6$
$h = 72$

e $2x + 1 = 17$ [subtract 1]
$2x = 16$ [divide by 2]
$x = 8$

f $3x - 4 = 5$ [add 4]
$3x = 9$ [divide by 3]
$x = 3$

g $\frac{k}{2} + 3 = 7$ [subtract 3]
$\frac{k}{2} = 4$ [multiply by 2]
$k = 8$

h $\frac{d-4}{3} = -9$ [multiply by 3]
$d - 4 = -27$ [add 4]
$d = -23$

4 a $5x + 8 = 29 - 2x$
Substitute $x = 3$,
LHS $= 5 \times 3 + 8$
$= 23$
RHS $= 29 - 2 \times 3$
$= 23$
LHS = RHS True

b $\frac{x}{3} + \frac{x}{4} = 7$
Substitute $x = 12$ in LHS
$\frac{12}{3} + \frac{12}{4} = 4 + 3$
$= 7$
LHS = RHS True

c $4(2x + 1) - 3(x + 5) = 28$
Substitute $x = 8$ in LHS
$4(2 \times 8 + 1) - 3(8 + 5)$
$= 4 \times 17 - 3 \times 13$
$= 29$
LHS ≠ RHS False

d $\frac{7x+1}{5} = \frac{2x-9}{3}$
Substitute $x = -4$
LHS $= \frac{7 \times -4 + 1}{5}$
$= -5\frac{2}{5}$
RHS $= \frac{2 \times -4 - 9}{3}$
$= -5\frac{2}{3}$
LHS ≠ RHS False

5 a $9p = 15 - p$
[add p to both sides]
$10p = 15$
[divide both sides by 10]
$p = 1.5$

b $7k + 5 = 2k + 25$
[subtract $2k$ from both sides]
$5k + 5 = 25$
$5k = 20$
$k = 4$

6 $A = 120, h = 12.5$
$V = Ah$
$= 120 \times 12.5$
$= 1500$

7 $l = 8, = 5$
$P = 2l + 2b$
$= 2 \times 8 + 2 \times 5$
$= 26$

8 $u = 28, a = -4, t = 5$
$v = u + at$
$= 28 + (-4) \times 5$
$= 28 - 20$
$= 8$

9 $a = 24, \ b = 45$
$c^2 = a^2 + b^2$
$= 24^2 + 45^2$
$= 2601$
$c = \sqrt{2601} \quad (c > 0)$
$= 51$

10 $r = 12$
$A = \pi r^2$
$= \pi \times 12^2$
$= 452.389\,3421\ldots$
$= 452.4$ (1 d.p.)

11 $x_1 = -2, x_2 = 1, y_1 = 8, y_2 = -4$
$m = \frac{y_2 - y_1}{x_2 - x_1}$
$= \frac{-4 - 8}{1 - (-2)}$
$= -4$

SUBSTITUTION, EQUATIONS AND FORMULAE
INTERMEDIATE TEST PAGE 31

1 $a = 5, b = -3$
$3a - b = 3 \times 5 - (-3)$
$= 15 + 3$
$= 18$
∴ [B] ✓ (1 mark)

2 $2(3x + 4) = 2x - 4$
$6x + 8 = 2x - 4$
$4x + 8 = -4$
$4x = -12$
$x = -3$
[Or try each option to see which one satisfies the equation.]
∴ [B] ✓ (1 mark)

3 If $x = -2$,
$3x^2 = 3 \times (-2)^2$
$= 3 \times 4$
$= 12$
∴ [C] ✓ (1 mark)

4 $A = (n - 2) \times 180°$
When $n = 15$,
$A = (15 - 2) \times 180°$
$= 2340°$
∴ [D] ✓ (1 mark)

5 $y = mx + b$
When $m = 2, \ b = -6$ and $y = 8$
$8 = 2x - 6$
$14 = 2x$
$7 = x$
∴ [D] ✓ (1 mark)

6 $v = u + at$
$u = v - at$ (A is not correct)
$v = u + at$
$at = v - u$
$a = \frac{v-u}{t}$ (B is correct)
$t = \frac{v-u}{a}$
(neither C nor D is correct)
∴ **[B]** ✓ (1 mark)

7 $6a + 9 - a = 8 - 3a + 1$
$5a + 9 = 9 - 3a$ ✓
$8a + 9 = 9$
$8a = 0$ ✓
$a = 0$ ✓ (3 marks)

8 $4(x-5) + 3(x-2) = 16$
$4x - 20 + 3x - 6 = 16$ ✓
$7x - 26 = 16$ ✓
$7x = 42$
$x = 6$ ✓ (3 marks)

9 $\frac{2t+4}{7} - 3 = 5$
$\frac{2t+4}{7} = 8$ ✓
$2t + 4 = 56$ ✓
$2t = 52$
$t = 26$ ✓
[Or multiply each term by 7 first;
$2t + 4 - 21 = 35$
$2t - 17 = 35$
$2t = 52$
$t = 26$] (3 marks)

10 $\frac{2m-5}{3} = \frac{3m+2}{5}$
$5(2m-5) = 3(3m+2)$ ✓
$10m - 25 = 9m + 6$ ✓
$m - 25 = 6$
$m = 31$ ✓ (3 marks)

11 **a** Between lines 1 and 2 ✓
because $3x - 5x = -2x$
(not $8x$) ✓
Between lines 3 and 4 ✓
because $x = 4 \div 8$
(not $8 \div 4$) ✓ (4 marks)

b $3x + 5 - 5x = 9$
$-2x + 5 = 9$
$-2x = 4$
$x = -2$ ✓ (1 mark)

12 $A = 20, y = 8$
$D = \frac{yA}{y+12}$
$= \frac{8 \times 20}{8+12}$ ✓
$= \frac{160}{20}$
$= 8$
The child's dose is 8 mL. ✓
(2 marks)
(Total: 25 marks)

SUBSTITUTION, EQUATIONS AND FORMULAE
ADVANCED TEST PAGE 32

1 **a** $5(x-3) + 2(x-7) = 6$
$5x - 15 + 2x - 14 = 6$ ✓
$7x - 29 = 6$
$7x = 35$
$x = 5$ ✓
(2 marks)

b $4(3k-2) = 3(1-2k)$
$12k - 8 = 3 - 6k$ ✓
$18k = 11$
$k = \frac{11}{18}$ ✓ (2 marks)

c $3(4p-5) = 2p - (7-p)$
$12p - 15 = 2p - 7 + p$ ✓
$9p = 8$
$p = \frac{8}{9}$ ✓ (2 marks)

d $11(7-2g) - 5(3g-8) = 8g$
$77 - 22g - 15g + 40 = 8g$ ✓
$117 = 45g$
$g = 2.6$ ✓
(2 marks)

e $\frac{3a}{4} + 7 = 2a + 1$
$3a + 28 = 8a + 4$ ✓
$24 = 5a$
$a = 4.8$ ✓ (2 marks)

f $\frac{9n-3}{7} = 4n + 5$
$9n - 3 = 28n + 35$ ✓
$-19n = 38$
$n = -2$ ✓ (2 marks)

g $\frac{5x-3}{4} = \frac{4x-2}{3}$
$3(5x-3) = 4(4x-2)$
$15x - 9 = 16x - 8$ ✓
$-x = 1$
$x = -1$ ✓ (2 marks)

h $\frac{m}{5} + \frac{m}{3} = 16$
$3m + 5m = 240$ ✓
$8m = 240$
$m = 30$ ✓ (2 marks)

i $\frac{3u+2}{5} - \frac{4u-1}{2} = \frac{2}{3}$
$30\left(\frac{3u+2}{5}\right) - 30\left(\frac{4u-1}{2}\right)$
$= 30 \times \frac{2}{3}$
$6(3u+2) - 15(4u-1) = 20$
$18u + 12 - 60u + 15 = 20$ ✓
$-42u = -7$
$u = \frac{1}{6}$ ✓
(2 marks)

2 **a** $x^2 = 0.25$
$x = \pm 0.5$ ✓ (1 mark)

b $9a^2 = 1$
$a^2 = \frac{1}{9}$
$a = \pm\frac{1}{3}$ ✓ (1 mark)

c $t^2 - 4 = 0$
$t^2 = 4$
$t = \pm 2$ ✓ (1 mark)

d $\frac{y^2}{2} - 5 = 3$
$\frac{y^2}{2} = 8$
$y^2 = 16$
$y = \pm 4$ ✓ (1 mark)

3 **a** $5x^2 - 2 = 7$
$5x^2 = 9$
$x^2 = 1.8$
$x = \sqrt{1.8}$ $(x > 0)$ ✓
$= 1.341\,6407\ldots$
$= 1.342$ (3 d.p.) ✓
(2 marks)

b It means there is only one answer instead of two. ✓
(1 mark)

4 **a** $5p - 3pq$
$= 5 \times -2 - 3 \times -2 \times -3$
$= -28$ ✓ (1 mark)

b $4p^2 = 4 \times (-2)^2$
$= 16$ ✓ (1 mark)

c $\frac{p+q}{p-q} = \frac{-2-3}{-2-(-3)}$
$= -5$ ✓ (1 mark)

d $pq^2 - p + q$
$= -2 \times (-3)^2 - (-2) + (-3)$
$= -2 \times 9 + 2 - 3$
$= -19$ ✓ (1 mark)

5 **a** $s = ut + \frac{1}{2}at^2$

When $u = 32$, $a = 9.8$ and $t = 7$,

$s = 32 \times 7 + \frac{1}{2} \times 9.8 \times 7^2$

$= 464.1$ ✓ (1 mark)

b $s = ut + \frac{1}{2}at^2$

When $s = 1599$, $u = 58$ and $t = 13$,

$1599 = 58 \times 13 + \frac{1}{2} \times a \times 13^2$

$1599 = 754 + 84.5a$ ✓

$84.5a = 845$

$a = 10$ ✓ (2 marks)

6 **a** $V = \frac{4}{3}\pi r^3$

When $r = 9$,

$V = \frac{4}{3} \times \pi \times 9^3$

$= 3053.62805\ldots$

$= 3053.63$ (2 d.p.) ✓ (1 mark)

b $V = \frac{4}{3}\pi r^3$

When $V = 222\,665$,

$222\,665 = \frac{4}{3}\pi r^3$

$r^3 = 222\,665 \div \frac{4}{3}\pi$

$r = \sqrt[3]{222\,665 \div \frac{4}{3}\pi}$ ✓

$= 37.599994\ldots$

$= 37.60$ (2 d.p.) ✓ (2 marks)

7 **a** $F = \frac{9C}{5} + 32$

When $C = 180$,

$F = \frac{9 \times 180}{5} + 32$

$= 356$

The oven should be set at a temperature of 356 °F. ✓ (1 mark)

b $F = \frac{9C}{5} + 32$

When $F = 68$,

$68 = \frac{9C}{5} + 32$

$36 = \frac{9C}{5}$

$9C = 180$

$C = 20$

The temperature was only 20 °C, which is not hot, so Tina was not correct. ✓ (1 mark)

8 **a** $d = \sqrt{(x_2 - x_1)^2 + (y_2 - y_1)^2}$

$d = 10$ when $x_1 = -3$, $y_1 = 2$ and $x_2 = 5$,

$10 = \sqrt{(5 - (-3))^2 + (y_2 - 2)^2}$

$10 = \sqrt{64 + (y_2 - 2)^2}$ ✓

$100 = 64 + (y_2 - 2)^2$

$36 = (y_2 - 2)^2$

So $y_2 - 2 = 6$ or $y_2 - 2 = -6$

$y_2 = 8$ or $y_2 = -4$ ✓ (2 marks)

b There are two points on the line $x = 5$ that are 10 units from the point (–3, 2). The points are (5, –4) and (5, 8). ✓ (1 mark)

(Total: 40 marks)

FURTHER EQUATIONS AND INEQUALITIES

SKILLS CHECK PAGE 34

1

a	$3 > 2$	True
b	$-3 > -2$	False
c	$6 < 15$	True
d	$-6 < -15$	False
e	$4 \geq -1$	True
f	$-4 \geq 1$	False
g	$-2 \leq 2$	True
h	$2 \leq -2$	False

2 **a** $-3 \geq 2$ False

b $5 < 2 \times 3$

$5 < 6$ True

c $12 + 8 \neq 35 - 5 \times 3$

$20 \neq 35 - 15$

$20 \neq 20$ False

d $19 > 17 - 3 \times -2$

$19 > 17 + 6$

$19 > 23$ False

e $6^2 \leq 2^6$

$36 \leq 64$ True

f $8 + 5 \not> 5 + 8$

$13 \not> 13$ True

3 **a** $5a - 4 < 2a$

Substitute $a = -2$

LHS $= 5 \times -2 - 4$

$= -14$

RHS $= 2 \times -2$

$= -4$

$-14 < -4$ True

b $\frac{9e - 1}{11} \geq 7$

Substitute $e = 5$

LHS $= \frac{9 \times 5 - 1}{11}$

$= 4$

$4 < 7$ (not $\geq$) False

c $3p + 1 \leq 3 - 3p$

Substitute $p = -1$

LHS $= 3 \times -1 + 1$

$= -3 + 1$

$= -2$

RHS $= 3 - 3 \times -1$

$= 3 - (-3)$

$= 6$

$-2 \leq 6$ True

d $\frac{x}{3} + 8 \geq -2$

Substitute $x = -30$

LHS $= \frac{-30}{3} + 8$

$= -10 + 8$

$= -2$

$-2 \geq -2$ True

4 **a** $-4p = 8$ [divide by –4]

$p = -2$

b $4p < -8$

$p < -2$

c $-4p > 8$

$p < -2$ [dividing both sides by –4, the inequality sign must be reversed]

d $3x + 17 = 74$

$3x = 57$

$x = 19$

e $2n - 1 > 15$

$2n > 16$

$n > 8$

f $5x \leq x + 8$

$4x \leq 8$ [subtracting x from both sides]

$x \leq 2$

g $\frac{a}{4} + 7 \geq -2$

$\frac{a}{4} \geq -9$

$a \geq -36$

h $\frac{t-3}{2} > -1$

$t - 3 > -2$

$t > 1$

5 **a** $x + 4 \le -2$

$x \le -6$

b $a - 5 > 5$

$a > 10$

c $4e < -2$

$e < -\frac{1}{2}$ $[\frac{-2}{4}]$

d $\frac{n}{3} \ge 6$

$n \ge 18$ $[6 \times 3]$

6 **a** $2x + 5 \ge 11$

$2x \ge 11 - 5$

$2x \ge 6$

$x \ge 3$

b $3a - 1 < -2$

$3a < -2 + 1$

$3a < -1$

$a < -\frac{1}{3}$

c $\frac{m+2}{3} \le 4$

$m + 2 \le 4 \times 3$

$m + 2 \le 12$

$m \le 12 - 2$

$m \le 10$

7 **a** $x \ge 1$

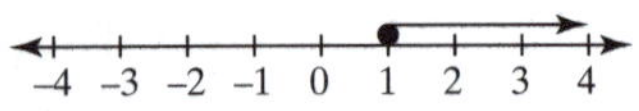

b $x < 2$

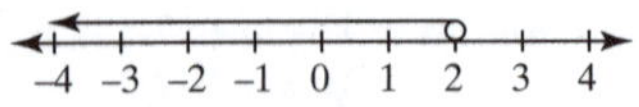

c $x \le -2$

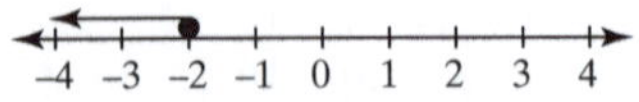

8 If $x = -1$,

$5(x + 1) - 2(3 - 2x)$

$= 5(-1 + 1) - 2(3 - 2 \times -1)$

$= -10$ $(\ne 10)$ False

9 If $x = -1$,

$7x - 3 = 7 \times -1 - 3$

$= -10$

$x - 8 = -1 - 8$

$= -9$ False

[because $-10 \not> -9$]

10 **a** $A = 36, b = 8$

$A = \frac{1}{2}bh$

$36 = \frac{1}{2} \times 8 \times h$

$= 4h$

$h = 36 \div 4$

$= 9$

b $A = 84, h = 7$

$A = \frac{1}{2}bh$

$84 = \frac{1}{2} \times b \times 7$

$= 3.5b$

$b = 84 \div 3.5$

$= 24$

The base is 24 cm long.

FURTHER EQUATIONS AND INEQUALITIES
INTERMEDIATE TEST PAGE 35

1 $11 < 2x - 15$ [add 15]

$26 < 2x$ [divide by 2]

$13 < x$

$x > 13$

None of the answers in A, B or C are greater than 13. [Instead of solving you could substitute each answer to see if it satisfies the inequality.]

∴ **[D]** ✓ (1 mark)

2 $\frac{7m}{3} - 8 \le 9 - \frac{m}{2}$

Substitute $m = -2$, $-12\frac{2}{3} < 10$ (True)

Substitute $m = 0$, $-8 \le 9$ (True)

Substitute $m = 6$, $6 \le 6$ (True)

Substitute $m = 9$, $13 \le 4\frac{1}{2}$ (False)

[Or $\frac{7m}{3} - 8 \le 9 - \frac{m}{2}$

(multiply both sides by 6)

$14m - 48 \le 54 - 3m$

$17m \le 102$

$m \le 6$

The only value of m not less than or equal to 6 is $m = 9$.]

∴ **[D]** ✓ (1 mark)

3 $-2x > -6$ [divide by –2]

$x < 3$

[The sign must be reversed because we divide by a negative number.]

∴ **[B]** ✓ (1 mark)

4 $3p - 17 = 5p + 19$ [add 17]

$3p = 5p + 36$ [subtract $5p$]

$-2p = 36$ [divide by –2]

$p = -18$

∴ **[A]** ✓ (1 mark)

5 Let the number be x.

Three times the number

$= 3 \times x$

$= 3x$

Subtracting 15 from $3x$ gives

$3x - 15$

Twice the number $= 2 \times x = 2x$

8 more than twice the number

$= 2x + 8$

$3x - 15 = 2x + 8$

∴ **[A]** ✓ (1 mark)

6 The number 8 more than $2x - 5$

$= 2x - 5 + 8$

$= 2x + 3$

∴ **[A]** ✓ (1 mark)

7 **a** Let x be the price of a bottle of soft drink.

$7x + \$2.75 = \$15.98$ ✓

(1 mark)

b $7x + \$2.75 = \$15.98$

$7x = \$13.23$

$x = \$1.89$

One bottle of soft drink costs \$1.89. ✓

The cost of 5 bottles

$= 5 \times \$1.89$

$= \$9.45$ ✓ (2 marks)

8 $5(2x + 3) \le 20$

$10x + 15 \le 20$

$10x \le 5$

$x \le \frac{1}{2}$ ✓

[Or $5(2x + 3) \le 20$

$2x + 3 \le 4$

$2x \le 1$

$x \le \frac{1}{2}$] (1 mark)

9 $9n + 4 > 2n - 17$

$7n + 4 > -17$ ✓

$7n > -21$

$n > -3$ ✓ (2 marks)

10 $3 - 2a > 7$
$-2a > 4$ ✓
$a < -2$ ✓

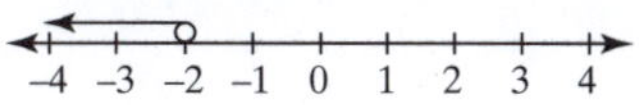

✓ (3 marks)

11 $3m + 17 \le 8m$
$17 \le 5m$ ✓
$3.4 \le m$
$m \ge 3.4$ ✓
[Or: $3m + 17 \le 8m$
$3m \le 8m - 17$
$-5m \le -17$
$m \ge 3.4$] (2 marks)

12 $2 - \frac{u}{7} \ge 3$
$-\frac{u}{7} \ge 1$ ✓
$u \le -7$ ✓
[Or multiply each term by 7 first.
$14 - u \ge 21$
$-u \ge 7$
$u \le -7$] (2 marks)

13 Let x be the number.
$5(x - 15) = 3x + 7$ ✓
$5x - 75 = 3x + 7$
$2x - 75 = 7$ ✓
$2x = 82$
$x = 41$
The number is 41. ✓ (3 marks)

14 Let the width be x cm.

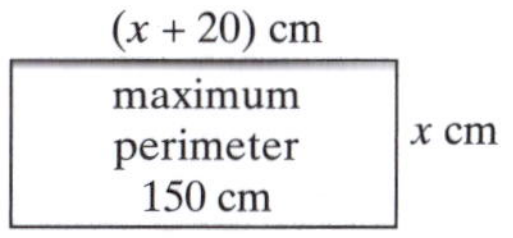

The length is $(x + 20)$ cm
$P = 2(x + 20) + 2x$
$= 2x + 40 + 2x$
$= 4x + 40$ ✓
Now $4x + 40 \le 150$ ✓
$4x \le 110$
$x \le 27.5$
If $x = 27.5$, $x + 20 = 27.5 + 20$
$= 47.5$
The largest rectangle is 27.5 cm wide and 47.5 cm long. ✓
(3 marks)
(Total: 25 marks)

FURTHER EQUATIONS AND INEQUALITIES
ADVANCED TEST PAGE 36

1 **a** $x \ge 3$

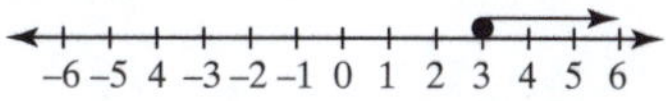

✓ (1 mark)

b $x < -1$ or $x > 4$

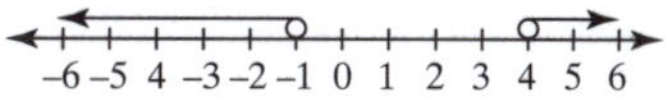

✓ (1 mark)

c $-2 \le x < 5$

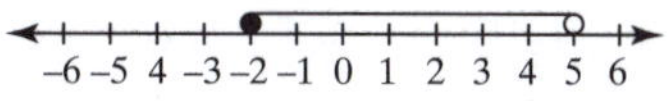

✓ (1 mark)

2 **a** $5x + 1 \ge 9 - 3x$
$8x \ge 8$
$x \ge 1$ ✓

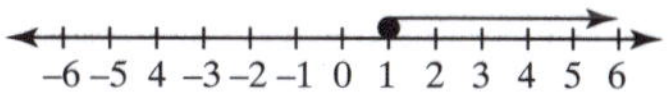

✓ (2 marks)

b $3(p + 5) - 1 < 11$
$3p + 15 - 1 < 11$
$3p < -3$
$p < -1$ ✓

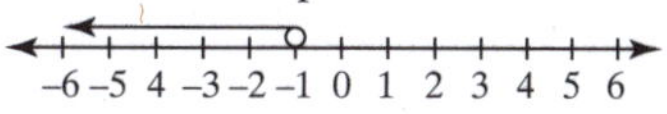

✓ (2 marks)

c $9 - 2n > 5$
$-2n > -4$
$n < 2$ ✓

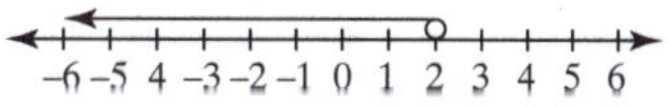

✓ (2 marks)

3 **a** $3(2m - 5) \le 2(m + 7)$
$6m - 15 \le 2m + 14$ ✓
$4m \le 29$
$m \le 7.25$ ✓(2 marks)

b $\frac{5k}{3} - 2 < 4 - k$
$5k - 6 < 12 - 3k$ ✓
$8k < 18$
$k < 2.25$ ✓ (2 marks)

c $\frac{6 - a}{2} > 4$
$6 - a > 8$
$-a > 2$ ✓
$a < -2$ ✓ (2 marks)

d $7(4q - 1) - 3(2q + 5) \ge 0$
$28q - 7 - 6q - 15 \ge 0$ ✓
$22q \ge 22$
$q \ge 1$ ✓
(2 marks)

e $12 - \frac{3d}{4} \le 18$
$48 - 3d \le 72$
$-3d \le 24$ ✓
$d \ge -8$ ✓ (2 marks)

f $5(2y - 3) - 4(3y - 7) > 2(y + 5)$
$10y - 15 - 12y + 28 > 2y + 10$
✓
$-2y + 13 > 2y + 10$
$-4y > -3$
$y < \frac{3}{4}$ ✓
(2 marks)

4 Jo divided by x without considering whether $x = 0$ is a solution to the equation. ✓
$x(x + 5) = x(x + 3)$
$x^2 + 5x = x^2 + 3x$
$5x = 3x$
$2x = 0$
$x = 0$ ✓ (2 marks)

5 **a** $y = mx + b$
$b = y - mx$ ✓ (1 mark)

b $A = \frac{1}{2}h(a + b)$
$2A = h(a + b)$
$a + b = \frac{2A}{h}$ ✓
$b = \frac{2A}{h} - a$ ✓ (2 marks)

c $c = \sqrt{a^2 + b^2}$
$c^2 = a^2 + b^2$
$b^2 = c^2 - a^2$ ✓
$b = \pm\sqrt{c^2 - a^2}$ ✓ (2 marks)

6 **a** $V = \pi r^2 h$
$r^2 = \frac{V}{\pi h}$
$r = \pm\sqrt{\frac{V}{\pi h}}$ ✓ (1 mark)

b There are no restrictions on the variables before the formula is rearranged, although in practical situations, because it is the formula for the volume of a cylinder, V, r and h would all be greater than 0. ✓
After it is rearranged, $h \ne 0$ and $\frac{V}{h} > 0$. ✓ (2 marks)

c As $r > 0$, $r = \sqrt{\dfrac{V}{\pi h}}$

When $V = 179$ and $h = 8.7$,

$r = \sqrt{\dfrac{179}{\pi \times 8.7}}$

$= 2.559\,127\ldots$

$= 2.6$ (1 d.p.) ✓ (1 mark)

7 **a** $A = \dfrac{1}{2}bh$

$27.3 = \dfrac{1}{2} \times 19.5 \times h$

$9.75h = 27.3$

$h = \dfrac{27.3}{9.75}$

$= 2.8$

The perpendicular height is 2.8 m. ✓ (1 mark)

b Let x m be the length of the hypotenuse.

$x^2 = 19.5^2 + 2.8^2$

$= 388.09$

$x = \sqrt{388.09}$ $(x > 0)$

$= 19.7$

The hypotenuse is 19.7 m long. ✓ (1 mark)

c The length of any side of a triangle cannot be negative. It makes no sense for any length to be negative. ✓ (1 mark)

8 **a** When $x = -23$,

$\sqrt{3 - 2x} = \sqrt{3 - 2 \times (-23)}$

$= 7$ ✓ (1 mark)

b $3 - 2x \geq 0$

$-2x \geq -3$

$x \leq 1.5$ ✓ (1 mark)

c $\sqrt{3 - 2x} = 5$

$3 - 2x = 25$

$-2x = 22$

$x = -11$ ✓ (1 mark)

d **i** $\sqrt{3 - 2x} > 5$ when $x < -11$ ✓ (1 mark)

[From part **c** we can see that the solution is either $x < -11$ or $x > -11$; from part **a** we can see that values of x less than -11 give values of the square root greater than 5.]

ii $\sqrt{3 - 2x} < 5$ when $-11 < x \leq 1.5$ ✓ (1 mark)

[Values of x greater than -11 will give values of the square root less than 5 but from part **b** we know that x cannot be greater than 1.5.]

(Total: 40 marks)

QUADRATIC EQUATIONS
SKILLS CHECK

PAGE 38

1 **a** If $x = 3$, $x^2 = 3^2$
$= 9$

b If $x = -3$, $x^2 = (-3)^2$
$= 9$

c If $x = 7$, $x^2 = 7^2$
$= 49$

d If $x = -7$, $x^2 = (-7)^2$
$= 49$

e If $x = -5$, $x^2 = (-5)^2$
$= 25$

f If $x = 5$, $x^2 = 5^2$
$= 25$

g If $x = 1$, $x^2 = 1^2$
$= 1$

h If $x = -1$, $x^2 = (-1)^2$
$= 1$

2 **a** $(-6)^2 = 36$
The other number is -6.

b $10^2 = 100$ and $(-10)^2 = 100$
The numbers are 10 and -10.

3 **a** $x^2 = 9$
$x = 3$ or $x = -3$

b $x^2 = 49$
$x = 7$ or $x = -7$

c $x^2 = 25$
$x = 5$ or $x = -5$

d $x^2 = 1$
$x = 1$ or $x = -1$

4 **a** If $p = 2$, $2p^2 = 2 \times 2^2$
$= 2 \times 4$
$= 8$

b If $p = -2$, $2p^2 = 2 \times (-2)^2$
$= 2 \times 4$
$= 8$

c If $p = 6$, $2p^2 = 2 \times 6^2$
$= 2 \times 36$
$= 72$

d If $p = -6$, $2p^2 = 2 \times (-6)^2$
$= 2 \times 36$
$= 72$

5 **a** $2p^2 = 8$
$p^2 = 4$
$p = 2$ or $p = -2$

b $2p^2 = 72$
$p^2 = 36$
$p = 6$ or $p = -6$

6 **a** $x^2 = 16$
$x = 4$ or $x = -4$

b $3e^2 = 243$
$e^2 = 81$
$e = 9$ or $e = -9$

c $4k^2 = 1$
$k^2 = \dfrac{1}{4}$
$k = \dfrac{1}{2}$ or $k = -\dfrac{1}{2}$

7 **a** $x^2 = 60$
$x = \pm\sqrt{60}$
$= \pm 7.745\,966\,69\ldots$
$= \pm 7.75$ (2 d.p.)

b $x^2 = 12.5$
$x = \pm\sqrt{12.5}$
$= \pm 3.535\,533\,906\ldots$
$= \pm 3.54$ (2 d.p.)

c $7x^2 = 45$
$x^2 = 6.428\,5714\ldots$
$x = \pm\sqrt{6.428\,5714\ldots}$
$= \pm 2.535\,462\,764\ldots$
$= \pm 2.54$ (2 d.p.)

8 **a** $5x = 0$ [divide both sides by 5]
$x = 0$

b $3a = 0$
$a = 0$

c $12t = 0$
$t = 0$

d $-37m = 0$
[divide both sides by -37]
$m = 0$

9 **a** $3(x - 7) = 0$
$x - 7 = 0$
$x = 7$

b $4(2x + 5) = 0$
$2x + 5 = 0$
$2x = -5$
$x = -2.5$

c $6(3x - 2) = 0$
$3x - 2 = 0$
$3x = 2$
$x = \dfrac{2}{3}$

10 a $x(x+4)=0$
$x=0$ or $x+4=0$
$x=0$ or $x=-4$

b $(x-7)(x+1)=0$
$x-7=0$ or $x+1=0$
$x=7$ or $x=-1$

c $(2a+5)(3a-2)=0$
$2a+5=0$ or $3a-2=0$
$2a=-5$ or $3a=2$
$a=-2\frac{1}{2}$ or $a=\frac{2}{3}$

11 a $x^2-9x=x(x-9)$

b $x^2+8x+15=(x+5)(x+3)$
[What 2 numbers add to 8 and multiply to 15?]

c $t^2+3t-10=(t+5)(t-2)$
[What 2 numbers add to 3 and multiply to −10?]

d $m^2-7m+12$
$=(m-3)(m-4)$
[Or $(m-4)(m-3)$]

e $n^2-4n-60$
$=(n-10)(n+6)$
[Or $(n+6)(n-10)$]

f $a^2-2a+1=(a-1)(a-1)$
$=(a-1)^2$

12 a When $x=3$,
$x^2+7x+12$
$=3^2+7\times3+12$
$=42\ (\neq 0)$ False

b When $y=-3$,
$y^2-4y-21$
$=(-3)^2-4\times-3-21$
$=0$ True

13 a $n^2-3n=0$
$n(n-3)=0$
$n=0$ or $n-3=0$
$n=0$ or $n=3$

b $x^2+6x+8=0$
$(x+2)(x+4)=0$
$x+2=0$ or $x+4=0$
$x=-2$ or $x=-4$

c $a^2-4a-45=0$
$(a-9)(a+5)=0$
$a-9=0$ or $a+5=0$
$a=9$ or $a=-5$

QUADRATIC EQUATIONS
INTERMEDIATE TEST PAGE 39

1 $4a^2=16$
$a^2=4$
$a=\pm2$
Both $a=2$ and $a=-2$ are correct.
$\therefore$ [D] ✓ (1 mark)

2 $p^2+8p-9=0$
$(p+9)(p-1)=0$
$p+9=0$ or $p-1=0$
$p=-9$ or $p=1$
$p=-9$ is a solution.
[Or substitute each value into the equation to see which solution satisfies it.]
$\therefore$ [D] ✓ (1 mark)

3 $x^2-11x+28=0$
$(x-4)(x-7)=0$
$x-4=0$ or $x-7=0$
$x=4$ or $x=7$
$\therefore$ [A] ✓ (1 mark)

4 $x=3$ or $x=-2$
$x-3=0$ or $x+2=0$
$(x-3)(x+2)=0$
$x^2+2x-3x-6=0$
$x^2-x-6=0$
$\therefore$ [D] ✓ (1 mark)

5 $(x-5)(x-2)=28$
$x^2-2x-5x+10=28$
$x^2-7x-18=0$
$(x+2)(x-9)=0$
$x+2=0$ or $x-9=0$
$x=-2$ or $x=9$
$\therefore$ [C] ✓ (1 mark)
[Or substitute to see which solutions work.]

6 $100x^2=1$
$x^2=0.01$
$x=\pm\sqrt{0.001}$
$x=\pm0.1$ ✓✓
[Or $\pm\frac{1}{10}$] (2 marks)

7 $4x^2-100=0$
$4x^2=100$
$x^2=25$
$x=\pm5$ ✓✓ (2 marks)

8 a $x^2+11x=0$
$x(x+11)=0$
$x=0$ or $x+11=0$
$x=0$ or $x=-11$
✓✓ (2 marks)

b $x^2+2x-80=0$
$(x+10)(x-8)=0$
$x+10=0$ or $x-8=0$
$x=-10$ or $x=8$
✓✓ (2 marks)

c $x^2-18x+81=0$
$(x-9)^2=0$ ✓
$x-9=0$
$x=9$ ✓
(2 marks)

d $x^2+14x+48=0$
$(x+6)(x+8)=0$
$x+6=0$ or $x+8=0$
$x=-6$ or $x=-8$
✓✓ (2 marks)

9 $(2x+3)^2=25$
$2x+3=5$ or $2x+3=-5$
$2x=2$ or $2x=-8$
$x=1$ or $x=-4$
✓✓ (2 marks)

10 a Because the two factors multiply to 35 it doesn't mean that one must be 35. ✓
The rule only works if the product is zero. If one of the factors was 35 the other would have to be 1 and that is not possible. ✓ (2 marks)

b $(x+3)(x+5)=35$
$x^2+5x+3x+15=35$
$x^2+8x-20=0$ ✓
$(x+10)(x-2)=0$
$x+10=0$ or $x-2=0$
$x=-10$ or $x=2$
✓✓ (3 marks)

11 a Let the width be x m.
The length $=(x+12)$ m ✓
$A=lb$
$=x(x+12)$ ✓
But the area is 45 m²
$x(x+12)=45$
$x^2+12x-45=0$ ✓
The width, x m, can be found by solving that equation. (3 marks)

b $x^2 + 12x - 45 = 0$
$(x - 3)(x + 15) = 0$
$x - 3 = 0$ or $x + 15 = 0$
$x = 3$ or $x = -15$ ✓
But $x > 0$, [because it is the width of the rectangle]
$\therefore x = 3$ ✓
$l = x + 12$
$= 3 + 12$
$= 15$
The length of the rectangle is 15 m and the width is 3 m. ✓ (3 marks)
(Total: 30 marks)

QUADRATIC EQUATIONS ADVANCED TEST PAGE 40

1 **a** $a^2 = \frac{1}{16}$
$a = \pm\frac{1}{4}$ ✓ (1 mark)

b $4t^2 = 25$
$t^2 = 6.25$
$t = \pm 2.5$ ✓ (1 mark)

c $12d^2 - 27 = 0$
$12d^2 = 27$
$d^2 = 2.25$
$d = \pm 1.5$ ✓ (1 mark)

2 If $x^2 + 1 = 0$ then $x^2 = -1$
The square of a real number is always positive so there is no real solution. ✓ (1 mark)

3 **a** $x^2 + 9x + 14 = 0$
$(x + 7)(x + 2) = 0$
$x + 7 = 0$ or $x + 2 = 0$
$x = -7$ or $x = -2$
✓ (1 mark)

b $a^2 - 2a - 48 = 0$
$(a - 8)(a + 6) = 0$
$a - 8 = 0$ or $a + 6 = 0$
$a = 8$ or $a = -6$
✓ (1 mark)

c $54 - 3x - x^2 = 0$
$(9 + x)(6 - x) = 0$
$9 + x = 0$ or $6 - x = 0$
$x = -9$ or $x = 6$ ✓
(1 mark)

4 **a** $x^2 = 5x$
$x^2 - 5x = 0$
$x(x - 5) = 0$ ✓
$x = 0$ or $x = 5$ ✓ (2 marks)

b $n^2 = 12n + 64$
$n^2 - 12n - 64 = 0$
$(n - 16)(n + 4) = 0$ ✓
$n = 16$ or $n = -4$
✓ (2 marks)

c $s^2 + 5s = 24$
$s^2 + 5s - 24 = 0$
$(s - 3)(s + 8) = 0$ ✓
$s = 3$ or $s = -8$ ✓
(2 marks)

5 **a** $x^2 = 7$
$x = \pm\sqrt{7}$ ✓
$= \pm 2.64575\ldots$
$= \pm 2.65$ (2 d.p.) ✓
(2 marks)

b $3y^2 = 11$
$y^2 = \frac{11}{3}$
$y = \pm\sqrt{\frac{11}{3}}$ ✓
$= \pm 1.914854\ldots$
$= \pm 1.91$ (2 d.p.) ✓
(2 marks)

c $5m^2 - 13 = 0$
$5m^2 = 13$
$m^2 = 2.6$
$m = \pm\sqrt{2.6}$ ✓
$= \pm 1.612451\ldots$
$= \pm 1.61$ (2 d.p.)
✓ (2 marks)

6 **a** $2x^2 - 28x + 66 = 0$
$x^2 - 14x + 33 = 0$
$(x - 3)(x - 11) = 0$ ✓
$x = 3$ or $x = 11$
✓ (2 marks)

b $3m^2 + 30m - 72 = 0$
$m^2 + 10m - 24 = 0$
$(m + 12)(m - 2) = 0$ ✓
$m = -12$ or $m = 2$
✓ (2 marks)

c $5y^2 - 5y - 10 = 0$
$y^2 - y - 2 = 0$
$(y - 2)(y + 1) = 0$ ✓
$y = 2$ or $y = -1$
✓ (2 marks)

7 **a** $y^3 + 7y^2 = 0$
$y^2(y + 7) = 0$
$y^2 = 0$ or $y + 7 = 0$ ✓
$y = 0$ or $y = -7$ ✓
(2 marks)

b $a^3 - 9a = 0$
$a(a^2 - 9) = 0$
$a = 0$ or $a^2 = 9$ ✓
$a = 0$ or $a = \pm 3$ ✓
(2 marks)

c $x^4 - 5x^2 + 4 = 0$
$(x^2 - 4)(x^2 - 1) = 0$
$x^2 = 4$ or $x^2 = 1$ ✓
$x = \pm 2$ or $x = \pm 1$
✓ (2 marks)

8 **a** $(x - 5)(x - 2) = -2$
$x^2 - 7x + 10 = -2$
$x^2 - 7x + 12 = 0$ ✓
$(x - 3)(x - 4) = 0$ ✓
$x = 3$ or $x = 4$ ✓
(2 marks)

b $(x + 6)(x - 7) = x + 21$
$x^2 - x - 42 = x + 21$
$x^2 - 2x - 63 = 0$ ✓
$(x - 9)(x + 7) = 0$
$x = 9$ or $x = -7$ ✓
(2 marks)

c $2x^2 = (x - 5)(x - 12)$
$2x^2 = x^2 - 17x + 60$
$x^2 + 17x - 60 = 0$ ✓
$(x + 20)(x - 3) = 0$
$x = -20$ or $x = 3$ ✓
(2 marks)

9 $(x + 4)(x + 6) + (x + 8)(x - 3)$
$= (x - 12)(x + 2)$
$x^2 + 10x + 24 + x^2 + 5x - 24$
$= x^2 - 10x - 24$ ✓
$x^2 + 25x + 24 = 0$ ✓
$(x + 24)(x + 1) = 0$
$x = -24$ or $x = -1$ ✓
(3 marks)
(Total: 40 marks)

SIMULTANEOUS EQUATIONS SKILLS CHECK PAGE 42

1 **a** $3a + 11 = -1$
$3a = -12$ [subtracting 11]
$a = -4$ [dividing by 3]

b $4x + 3(6x - 5) - 2 = 16$
$4x + 18x - 15 - 2 = 16$
$22x - 17 = 16$
$22x = 33$
$x = 1.5$

2 **a** $p + q = 5$
When $q = -2$, $p - 2 = 5$
$p = 7$

b $3p - q = 17$
When $q = -2$, $3p + 2 = 17$
$3p = 15$
$p = 5$

c $4p + 3q - 10 = 0$
When $q = -2$,
$4p + 3 \times -2 - 10 = 0$
$4p - 16 = 0$
$4p = 16$
$p = 4$

3 **a**

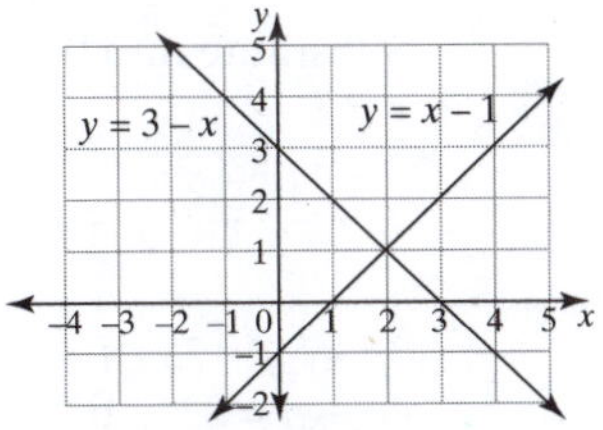

The lines intersect when $x = 2$ and $y = 1$.
[or at the point (2, 1)]

b

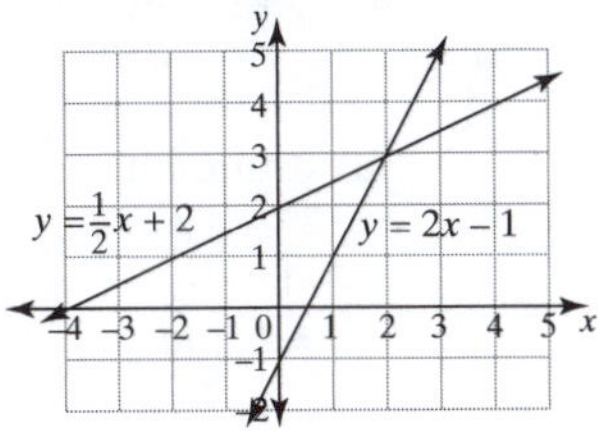

The lines intersect when $x = 2$ and $y = 3$.
[or at the point (2, 3)]

c

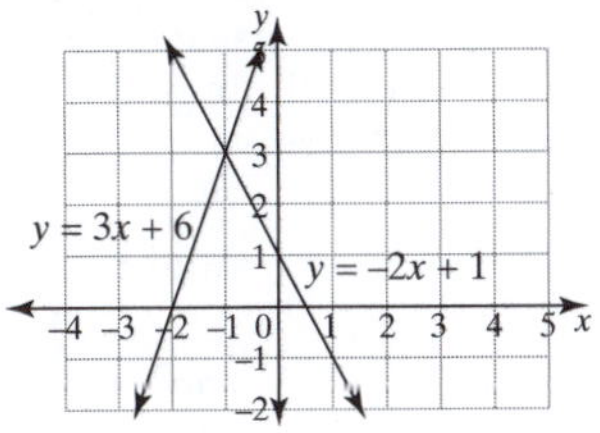

The lines intersect when $x = -1$ and $y = 3$.
[or at the point (–1, 3)]

4 $y = 2x - 5$

x	1	2	3	4	5	6	7
y	–3	–1	1	3	5	7	9

[continuing the pattern]
$y = x + 1$

x	1	2	3	4	5	6	7
y	2	3	4	5	6	7	8

The values of x and y that satisfy both equations are $x = 6$ and $y = 7$.

5 **a** $y = 3x - 8$

x	0	1	2	3	4	5	6
y	–8	–5	–2	1	4	7	10

b $x = y$ when $x = 4$ and $y = 4$

c $2x + y = 7$ when $x = 3$ and $y = 1$

6 **a** $7x - y = 10$
$-y = -7x + 10$
$y = 7x - 10$

b $5x + y - 2 = 0$
$y = -5x + 2$

c $6x + 2y - 1 = 0$
$2y = -6x + 1$
$y = -3x + \frac{1}{2}$

7 **a** $2a + b = 5$ [i]
$3a - b = 5$ [ii]
[i] + [ii] $5a = 10$

b $8x + 5y - 2 = 0$ [i]
$3x - 5y + 13 = 0$ [ii]
[i] + [ii] $11x + 11 = 0$

c $7p - 3q = 15$ [i]
$2p + 3q = 18$ [ii]
[i] + [ii] $9p = 33$

8 **a** $6x + 7y = 9$ [i]
$2x + 7y = 13$ [ii]
[i] – [ii] $4x = -4$

b $11a - 3b = 8$ [i]
$5a - 3b = 2$ [ii]
[i] – [ii] $6a = 6$

c $5a - 3b - 1 = 0$ [i]
$5a + 6b + 10 = 0$ [ii]
[i] – [ii] $-9b - 11 = 0$

9 **a** Adding equation [i] to equation [ii] will eliminate b [giving $10a = 40$].

b Adding or subtracting.
Adding equation [i] to equation [ii] will eliminate y [giving $16x = 34$].
Subtracting equation [ii] from equation [i] will eliminate x [giving $4y = 4$].

c Subtracting one equation from the other will eliminate q.
[Subtracting [ii] from [i] will give $3p + 22 = 0$.]

10 **a** The number of blue pens plus the number of red pens equals 15.
$x + y = 15$

b The number of blue pens is five more than the number of red pens.
$x = y + 5$

c The number of blue pens is equal to two times the number of red pens.
$x = 2y$

SIMULTANEOUS EQUATIONS
INTERMEDIATE TEST PAGE 43

1 The point of intersection is where both $x = 3$ and $y = 4$.
i.e. at the point (3, 4)
∴ [A] ✓ (1 mark)

2 $2x + y = 5$ [i]
$x - 3y = 13$ [ii]
If $x = 4$ and $y = 3$
$2x + y = 8 + 3$
$= 13 \ (\neq 5)$
If $x = 4$ and $y = -3$
$2x + y = 8 - 3$
$= 5$
$x - 3y = 4 - 3 \times -3$
$= 13$
[Also check that C and D do not work.]
∴ [B] ✓ (1 mark)

3 $y = 4x + 7$ and $y = -2x - 5$
$4x + 7 = -2x - 5$
$6x = -12$
$x = -2$
Substitute: $y = 4 \times -2 + 7$
$= -1$
∴ [A] ✓ (1 mark)

4 $y = x - 6$ and $y = -12x - 6$ are the only two equations with the same y-intercept.
So they have a simultaneous solution when $x = 0$.
∴ [C] ✓ (1 mark)

5 **a**

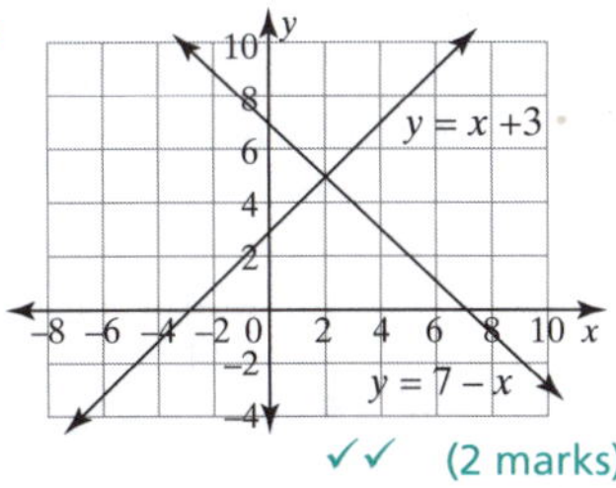

✓✓ (2 marks)

b From the graph, $x = 2$ ✓ and $y = 5$ ✓
[Or the point (2, 5)]
(2 marks)

6 $7x - 2y = 5$ [i]
$y = 3x$ [ii]
Substitute [ii] into [i]
$7x - 2(3x) = 5$
$7x - 6x = 5$
$x = 5$ ✓
Substitute into [ii]
$y = 3 \times 5$
$= 15$ ✓
$\therefore x = 5$ and $y = 15$
[Check in (i),
$7 \times 5 - 2 \times 15 = 5$] (2 marks)

7 $7a + 2b = 16$ [i]
$3a - 2b = 24$ [ii]
[i] + [ii] $10a = 40$
$a = 4$ ✓
Substitute in [i] [or into [ii]]
$28 + 2b = 16$
$2b = -12$
$b = -6$ ✓
$\therefore a = 4$ and $b = -6$
[Check in [ii], $12 - (-12) = 24$]
(2 marks)

8 **a** $2x + y - 5 = 0$
$y = -2x + 5$ ✓
(1 mark)

b $2x + y - 5 = 0$ [i]
$3x - 4y + 42 = 0$ [ii]
From **a** $y = -2x + 5$ [iii]
Substitute into [ii]
$3x - 4(-2x + 5) + 42 = 0$
$3x + 8x - 20 + 42 = 0$
$11x + 22 = 0$
$11x = -22$
$x = -2$ ✓
Substitute into [iii]
$y = -2 \times -2 + 5$
$= 9$ ✓
$\therefore x = -2$ and $y = 9$ (2 marks)

9 $9m + 4p = 41$ [i]
$5m + 2p = 23$ [ii]
[ii] × 2 $10m + 4p = 46$ [iii]
[iii] – [i] $m = 5$ ✓
Substitute in [i]
$9 \times 5 + 4p = 41$
$45 + 4p = 41$
$4p = -4$
$p = -1$ ✓
$\therefore m = 5$ and $p = -1$ (2 marks)

10 $2x - y + 8 = 0$ [i]
$3x - y - 2 = 0$ [ii]
[ii] – [i] $x - 10 = 0$
$x = 10$ ✓
Substitute in [i]
$2 \times 10 - y + 8 = 0$
$28 - y = 0$
$y = 28$ ✓
$\therefore x = 10$ and $y = 28$
[Or from [i] $y = 2x + 8$
Substitute into [ii]
$3x - (2x + 8) - 2 = 0$
$x - 10 = 0$] (2 marks)

11 $5a + 3b = 8$ [i]
$3a - 4b = 28$ [ii]
[i] × 4 $20a + 12b = 32$ [iii]
[ii] × 3 $9a - 12b = 84$ [iv] ✓
[iii] + [iv] $29a = 116$
$a = 4$ ✓
Substitute into [i]
$5 \times 4 + 3b = 8$
$20 + 3b = 8$
$3b = -12$
$b = -4$ ✓ (3 marks)

12 Let x be the number of cans of dog food and y be the number of cans of cat food.
$x + y = 24$ [i]
$140x + 90y = 2910$ [ii] ✓
From [i] $y = 24 - x$ [iii]
Substitute into [ii]
$140x + 90(24 - x) = 2910$
$140x + 2160 - 90x = 2910$
$50x = 750$
$x = 15$ ✓
Substitute into [iii]
$y = 24 - 15$
$= 9$ ✓
There were 15 cans of dog food and 9 cans of cat food. ✓
(4 marks)
(Total: 26 marks)

SIMULTANEOUS EQUATIONS ADVANCED TEST PAGE 44

1 **a** $y = 5x - 6$ [i]
$y = 3x + 2$ [ii]
$5x - 6 = 3x + 2$
$2x = 8$
$x = 4$ ✓
Substitute into [i]:
$y = 5 \times 4 - 6$
$= 14$
$\therefore x = 4$ and $y = 14$ ✓
(2 marks)

b $y = 7 - 2x$ [i]
$x + 3y = 11$ [ii]
Substitute [i] into [ii]
$x + 3(7 - 2x) = 11$
$x + 21 - 6x = 11$
$-5x = -10$
$x = 2$ ✓
Substitute into [i]
$y = 7 - 2 \times 2$
$= 3$
$\therefore x = 2$ and $y = 3$ ✓
(2 marks)

c $3p - 5q = 23$ [i]
$4p + q = 23$ [ii]
From [ii] $q = 23 - 4p$
Substitute into [i]
$3p - 5(23 - 4p) = 23$
$3p - 115 + 20p = 23$
$23p = 138$
$p = 6$ ✓
Substitute into [ii]
$q = 23 - 4 \times 6$
$= -1$
$\therefore p = 6$ and $q = -1$ ✓
(2 marks)

2 **a** $a + b = 4$ [i]
$a - b = -10$ [ii]
[i] + [ii] $2a = -6$
$a = -3$ ✓
Substitute into [i]
$-3 + b = 4$
$b = 7$
$\therefore a = -3$ and $b = 7$ ✓
(2 marks)

b $8x - 3y = 30$ [i]
$5x - 3y = 3$ [ii]
[i] – [ii] $3x = 27$
$x = 9$ ✓
Substitute into [i]:
$8 \times 9 - 3y = 30$
$3y = 42$
$y = 14$
$\therefore x = 9$ and $y = 14$ ✓
(2 marks)

c $9m + 5n - 3 = 0$ [i]
$9m - 7n + 69 = 0$ [ii]
[i] – [ii] $12n - 72 = 0$
$12n = 72$
$n = 6$ ✓
Substitute into [i]
$9m + 5 \times 6 - 3 = 0$
$9m = -27$
$m = -3$
$\therefore m = -3$ and $n = 6$ ✓
(2 marks)

d $4p + 3q = 27$ [i]
$3p + q = 14$ [ii]
[ii] × 3: $9p + 3q = 42$ [iii]
[iii] – [i] $5p = 15$
$p = 3$ ✓
Substitute into [ii]
$3 \times 3 + q = 14$
$q = 5$
$\therefore p = 3$ and $q = 5$ ✓
(2 marks)

e $9a - 4b = -1$ [i]
$5a + 2b = -9$ [ii]
[ii] × 2: $10a + 4b = -18$ [iii]
[i] + [iii] $19a = -19$
$a = -1$ ✓
Substitute into [i]:
$9 \times -1 - 4b = -1$
$4b = -8$
$b = -2$
$\therefore a = -1$ and $b = -2$ ✓
(2 marks)

f $6k - 5m = 19$ [i]
$7k - 3m = 42$ [ii]
[i] × 3: $18k - 15m = 57$ [iii]
[ii] × 5: $35k - 15m = 210$ [iv]
[iv] – [iii] $17k = 153$
$k = 9$ ✓
Substitute into [i]
$6 \times 9 - 5m = 19$
$5m = 35$
$m = 7$
$\therefore k = 9$ and $m = 7$ ✓
(2 marks)

3 **a**

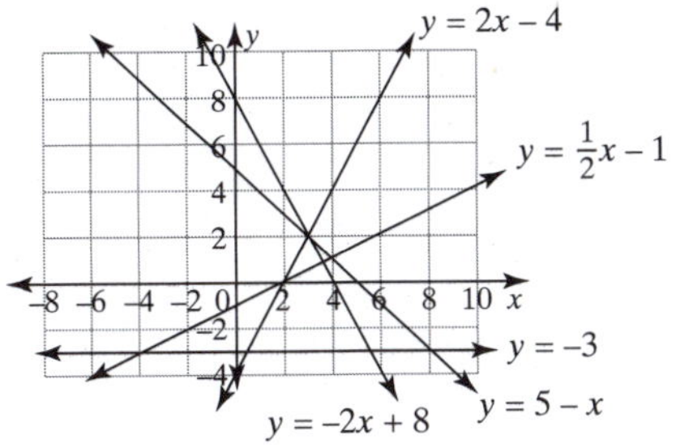

✓✓✓

b **i** (2, 0) ✓ (1 mark)
ii (4, 1) ✓ (1 mark)
iii (–4, –3) ✓ (1 mark)

c It is the point where $y = -2x + 8$, $y = 5 - x$ and $y = 2x - 4$ all intersect. ✓
(1 mark)

4 **a** $x = x^2$
$x - x^2 = 0$
$x(1 - x) = 0$
$x = 0$ or $x = 1$ ✓(1 mark)

b When $x = 0$, $y = 0$ and when $x = 1$, $y = 1$. ✓ (1 mark)

c Two places; at (0, 0) and (1, 1). ✓ (1 mark)

d Two places. ✓ (1 mark)

e $x^2 = 4$
$x = \pm 2$
When $x = -2$, $y = 4$ and when $x = 2$, $y = 4$. ✓(1 mark)

5 **a** $y = x^2 + 5x$ [i]
$y = x + 12$ [ii]
$x^2 + 5x = x + 12$
$x^2 + 4x - 12 = 0$ ✓
$(x + 6)(x - 2) = 0$
$x = -6$ or $x = 2$
Substitute into [ii]
When $x = -6$, $y = 6$ ✓
When $x = 2$, $y = 14$ ✓
(3 marks)

b $y = x^2 + x - 10$ [i]
$y = 2x - 4$ [ii]
$x^2 + x - 10 = 2x - 4$
$x^2 - x - 6 = 0$ ✓
$(x - 3)(x + 2) = 0$
$x = 3$ or $x = -2$
Substitute into [ii]
When $x = 3$, $y = 2 \times 3 - 4$
$= 2$ ✓
When $x = -2$, $y = 2 \times -2 - 4$
$= -8$ ✓
(3 marks)

6 $y = x^2 + 8x + 3$ [i]
$y = x^2 - 2x - 7$ [ii]
$x^2 + 8x + 3 = x^2 - 2x - 7$
$10x = -10$
$x = -1$ ✓
Substitute into [i]
$y = (-1)^2 + 8 \times -1 + 3$
$= -4$
$\therefore x = -1$ and $y = -4$ ✓
(2 marks)

7 **a** You can immediately see the point of intersection once the graph is drawn, and easily see how many points of intersection there are. ✓ (1 mark)

b Sketching a graph can be time-consuming and not always accurate. Solving equations algebraically will give the exact values. ✓
(1 mark)
(Total: 40 marks)

PARALLEL AND PERPENDICULAR LINES
SKILLS CHECK PAGE 46

1 **a**

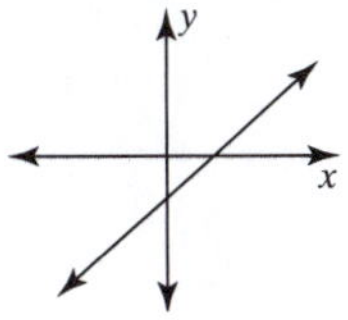

positive

b

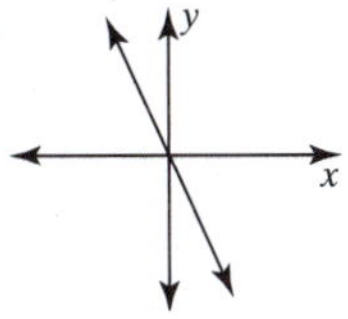

negative

c

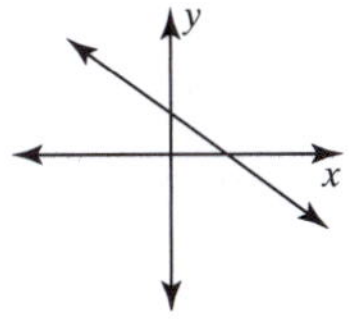

negative

d

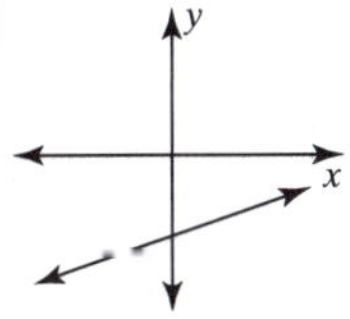

positive

2 **a** $m = \frac{\text{rise}}{\text{run}}$
$= \frac{2}{3}$

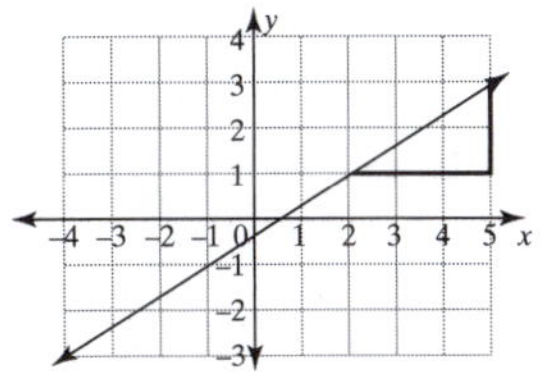

b $m = \frac{\text{rise}}{\text{run}}$
$= \frac{-1}{2}$
$= -\frac{1}{2}$

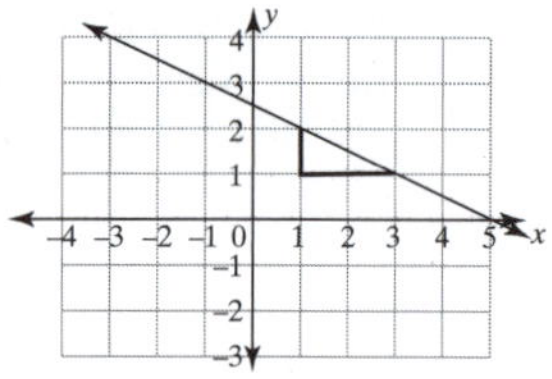

c $m = \dfrac{\text{rise}}{\text{run}}$
$= \dfrac{2}{1}$
$= 2$

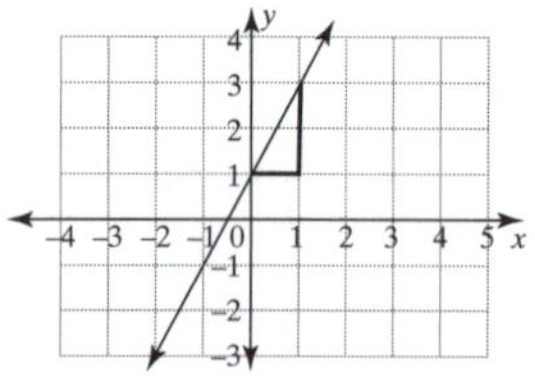

d $m = \dfrac{\text{rise}}{\text{run}}$
$= \dfrac{-1}{1}$
$= -1$

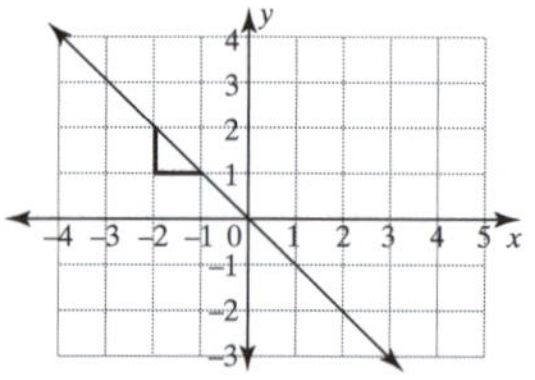

3 $(1, 2), (5, 0)$
$m = \dfrac{y_2 - y_1}{x_2 - x_1}$
$= \dfrac{0 - 2}{5 - 1}$
$= -\dfrac{1}{2}$
[Or plot the points and find the rise and run.]

4 a $y = 2x + 3$
$m = 2,\ b = 3$

b $y = 4x - 1$
$m = 4,\ b = -1$

c $y = -6x$ $[y = -6x + 0]$
$m = -6, b = 0$

d $y = x - 5$ $[y = 1x - 5]$
$m = 1, b = -5$

e $y = \dfrac{1}{3}x - 2$
$m = \dfrac{1}{3},\ b = -2$

f $y = \dfrac{x}{4}$ $[y = \dfrac{1}{4}x + 0]$
$m = \dfrac{1}{4}, b = 0$

g $y = \dfrac{-2x}{3} + \dfrac{1}{2}$
$m = \dfrac{-2}{3}, b = \dfrac{1}{2}$

h $y = 5 - 2x$ $[y = -2x + 5]$
$m = -2, b = 5$

5 a $m = 3, b = 2$
$y = 3x + 2$

b $m = -2, b = 0$
$y = -2x$

c $m = \dfrac{1}{2}, b = -\dfrac{1}{3}$
$y = \dfrac{1}{2}x - \dfrac{1}{3}$

6 a If the gradients of two lines are equal then the lines are **parallel**.

b If two lines are parallel then their gradients are **equal**.

7 a If $m_1 = m_2$ the two lines are **parallel**.

b If $m_1m_2 = -1$ the two lines are **perpendicular**.

8 a $-\dfrac{1}{2}$
[Parallel lines have the same gradient.]

b 2
[The negative reciprocal of $-\dfrac{1}{2}$.]

9 a Both lines have gradient 3 so they are parallel. True.

b Both lines have gradient $\dfrac{1}{2}$ so they are parallel. True.

c Both lines have gradient –1 so they are parallel. True.

PARALLEL AND PERPENDICULAR LINES
INTERMEDIATE TEST PAGE 47

1 $y = 3x - 4$ has gradient 3.
Parallel lines have gradient 3.
The line parallel to $y = 3x - 4$ is $y = 3x + 7$.
∴ [D] ✓ (1 mark)

2 $y = 2x - 5$
Gradient = 2 [coefficient of x]
Any perpendicular line has gradient $-\dfrac{1}{2}$.
[the negative reciprocal of 2]
∴ [D] ✓ (1 mark)

3 $y = 2 - x$
$y = -x + 2$
Gradient = –1
Any parallel line has $m = -1$.
Any line that passes through the origin has y-intercept 0.
Equation is $y = -x$
∴ [B] ✓ (1 mark)

4 a gradient $= \dfrac{\text{rise}}{\text{run}}$
$= \dfrac{2}{5}$ ✓ (1 mark)

b y-intercept = 2 ✓ (1 mark)

c The line $y = \dfrac{2}{5}x + 4$ is parallel to the given line with y-intercept 4.

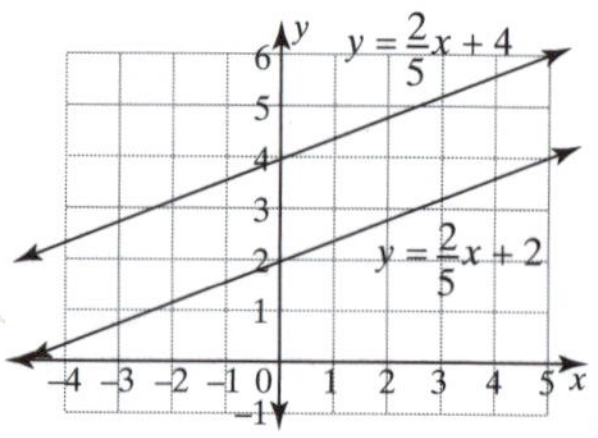

✓✓
(2 marks)

5

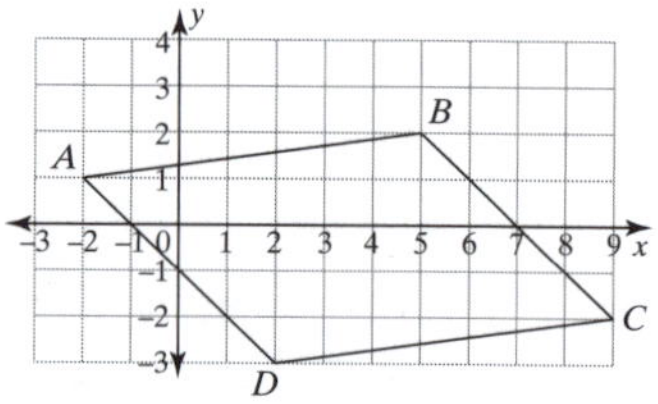

a $A(-2, 1), B(5, 2)$
$m = \dfrac{\text{rise}}{\text{run}}$
$= \dfrac{1}{7}$ ✓
$D(2, -3), C(9, -2)$
$m = \dfrac{\text{rise}}{\text{run}}$
$= \dfrac{1}{7}$
$m_{AB} = m_{DC}$
∴ AB is parallel to DC ✓
(2 marks)

b $A(-2, 1), D(2, -3)$

$m = \frac{\text{rise}}{\text{run}}$

$= \frac{-4}{4}$

$= -1$ ✓

$B(5, 2), C(9, -2)$

$m = \frac{\text{rise}}{\text{run}}$

$= \frac{-4}{4}$

$= -1$

$m_{AD} = m_{BC}$

$\therefore AD$ is parallel to BC ✓ (2 marks)

c $ABCD$ is a parallelogram. ✓ [Both pairs of opposite sides are parallel.] (1 mark)

6

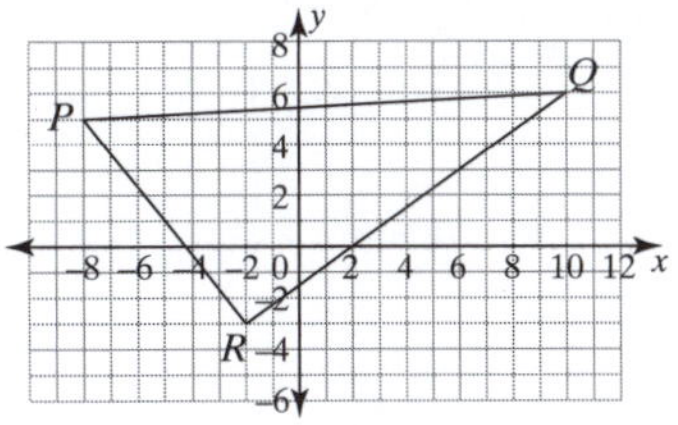

[From the diagram if the triangle has a right angle it will be at R.]

$m_{PR} = \frac{\text{rise}}{\text{run}}$

$= \frac{-8}{6}$

$= -\frac{4}{3}$ ✓

$m_{QR} = \frac{\text{rise}}{\text{run}}$

$= \frac{9}{12}$

$= \frac{3}{4}$ ✓

$\therefore$ the triangle is right-angled ✓

PR and QR are perpendicular because the gradients are negative reciprocals. ✓ (4 marks)

7 **a** $y = -4x + 3$

The gradient of this line is -4. Line k is parallel to line l so it also has gradient -4. ✓

Line k has y-intercept -1.

So $m = -4$ and $b = -1$

The equation of line k is $y = -4x - 1$. ✓ (2 marks)

b Line m is perpendicular to line l so its gradient is $\frac{1}{4}$. ✓

Lines l and m meet on the y-axis. The y-intercept of line l is 3 so the y-intercept of line m is also 3. ✓

So $m = \frac{1}{4}$ and $b = 3$

The equation of line m is

$y = \frac{1}{4}x + 3$ ✓ (3 marks)

8 **a** $A(7, -8), B(1, 1)$

$m_{AB} = \frac{y_2 - y_1}{x_2 - x_1}$

$= \frac{1 - (-8)}{1 - 7}$

$= \frac{9}{-6}$

$= -\frac{3}{2}$ ✓ (1 mark)

b $B(1, 1), \quad C(-3, 7)$

$m_{BC} = \frac{y_2 - y_1}{x_2 - x_1}$

$= \frac{7 - 1}{-3 - 1}$

$= \frac{6}{-4}$

$= -\frac{3}{2}$ ✓ (1 mark)

c A, B and C are collinear. [They lie on the same straight line.] ✓

AB and BC have the same gradient and B is a common point. ✓ (2 marks)

(Total: 25 marks)

PARALLEL AND PERPENDICULAR LINES
ADVANCED TEST PAGE 48

1 **a** $y = 3x - 2$ has gradient 3

$x + 3y - 4 = 0$

$3y = -x + 4$

$y = -\frac{1}{3}x + \frac{4}{3}$

This line has gradient $-\frac{1}{3}$. ✓

Now $3 \times -\frac{1}{3} = -1$

So the lines are perpendicular. ✓ (2 marks)

b $2x + y - 3 = 0$

$y = -2x + 3$

gradient $= -2$ ✓

$x + 2y + 5 = 0$

$2y = -x - 5$

$y = -\frac{1}{2}x - 2.5$

gradient $= -\frac{1}{2}$

So the lines are neither parallel nor perpendicular. ✓ (2 marks)

c $y = \frac{3}{4}x - 5$

gradient $= \frac{3}{4}$

$3x - 4y - 2 = 0$

$4y = 3x - 2$

$y = \frac{3}{4}x - 0.5$

gradient $= \frac{3}{4}$ ✓

The gradients are equal so the lines are parallel. ✓ (2 marks)

2 **a** $l: y = 7 - x$

$= -x + 7$

Gradient $= -1$ ✓

Parallel line has equation

$y = -x + b$

It passes through $(2, 1)$.

So $1 = -2 + b$

$b = 3$

The equation is $y = -x + 3$. ✓ (2 marks)

b The two lines intersect on the y-axis so have the same y-intercept.

Line l has y-intercept 7

Perpendicular line has gradient 1. ✓

Equation is $y = x + 7$ ✓ (2 marks)

3 $P(1, 8), Q(-4, 5)$

a $m = \frac{y_2 - y_1}{x_2 - x_1}$

$= \frac{5 - 8}{-4 - 1}$

$= \frac{3}{5}$ ✓ (1 mark)

b $PQ \perp QR$

$\therefore$ gradient of $QR = -\frac{5}{3}$

At R, $y = 0$

So R is the point $(x, 0)$.

$$m = \frac{y_2 - y_1}{x_2 - x_1}$$

$$-\frac{5}{3} = \frac{5-0}{-4-x} \checkmark$$

$$-\frac{5}{3} = -\frac{5}{4+x}$$

$$4 + x = 3$$

$$x = -1$$

So R is the point $(-1, 0)$ ✓ (2 marks)

4 **a** $S(-10, -5)$, $T(-1, 7)$

$$m = \frac{y_2 - y_1}{x_2 - x_1}$$

$$m_{ST} = \frac{7-(-5)}{-1-(-10)}$$

$$= \frac{4}{3} \checkmark$$ (1 mark)

b $T(-1, 7)$, $U(15, -5)$

$$m = \frac{y_2 - y_1}{x_2 - x_1}$$

$$m_{TU} = \frac{-5-7}{15-(-1)}$$

$$= -\frac{3}{4} \checkmark$$

Now $\frac{4}{3} \times -\frac{3}{4} = -1$

So $ST \perp TU$

$\therefore \Delta STU$ is right-angled at T. ✓ (2 marks)

c $V(6, -17)$ ✓ (1 mark)

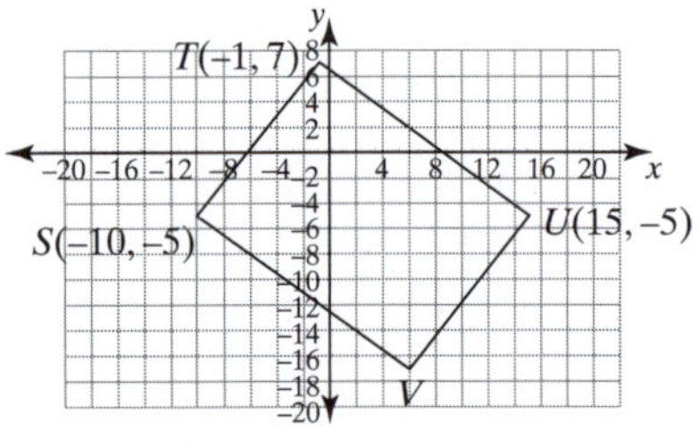

d $STUV$ is a rectangle.

(A parallelogram with one angle a right-angle)

Distance $SU = 25$ units

T is 12 units above line SU. ✓

$$A = 2 \times \frac{1}{2} \times 25 \times 12$$

$$= 300$$

The area is 300 units². ✓ (2 marks)

5 $A(-2, 4)$, $B(6, 2)$

a midpoint $= \left(\frac{-2+6}{2}, \frac{4+2}{2}\right)$

$= (2, 3)$ ✓ (1 mark)

b $m = \frac{y_2 - y_1}{x_2 - x_1}$

$= \frac{2-4}{6-(-2)}$

$= -\frac{1}{4}$ ✓ (1 mark)

c Perpendicular line has gradient 4.

$y - y_1 = m(x - x_1)$

$y - 3 = 4(x - 2)$

$y - 3 = 4x - 8$

$y = 4x - 5$ ✓ (1 mark)

d $A(-2, 4)$, $C(-4, 2)$

midpoint $= (-3, 3)$

$m = \frac{y_2 - y_1}{x_2 - x_1}$

$= \frac{2-4}{-4-(-2)}$

$= 1$

Perpendicular line has $m = -1$ ✓

$y - y_1 = m(x - x_1)$

$y - 3 = -1(x - (-3))$

$y - 3 = -x - 3$

$y = -x$ ✓ (2 marks)

e At point of intersection:

$4x - 5 = -x$

$5x = 5$

$x = 1$ ✓

When $x = 1$, $y = -1$

$\therefore$ the point of intersection is $(1, -1)$. ✓ (2 marks)

f $B(6, 2)$, $C(-4, 2)$

midpoint $= (1, 2)$

Now both B and C lie on the horizontal line $y = 2$

So the perpendicular line will be vertical.

The equation of the perpendicular bisector will be $x = 1$. ✓ (1 mark)

g All three bisectors will meet at the same point; $(1, -1)$. ✓ (1 mark)

6 **a** $ax + by + c = 0$

$by = -ax - c$

$y = -\frac{a}{b}x - \frac{c}{b}$ ✓ (1 mark)

b Parallel lines have the same gradient.

So $m = -\frac{a}{b}$ ✓ (1 mark)

c $5x - 3y + 7 = 0$

$m = -\frac{5}{-3}$

$= \frac{5}{3}$

So the gradient of any parallel line is $\frac{5}{3}$. ✓ (1 mark)

d $m = \frac{5}{3}$, y-intercept $= 0$

Equation is $y = \frac{5}{3}x$

$3y = 5x$

$5x - 3y = 0$ ✓ (1 mark)

e $m = \frac{b}{a}$

[The negative reciprocal.] ✓ (1 mark)

f $m = \frac{b}{a}$

$= \frac{2}{4}$

$= \frac{1}{2}$ ✓ (1 mark)

g $y = \frac{1}{2}x$

$2y = x$

$x - 2y = 0$ ✓ (1 mark)

(Total: 35 marks)

FURTHER GRAPHS
SKILLS CHECK PAGE 50

1 **a** parabola

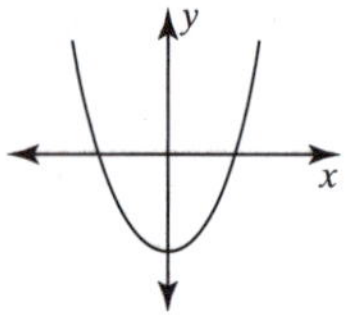

b straight line

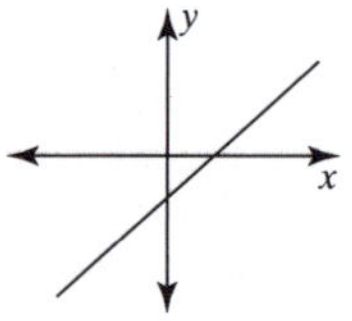

c exponential curve

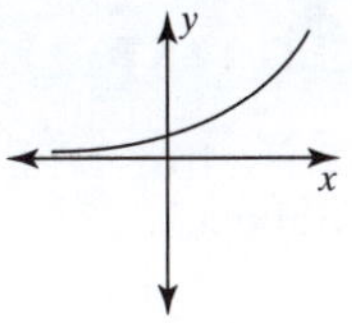

d parabola

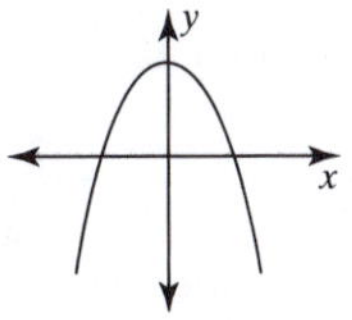

2

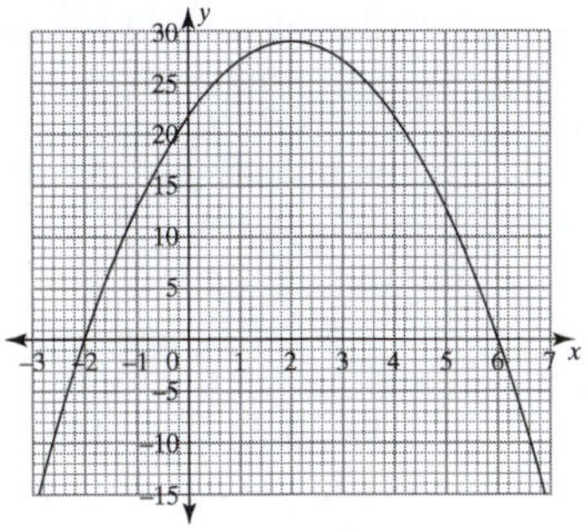

a y-intercept = 22

b When $x = 4$, $y = 22$

c The parabola cuts the x-axis at $x = -2$ and $x = 6$. So the points are $(-2, 0)$ and $(6, 0)$.

d The vertex is at $(2, 29)$.

e $y = 13$ when $x = -1$ and $x = 5$.

3 **a**

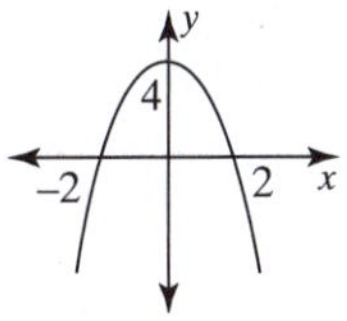

$y = 4 - x^2$
[parabola, 'upside down', y-intercept = 4]

b

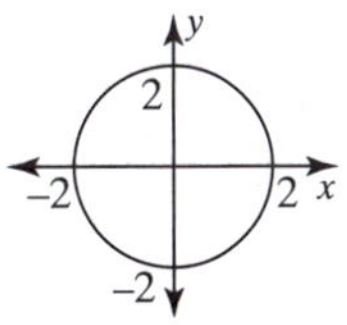

$x^2 + y^2 = 4$
[circle, centre $(0, 0)$, radius 2]

c

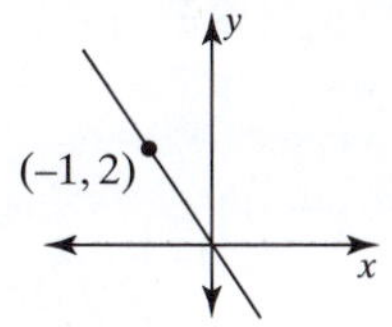

$y = -2x$
[gradient = −2, y-intercept = 0]

d

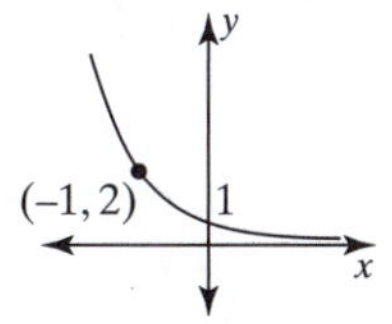

$y = 2^{-x}$
[exponential curve, when $x = -1$, $y = 2$]

e

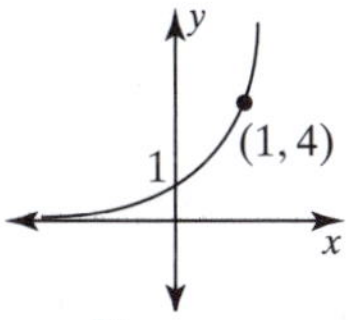

$y = 4^x$
[exponential curve, when $x = 1$, $y = 4$]

f

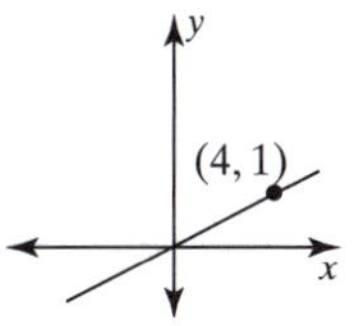

$y = \frac{x}{4}$

[gradient $-\frac{1}{4}$, y-intercept = 0]

g

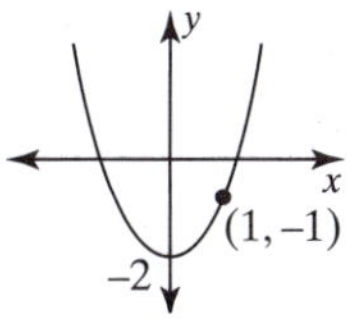

$y = x^2 - 2$
[parabola, y-intercept −2]

h

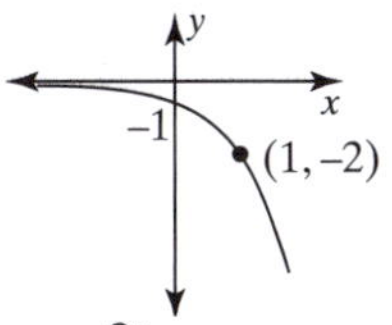

$y = -2^x$
[exponential curve, 'upside down']

4 **a** $y = 3 - x$ straight line
[$y = -x + 3$, gradient −1, y-intercept 3]

b $y = 3x^2$ parabola

c $y = 3^x$ exponential curve

d $y = x^3$ cubic curve

e $y = x^2 + 3$ parabola

f $x^2 + y^2 = 3$ circle
[centre $(0, 0)$, radius $\sqrt{3}$]

g $y = \frac{x}{3}$ straight line
[$y = \frac{1}{3}x$, gradient $\frac{1}{3}$, y-intercept 0]

h $y = -x^2 + 2$ parabola

i $y = 9 - x^2$ parabola

j $x + y = 9$ straight line
[$y = -x + 9$]

k $x^2 + y^2 = 9$ circle
[centre $(0, 0)$, radius 3]

l $y = 2^{-x}$ exponential curve

FURTHER GRAPHS
INTERMEDIATE TEST PAGE 51

1 Equations of straight lines can be put in the form $y = mx + b$.
$y = \frac{3}{x}$ is not of this form so it is not the equation of a straight line.
[The other equations can be rearranged:
A: $y = -x + 3$, B: $y = x$,
C: $y = \frac{1}{2}x + \frac{3}{2}$]
∴ **[D]** ✓ (1 mark)

2 $x^2 + y^2 = 50$
Try each option.
$(-1, -7)$: $(-1)^2 + (-7)^2 = 1 + 49 = 50$
$(7, -1)$: $7^2 + (-1)^2 = 49 + 1 = 50$
$(2, 6)$: $2^2 + 6^2 = 4 + 36 = 40$ (not 50)
$(5, 5)$: $5^2 + 5^2 = 25 + 25 = 50$
The point that does not lie on the circle is $(2, 6)$.
∴ **[C]** ✓ (1 mark)

3 $y = x^2 + x$

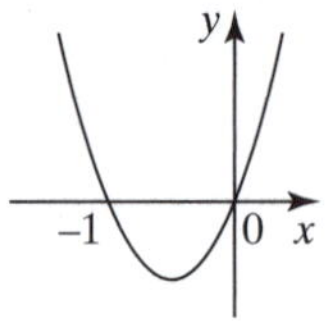

When $x = 0$, $y = 0^2 + 0$
$= 0$
When $x = -1$, $y = (-1)^2 + (-1)$
$= 0$
The only possible graph is B.
$\therefore$ **[B]** ✓ (1 mark)

4 The equation of a parabola is a quadratic equation.
$y = (x + 1)(x - 1)$
$= x^2 - 1$
This is the equation of a parabola.
$\therefore$ **[A]** ✓ (1 mark)

5 **a** From the graph there is one solution to the equation $x^3 - 1 = 1 - x^3$. ✓

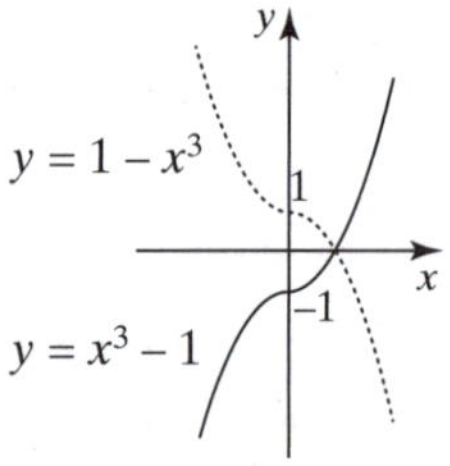

[The two graphs have only one point of intersection.]
(1 mark)

b $x^3 - 1 = 1 - x^3$
$2x^3 = 2$
$x^3 = 1$
$x = 1$ $[\sqrt[3]{1}]$ ✓
(1 mark)

6 $x^2 + y^2 = 36$
$x^2 + y^2 = 6^2$
This circle, centre O, has radius 6. ✓
The distance from O to P is 8 units. So, the point P will lie outside the circle because the distance from O to P is greater than the radius of the circle. ✓ (2 marks)

7 **a** The constant c moves the graph up or down the y-axis. ✓ (1 mark)

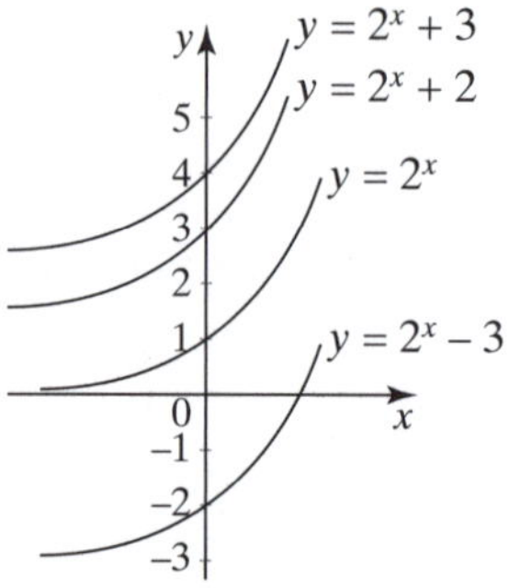

b $y = 2^x$ has y-intercept 1.
$y = 2^x + 7$ will have y-intercept 8. $[1 + 7]$ ✓
(1 mark)

c $y = 2^x - 1$ will pass through the origin. [The graph $y = 2^x$ needs to move down 1.]
$y = 2^x + c$ will pass through the origin when $c = -1$ ✓
(1 mark)

8 **a** The two graphs are exactly the same. ✓ (1 mark)

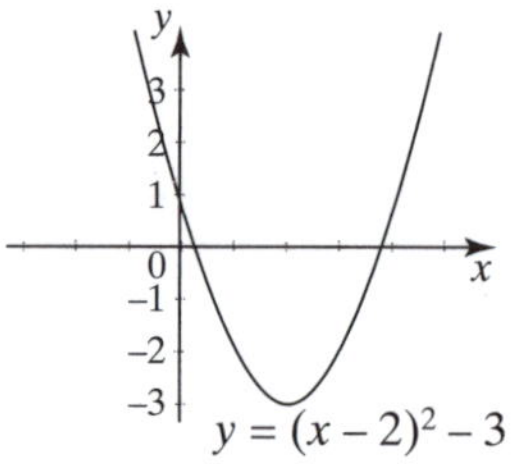

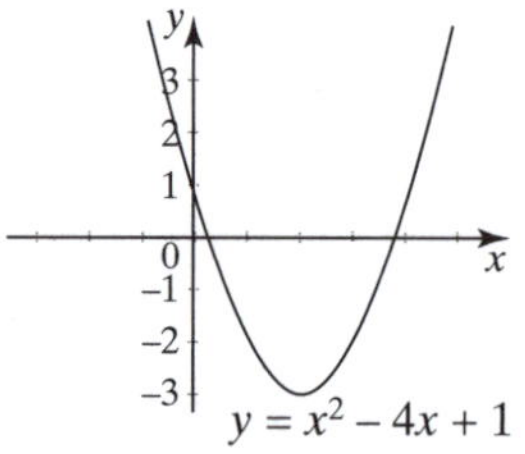

b $(x - 2)^2 - 3$
$= (x - 2)(x - 2) - 3$
$= x^2 - 2x - 2x + 4 - 3$ ✓
$= x^2 - 4x + 1$ ✓
The two expressions are in different forms but when simplified are the same so the graphs are also the same. ✓ (3 marks)
(Total: 15 marks)

FURTHER GRAPHS
ADVANCED TEST PAGE 52

1 **a** $y = x^2 - 10x + 21$ is a parabola. ✓ (1 mark)

b When $x = 0$, $y = 21$
The y-intercept is 21. ✓
(1 mark)

c When $y = 0$,
$x^2 - 10x + 21 = 0$
$(x - 3)(x - 7) = 0$
$x = 3$ or $x = 7$
The x-intercepts are 3 and 7. ✓✓ (2 marks)

d The equation of the axis of symmetry is $x = 5$.
[It passes through the midpoint of the x-intercepts.] ✓ (1 mark)

e When $x = 5$,
$y = 5^2 - 10 \times 5 + 21$
$= -4$
The minimum value is –4. ✓ (1 mark)

f

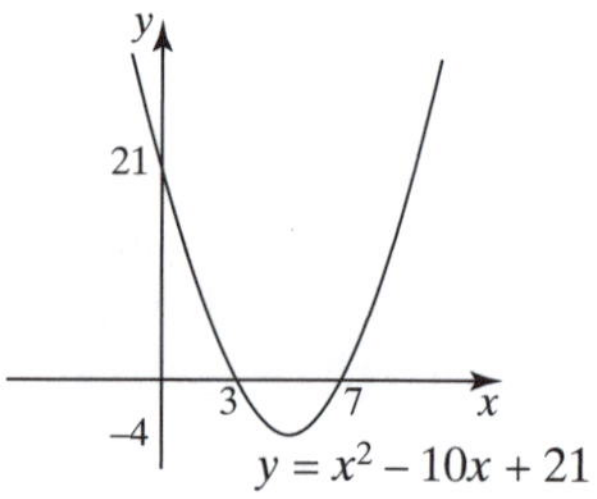

✓✓✓ (3 marks)

2 **a** Both curves are the same size and shape as $y = x^2$.
$y = x^2 + 2$ has moved up two units from the position of $y = x^2$ and $y = x^2 - 3$ has moved down 3 units. ✓
(1 mark)

b Both curves are the same size and shape as $y = x^2$.
$y = (x + 2)^2$ has moved two units across to the left from the position of $y = x^2$ and $y = (x - 3)^2$ has moved three units to the right. ✓ (1 mark)

c Both curves are the same size and shape but $y = -x^2$ is upside down (concave down). ✓ (1 mark)

d $y = 2x^2$ and $y = \frac{1}{2}x^2$ have the same axis of symmetry and minimum point as $y = x^2$ but $y = 2x^2$ is thinner than $y = x^2$ and $y = \frac{1}{2}x^2$ is wider. ✓ (1 mark)

3 **a** $y = (x + 2)^2 - 3$ has a minimum turning point at $(-2, -3)$ ✓✓ (2 marks)

b $y = 5 - (3 - x)^2$ has a maximum turning point at $(3, 5)$ ✓✓ (2 marks)

4 The curve $(x - 1)^2 + (y + 3)^2 = 16$ is a circle of radius 4 units. ✓
The centre is at $(1, -3)$. ✓
(2 marks)

5 **a** If n is even, $y = x^n$ will be symmetrical about the y-axis. The y-axis will be an axis of symmetry. ✓(1 mark)

b If n is odd, $y = x^n$ will have point symmetry about the origin. If rotated through 180° about (0, 0) it will be exactly the same. ✓(1 mark)

6 $y = (x - 2)(x - 4)(x + 1)$.

a When $x = 0$,
$y = (0 - 2)(0 - 4)(0 + 1)$
$= 8$ ✓ (1 mark)

b When $y = 0$,
$(x - 2)(x - 4)(x + 1) = 0$
$x = 2$ or $x = 4$ or $x = -1$ ✓
(1 mark)

c If x has a large positive value, y will also have a large positive value. ✓
(1 mark)

d If x has a large negative value, y will also have a large negative value. ✓
(1 mark)

e

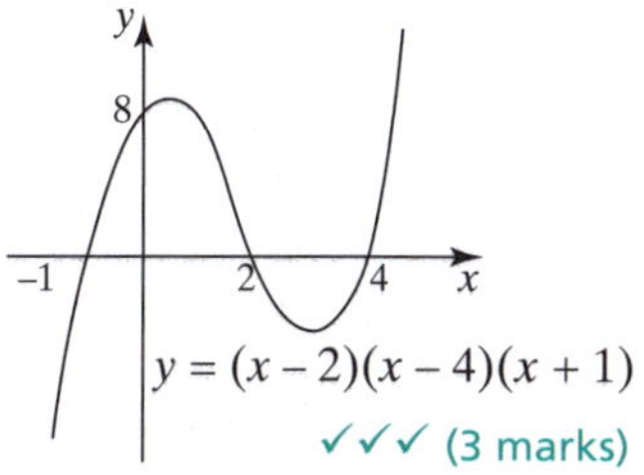

✓✓✓ (3 marks)

7 **a** $x^2 + y^2 = 25$ [i]
$y = 7 - x$ [ii]
Substitute [ii] into [i]
$x^2 + (7 - x)^2 = 25$
$x^2 + 49 - 14x + x^2 = 25$
$2x^2 - 14x + 24 = 0$
$x^2 - 7x + 12 = 0$
$(x - 3)(x - 4) = 0$
$x = 3$ or $x = 4$ ✓
Substitute into [ii]
When $x = 3$, $y = 7 - 3$
$= 4$
When $x = 4$, $y = 7 - 4$
$= 3$ ✓
(2 marks)

b

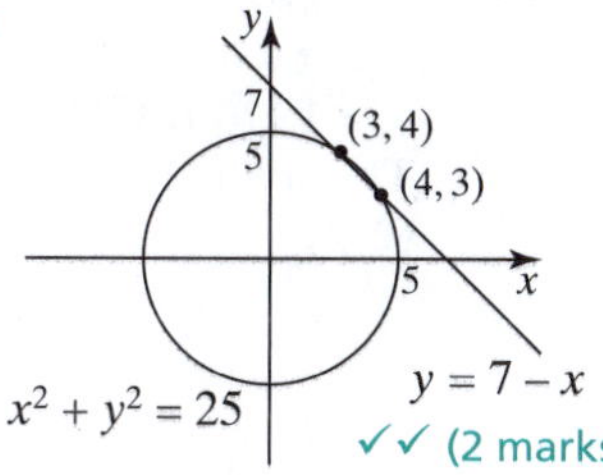

✓✓ (2 marks)

8

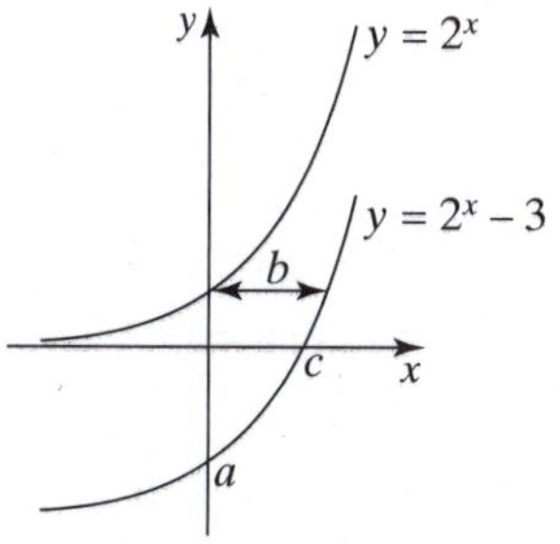

a $y = 2^x - 3$
When $x = 0$,
$y = 2^0 - 3$
$= -2$
So $a = -2$ ✓ (1 mark)

b $y = 2^x$ cuts the y-axis at $y = 1$
$y = 2^x - 3$
When $y = 1$,
$1 = 2^x - 3$
$4 = 2^x$
But $2^2 = 4$
So $x = 2$
$\therefore b = 2$ ✓ (1 mark)

c $y = 2^x - 3$ cuts the x-axis at $x = c$
When $x = c$ and $y = 0$
$2^c - 3 = 0$
$2^c = 3$
Now $2^1 = 2$ and $2^2 = 4$
So c lies between 1 and 2.
$2^{1.5} = 2.828\ldots$
$2^{1.6} = 3.031\ldots$
So c lies between 1.5 and 1.6
$2^{1.58} = 2.989\ldots$
$2^{1.59} = 3.010\ldots$
So c lies between 1.58 and 1.59
$2^{1.584} = 2.997\ldots$
$2^{1.585} = 3.000\ldots$
So c lies between 1.584 and 1.585
$2^{1.5845} = 2.999\ldots$
So $c = 1.585$ to three decimal places. ✓ (1 mark)
(Total: 35 marks)

SURFACE AREA AND VOLUME
SKILLS CHECK PAGE 54

1 A regular hexagon has 6 equal sides.
Perimeter $= 6 \times 26$ cm
$= 156$ cm

2 **a**

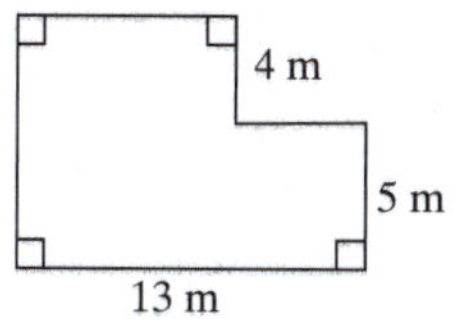

Total length at top = 13 m
[It must be the same as the bottom.]
Left side = total right side
Right side = 4 m + 5 m
= 9 m
$P = 2 \times 13 + 2 \times 9$
$= 44$
The perimeter is 44 m.

b

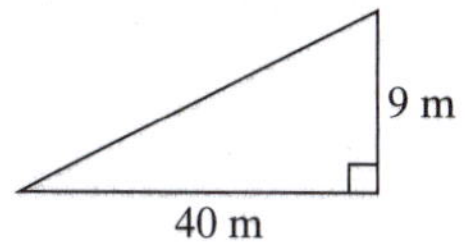

Let the length of the third side be x m.
$x^2 = 9^2 + 40^2$
$= 81 + 1600$
$= 1681$
$x = \sqrt{1681}$ $(x > 0)$
$= 41$
Perimeter
= 9 m + 40 m + 41 m
= 90 m

3 Radius = 19 cm

$C = 2\pi r$

$= 2 \times \pi \times 19$

$= 119.380\,5208\ldots$

The circumference is 119.4 cm (1 d.p.)

4 **a** Length = 3 m

Breadth = 70 cm = 0.7 m

$A = lb$

$= 3 \times 0.7$

$= 2.1$

The area is 2.1 m^2 [or 21 000 cm^2]

b Diameter = 16 m,

Radius = 8 m

$A = \pi r^2$

$= \pi \times 8^2$

$= 201.061\,9298\ldots$

$= 201.1$ (1 d.p.)

The area is 201.1 m^2, to 1 decimal place.

c $x = 23,\ y = 17$

$A = \frac{1}{2}xy$

$= \frac{1}{2} \times 23 \times 17$

$= 195.5$

The area is 195.5 cm^2.

5 **a**

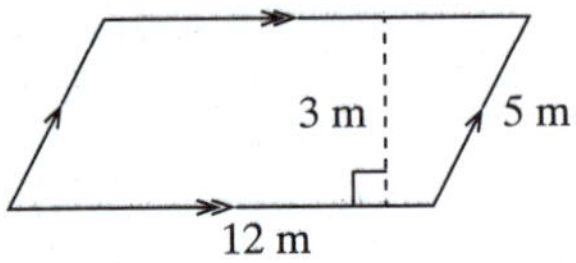

$A = bh$

$= 12 \times 3$

$= 36$

The area is 36 m^2.

b

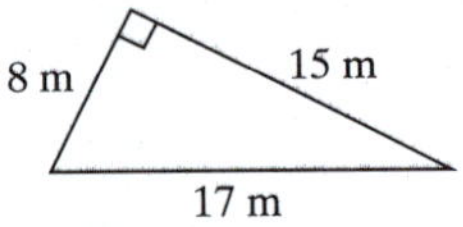

$A = \frac{1}{2}bh$

$= \frac{1}{2} \times 8 \times 15$

$= 60$

The area is 60 m^2.

c

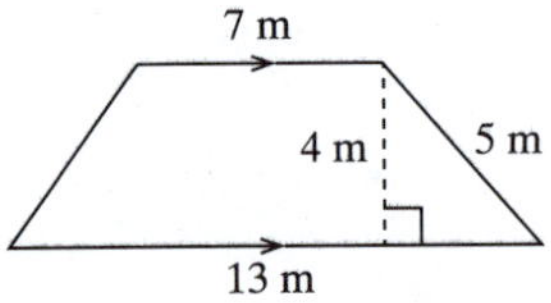

$A = \frac{1}{2}h(a + b)$

$= \frac{1}{2} \times 4 \times (7 + 13)$

$= 40$

The area is 40 m^2.

6 **a** $V = Ah$

$= 65 \times 7$

$= 455$

The volume is 455 cm^3.

b $V = lbh$

$= 8 \times 5 \times 3$

$= 120$

The volume is 120 m^3

c $V = \pi r^2 h$

$= \pi \times 0.6^2 \times 1$

$= 1.130\,973\,355\ldots$

The volume is 1.13 m^3 (2 d.p.)

7 **a** $A = 6s^2$

$= 6 \times 14^2$

$= 1176$

The surface area is 1176 cm^2.

b $l = 6, b = 4, h = 3$

$A = 2(lb + lh + bh)$

$= 2(6 \times 4 + 6 \times 3 + 4 \times 3)$

$= 108$

The surface area is 108 m^2.

c $r = 10, h = 15$

[radius = diameter ÷ 2]

$A = 2\pi r^2 + 2\pi rh$

$= 2 \times \pi \times 10^2 + 2 \times \pi \times 10 \times 15$

$= 1570.796\,327\ldots$

The surface area is 1570.8 cm^2. (1 d.p.)

8 **a**

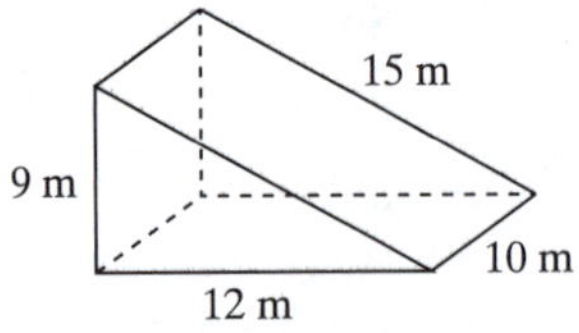

Area of triangular face

$= \frac{1}{2} \times 12 \times 9$

$= 54\ m^2$

$V = Ah$

$= 54 \times 10$

$= 540\ m^3$

b Surface area:

$A = 2 \times 54 + 12 \times 10 + 9 \times 10 + 15 \times 10$

$= 468\ m^2$

SURFACE AREA AND VOLUME

INTERMEDIATE TEST PAGE 55

1 $l = 6, b = 5, h = 4$

$A = 2(lb + lh + bh)$

$= 2 \times (6 \times 5 + 6 \times 4 + 5 \times 4)$

$= 148$

The surface area is 148 cm^2.

∴ [C] ✓ (1 mark)

2

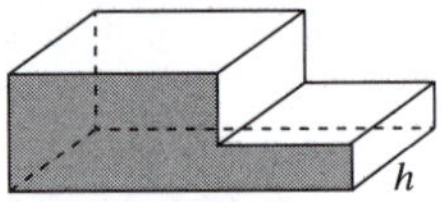

$V = Ah$ only in B.

[In A, the solid is a cone, $V = \frac{1}{3}Ah$. In C, h is not the perpendicular height, it is not perpendicular to the shaded face. In D, the shaded face is not the 'base' of the prism, (slices of the prism parallel to the shaded face get smaller).]

∴ [B] ✓ (1 mark)

3 $r = 50, h = 20$

$V = \pi r^2 h$

$= \pi \times 50^2 \times 20$

$= 157\,079.6327\ldots$

The volume is approximately 157 080 cm^3.

∴ [A] ✓ (1 mark)

4 Surface area = 13.5 m^2

Area of each face

$= (13.5 \div 6)\ m^2$

$= 2.25\ m^2$

Length of each side $= \sqrt{2.25}$ m

$= 1.5$ m

Volume $= (1.5\ m)^3$

$= 3.375\ m^3$

The closest option is 3 m^3.

∴ [D] ✓ (1 mark)

5 **a** Let h cm be the height of the triangle.

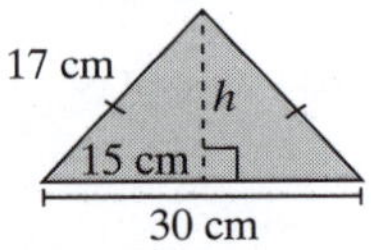

$17^2 = h^2 + 15^2$

$h^2 = 17^2 - 15^2$

$= 289 - 225$

$= 64$

$h = 8 \quad (h > 0)$

The perpendicular height of the shaded face is 8 cm. ✓

(1 mark)

b Area of the shaded face:

$A = \frac{1}{2}bh$

$= \frac{1}{2} \times 30 \times 8$

$= 120$ ✓

Volume of the prism:

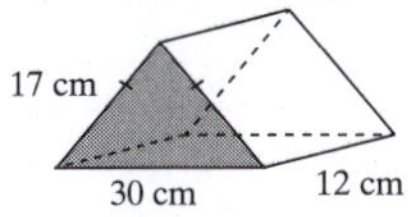

$V = AH$

$= 120 \times 12$

$= 1440$

The volume of the prism is $1440\ \text{cm}^3$. ✓ (2 marks)

c The area of the shaded face is $120\ \text{cm}^2$. (from part **b**)

Surface area:

$A = 2 \times 120 + 2 \times 17 \times 12 + 30 \times 12$ ✓

$= 1008$

The surface area is $1008\ \text{cm}^2$. ✓ (2 marks)

6 **a** Radius = 25 cm = 0.25 m

$V = Ah$

$= \frac{1}{2} \times \pi \times 0.25^2 \times 2.4$ ✓

$= 0.235\,6194\ldots$

$= 0.24$ (2 d.p.)

The volume is $0.24\ \text{m}^3$, to two decimal places. ✓ (2 marks)

b Capacity $\approx 0.24 \times 1000$ L

$= 240$ L ✓ (1 mark)

7 **a**

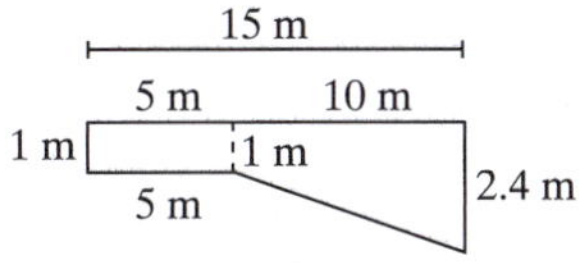

Area = area of rectangle + area of trapezium

$A = 5 \times 1 + \frac{1}{2} \times 10 \times (1 + 2.4)$ ✓

$= 5 + 17$

$= 22$

The area of the cross-section is $22\ \text{m}^2$. ✓ (2 marks)

b

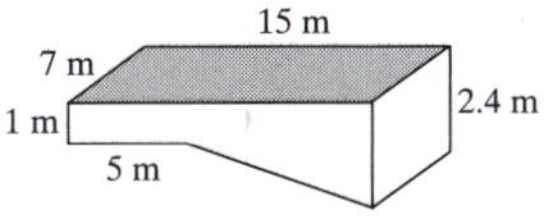

$V = Ah$

$= 22 \times 7$

$= 154$

The volume is $154\ \text{m}^3$. ✓

Capacity $= 154 \times 1000$ L

$= 154\,000$ L

Time

$= (154\,000 \div 5)$ seconds

$= 30\,800$ seconds ✓

$= 513.333\ldots$ minutes

$= 8.5555\ldots$ h

$= 8\frac{1}{2}$ h to the nearest half-hour ✓ (3 marks)

8 **a**

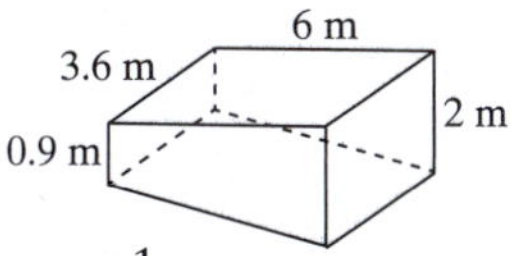

$A = \frac{1}{2}h(a + b)$

$= \frac{1}{2} \times 6 \times (0.9 + 2)$

$= 8.7$

The area of each trapezoidal wall is $8.7\ \text{m}^2$. ✓ (1 mark)

b Shallow end: $A = 0.9 \times 3.6$

$= 3.24$

The area of the shallow end wall is $3.24\ \text{m}^2$.

Deep end: $A = 2 \times 3.6$

$= 7.2$

The area of the deep end wall is $7.2\ \text{m}^2$. ✓

Total area

$= (2 \times 8.7 + 3.24 + 7.2)\ \text{m}^2$

$= 27.84\ \text{m}^2$ ✓ (2 marks)

c Let x m be the length of the bottom.

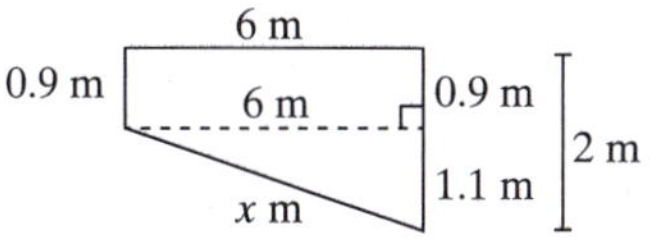

By Pythagoras' Theorem:

$x^2 = 6^2 + 1.1^2$

$= 37.21$

$x = \sqrt{37.21}$

$= 6.1$ ✓

The bottom of the pool is a rectangle 6.1 m long and 3.6 m wide.

$A = 6.1 \times 3.6$

$= 21.96$

The area of the bottom is $21.96\ \text{m}^2$. ✓ (2 marks)

d Area to be painted

$= (21.96 + 27.84)\ \text{m}^2$

$= 49.8\ \text{m}^2$

Total area (two coats of paint)

$= 2 \times 49.8\ \text{m}^2$

$= 99.6\ \text{m}^2$ ✓

The most tins of paint will be needed when each one covers the least area.

If the tins cover $19\ \text{m}^2$ then the number required

$= 99.6 \div 19$

$= 5.242\ldots$ ✓

So 6 tins would be needed. ✓ (3 marks)

(Total: 25 marks)

SURFACE AREA AND VOLUME ADVANCED TEST PAGE 56

1 **a**

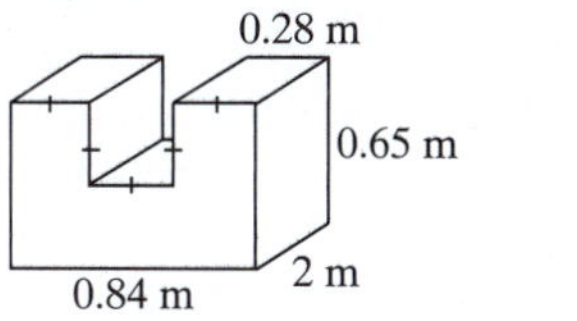

Area of cross-section

$= (0.84 \times 0.65 - 0.28^2)\ \text{m}^2$

$= 0.4676\ \text{m}^2$ ✓

Volume $= (0.4676 \times 2)\ \text{m}^3$

$= 0.9352\ \text{m}^3$

$= 0.94\ \text{m}^3$ (2 d.p.) ✓ (2 marks)

b

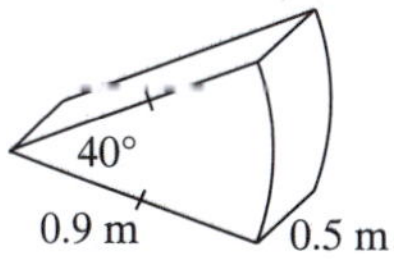

Area of cross-section

$= (\frac{40}{360} \times \pi \times 0.9^2)\ \text{m}^2$

$= 0.282\,74\ldots\ \text{m}^2$ ✓

Volume

$= (0.282\,74\ldots \times 0.5)\ \text{m}^3$

$= 0.141\,37\ldots\ \text{m}^3$

$= 0.14\ \text{m}^3$ (2 d.p.) ✓ (2 marks)

c

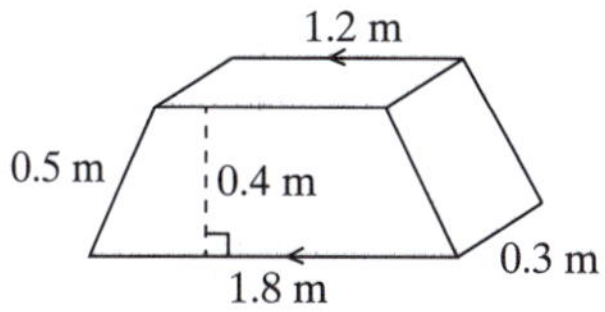

Area of cross-section

$= (\frac{0.4}{2}(1.2 + 1.8))$ m²

$= 0.6$ m² ✓

Volume

$= (0.6 \times 0.3)$ m³

$= 0.18$ m³ ✓ (2 marks)

d

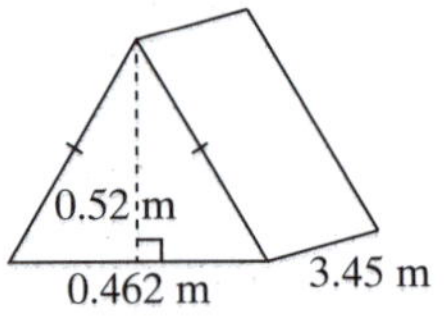

Area of cross-section

$= (\frac{1}{2} \times 0.462 \times 0.52)$ m²

$= 0.12012$ m² ✓

Volume

$= (0.12012 \times 3.45)$ m³

$= 0.414414$ m³

$= 0.41$ m³ (2 d.p.) ✓

(2 marks)

2 a Perimeter of cross-section

$= 2 \times (0.84 + 0.65 + 0.28)$ m

$= 3.54$ m ✓

Surface area

$= (3.54 \times 2 + 2 \times 0.4676)$ m²

$= 8.0152$ m² ✓ (2 marks)

b Perimeter

$= (\frac{40}{360} \times 2 \times \pi \times 0.9$

$+ 2 \times 0.9)$ m

$= 2.4283\ldots$ m ✓

Surface area

$= (2 \times 0.28274\ldots$

$+ 2.4283\ldots \times 0.5)$ m²

$= 1.77964\ldots$ m²

$= 1.78$ m² (2 d.p.) ✓

(2 marks)

c By Pythagoras' theorem we can see that the unknown side is also 0.5 m long. ✓

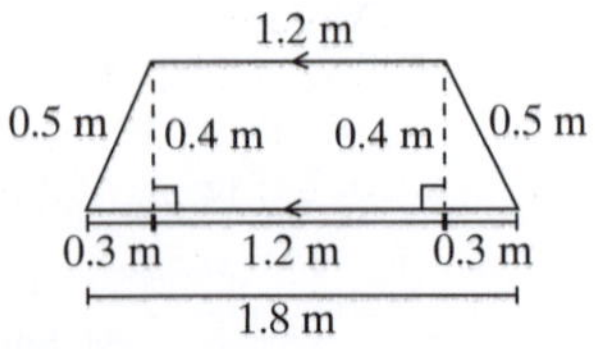

Surface area

$= (2 \times 0.6 + (1.2 + 1.8$

$+ 2 \times 0.5) \times 0.3)$ m²

$= 2.4$ m² ✓ (2 marks)

d Let x m be the length of the hypotenuse.

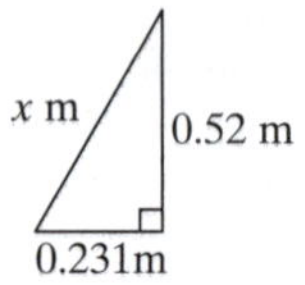

$x^2 = 0.52^2 + 0.231^2$

$= 0.323761$

$x = \sqrt{0.323761}$ $(x > 0)$

$= 0.569$ ✓

Surface area

$= (2 \times 0.12012$

$+ (0.462 + 2 \times 0.569)$

$\times 3.45)$ m²

$= 5.76024$ m² ✓ (2 marks)

3 a $V = Ah$

$= 697 \times 19$

$= 13243$ cm³ ✓ (1 mark)

b Surface area

$= (2 \times 697 + 96 \times 19)$ cm²

$= 3218$ cm² ✓ (1 mark)

4 a Area of cross-section

$= \frac{1}{2} \times 10.07 \times 12.24$

$- 7.35 \times 1.85$

$= 48.0309$ m² ✓

$V = Ah$

$= 48.0309 \times 23.45$

$= 1126.324605$

$= 1126.32$ m² (2 d.p.) ✓

(2 marks)

b

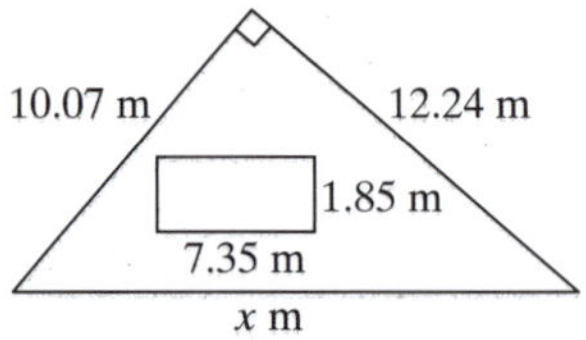

Let x m be the length of the hypotenuse.

$x^2 = 10.07^2 + 12.24^2$

$= 251.2225$

$x = \sqrt{251.2225}$ $(x > 0)$

$= 15.85$ ✓

Area of exterior rectangular faces

$= ((10.07 + 12.24 + 15.85)$

$\times 23.45)$ m²

$= 894.852$ m²

Area of faces of hole

$= (2 \times (7.35 + 1.85)$

$\times 23.45)$ m²

$= 431.48$ m² ✓

Total surface area

$= (2 \times 48.0309 + 894.852$

$+ 431.48)$ m²

$= 1422.3938$ m²

$= 1422.39$ m² (2 d.p.) ✓

(3 marks)

5 a Diameter of semicircle

$= 72$ cm

∴ radius $= 36$ cm

Width of rectangle

$= (87 - 36)$ cm

$= 51$ cm ✓

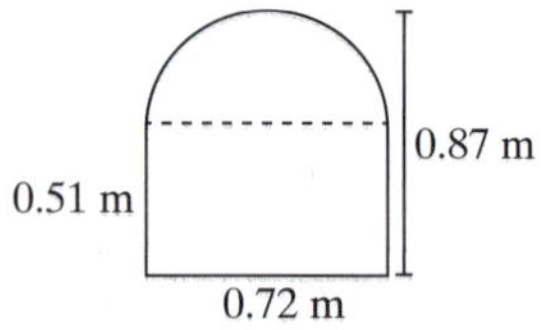

Area of cross-section

$= (0.72 \times 0.51$

$+ \frac{1}{2} \times \pi \times 0.36^2)$ m²

$= 0.570775\ldots$ m² ✓

Volume

$= (0.570775\ldots \times 1.4)$ m³

$= 0.799085\ldots$ m³

$= 0.80$ m³ (2 d.p.) ✓

(3 marks)

b Perimeter of cross-section

$= (\frac{1}{2} \times 2 \times \pi \times 0.36$

$+ 2 \times 0.51 + 0.72)$ m

$= 2.87097\ldots$ m ✓

Surface area

$= (2 \times 0.570775\ldots$

$+ 2.87097\ldots \times 1.4)$ m² ✓

$= 5.160913\ldots$ m²

$= 5.16$ m² (2 d.p.) ✓

(3 marks)

6 a Diagonal of square is diameter of the circle; $2r$ units

A square is also a rhombus so the area of the square is half the product of the diagonals.

Area of square

$= (\frac{1}{2} \times 2r \times 2r)$ units²

$= 2r^2$ units² ✓

Shaded area

= area of circle

− area of square

$= (\pi r^2 - 2r^2)$ units²

$= r^2(\pi - 2)$ units² ✓

(2 marks)

b Diameter = 8 cm; $r = 4$
Area = $4^2(\pi - 2)$ cm^2
$= 16(\pi - 2)$ cm^2 ✓
(1 mark)

c $V = Ah$
$= 16(\pi - 2) \times 6$
$= 109.59289\ldots$
$= 109.6$ cm^3 (1 d.p.) ✓
(1 mark)

d Curved surface:
$A = 2\pi rh$
$= 2 \times \pi \times 4 \times 6$
$= 48\pi$ units^2 ✓
Total surface area
$= 48\pi + 2 \times 16(\pi - 2)$
$= 48\pi + 32\pi - 64$
$= (80\pi - 64)$ cm^2 ✓
(2 marks)

7 **a** $A = 2\pi rh + 2\pi r^2$
$2287 = 2\pi r \times 15 + 2\pi r^2$
$2\pi r^2 + 30\pi r - 2287 = 0$ ✓
[divide by 2π]
$r^2 + 15r - 363.987\ldots = 0$
So a solution to the equation $r^2 + 15r - 364 = 0$ will give an approximate value of r. ✓ (2 marks)

b $r^2 + 15r - 364 = 0$
$(r + 28)(r - 13) = 0$
So $r = -28$ or $r = 13$ ✓
But $r > 0$
$\therefore r = 13$
An approximate value for the length of the radius is 13 cm. ✓ (2 marks)

c $V = \pi r^2 h$
$\approx \pi \times 13^2 \times 15$
$= 7963.937\ldots$
So, the volume is approximately 7964 cm^3. ✓
(1 mark)
(Total: 40 marks)

ANGLE PROPERTIES
SKILLS CHECK PAGE 58

1 **a**

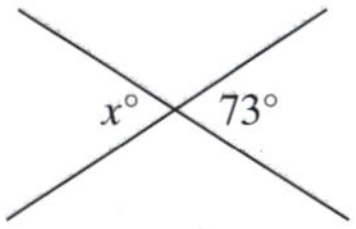

$x = 73$
(vertically opposite angles)

b

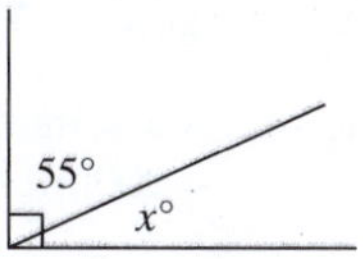

$x + 55 = 90$
(angles in a right angle)
$x = 35$

c

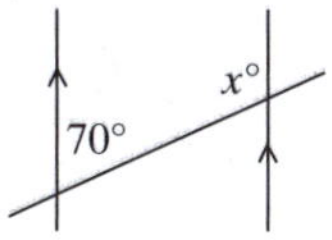

$x + 70 = 180$
(co-interior angles, parallel lines)
$x = 110$

d

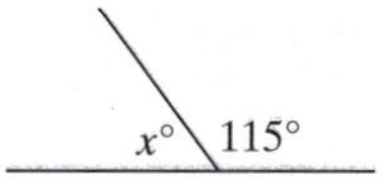

$x + 115 = 180$
(angles in a straight line)
$x = 65$

e

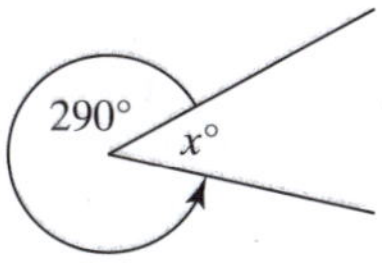

$x + 290 = 360$
(angles in a revolution)
$x = 70$

f

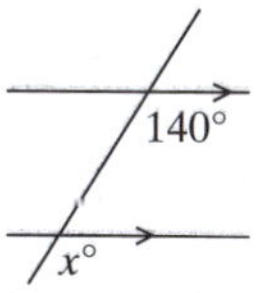

$x = 140$ (corresponding angles, parallel lines)

g

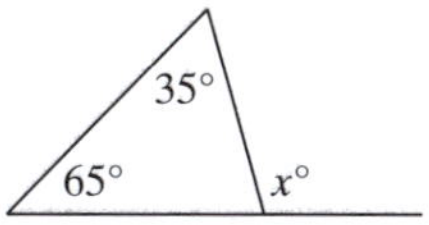

$x = 35 + 65$
(exterior angle of a Δ)
$x = 100$

h

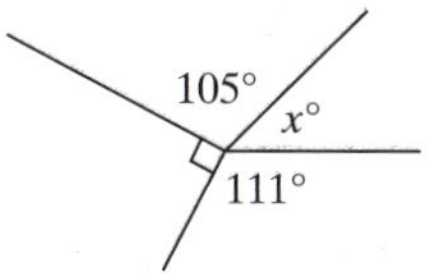

$x + 111 + 90 + 105 = 360$
(angles in a revolution)
$x + 306 = 360$
$x = 54$

i

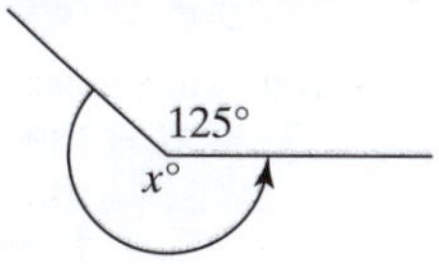

$x + 125 = 360$
(angles in a revolution)
$x = 235$

j

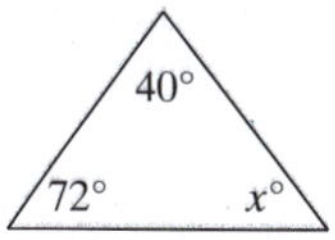

$x + 40 + 72 = 180$
(angle sum of a Δ)
$x + 112 = 180$
$x = 68$

k

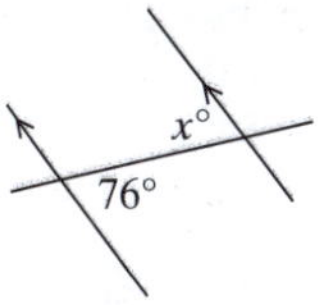

$x = 76$
(alternate angles, parallel lines)

l

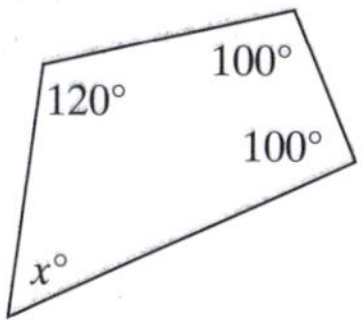

$x + 120 + 100 + 100 = 360$
(angle sum of a quad.)
$x + 320 = 360$
$x = 40$

2 **a** $x + 22 = 69$
(exterior angle of Δ)
$x = 47$

b $x = 67$ (angles opposite equal sides, isosceles Δ)

c $x = 60$ (equilateral Δ)

d

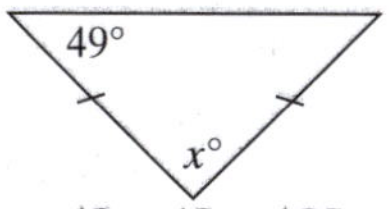

$x + 49 + 49 = 180$
(angle sum of isosceles Δ)
$x + 98 = 180$
$x = 82$

e

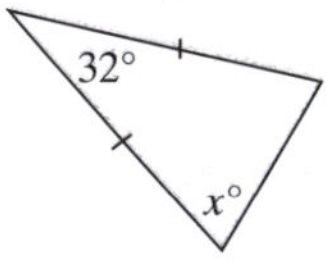

$x + x + 32 = 180$
(angle sum isosceles Δ)
$2x + 32 = 180$
$2x = 148$
$x = 74$

3 a $x = 79$ (opposite angles of parallelogram)

b $x = 90$ (diagonals of a rhombus meet at right-angles)

c 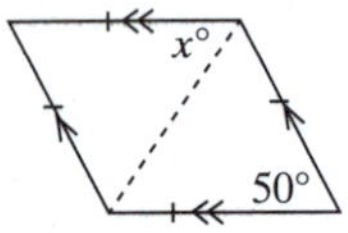

The quadrilateral is a rhombus.The diagonals bisect the angles.
$2x + 50 = 180$
(co-interior angles, parallel lines)
$2x = 130$
$x = 65$

d The quadrilateral is a parallelogram [both pairs of opposite sides are equal].

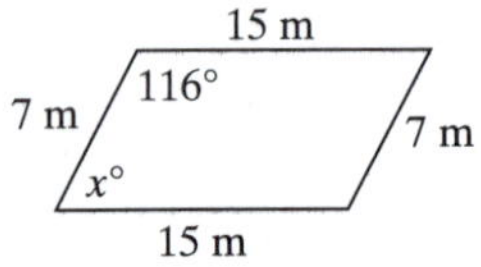

$x + 116 = 180$
(co-interior angles, parallel lines)
$x = 64$

e The quadrilateral is a square [a rhombus with one angle 90°].

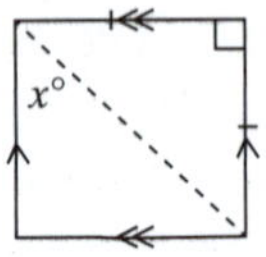

$x = 45$ [The diagonals bisect the angles. Each angle is a right angle.]

4 a A **hexagon** has 6 sides.

b A **decagon** has 10 sides.

c A **dodecagon** has 12 sides.

5 a A pentagon has **5** sides.

b An octagon has **8** sides.

c A heptagon has **7** sides.

6 Angle sum $= 5 \times 180°$
$= 900°$

ANGLE PROPERTIES
INTERMEDIATE TEST PAGE 59

1 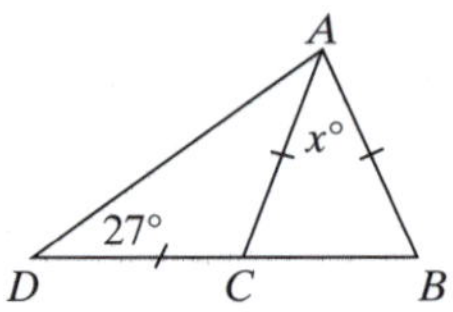

$\angle DAC = \angle ADC$
$= 27°$
($DC = AC$, angles opposite equal sides)
$\angle ACB = 27° + 27°$
(exterior angle)
$= 54°$
$\angle ABC = \angle ACB = 54°$
($AC = AB$, angles opposite equal sides)
$x° + 54° + 54° = 180°$
(angle sum ΔABC)
$x + 108 = 180$
$x = 72$
$\therefore$ [C] ✓ (1 mark)

2 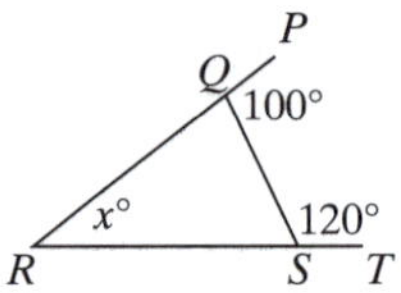

$\angle RQS = 80°$
($\angle RQP$ is a straight line)
$\angle QRS + \angle RQS = \angle QST$
(exterior angle ΔRQS)
$x + 80 = 120$
$x = 40$
$\therefore$ [C] ✓ (1 mark)

3 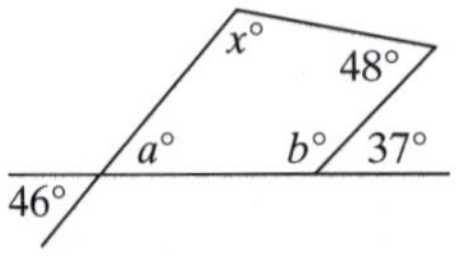

$a = 46$ (vertically opposite)
$b = 143$
(angles in a straight line)
$x + 48 + 143 + 46 = 360$
(angle sum of a quadrilateral)
$x + 237 = 360$
$x = 123$
$\therefore$ [A] ✓ (1 mark)

4 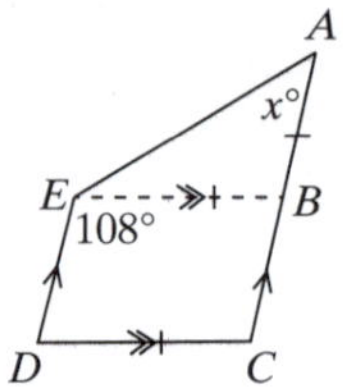

$\angle ABE = 108°$
(alternate, $AC \parallel DE$)
$EB = DC$
(opposite sides of parallelogram)
but $AB = DC$ (given)
$\therefore EB = AB$
$\therefore \angle AEB = \angle EAB$
(angles opposite equal sides)
$2x + 108 = 180$
(angle sum ΔAEB)
$2x = 72$
$x = 36$
$\therefore$ [A] ✓ (1 mark)

5 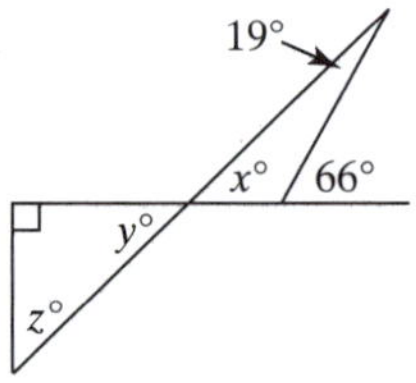

a $x + 19 = 66$ (exterior angle Δ)
$x = 47$ ✓ (1 mark)

b $x = y$
(vertically opposite angles)
$y = 47$ ✓ (1 mark)

c $y + z + 90 = 180$
(angle sum Δ)
$47 + z + 90 = 180$
$z + 137 = 180$
$z = 43$ ✓ (1 mark)

6 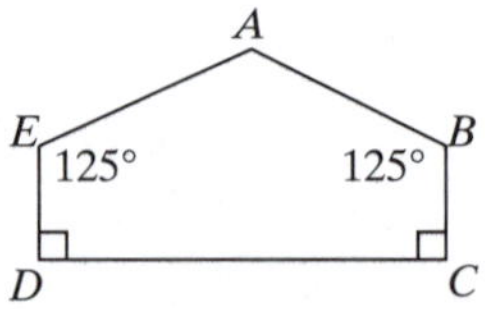

a Pentagon has 5 sides
$A = (n - 2) \times 180°$
$= (5 - 2) \times 180°$ ✓
$= 3 \times 180°$
$= 540°$ ✓
The angle sum of a pentagon is 540°. (2 marks)

b $\angle EAB + \angle ABC + \angle BCD$
$+ \angle CDE + \angle DEA = 540°$
(angle sum of a pentagon) ✓
$\angle EAB + 125° + 90°$
$+ 90° + 125° = 540°$
$\angle EAB + 430° = 540°$
$\angle EAB = 110°$ ✓
(2 marks)

7 An octagon has 8 sides: $n = 8$
$A = (n - 2) \times 180°$
$= (8 - 2) \times 180°$
$= 6 \times 180°$
$= 1080°$ ✓
The angle sum of an octagon is 1080°.
Each angle of a regular octagon
$= 1080° \div 8$
$= 135°$ ✓ (2 marks)

8 **a** Decagon has 10 sides.
Exterior angles add to 360°. ✓
Each exterior angle
$= 360° \div 10$
$= 36°$ ✓ (2 marks)

b Each interior angle
$= 180° - 36°$
(angles in a straight line) ✓
$= 144°$ ✓

c $A = 144° \times 10$
$= 1440°$ ✓ (3 marks)

9

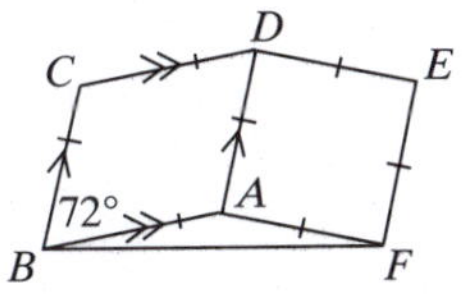

$ABCD$ is a rhombus
(parallelogram with adjacent sides equal)
$\angle DAB + \angle ABC = 180°$ (co-interior, $BC \parallel AD$)
$\angle DAB = 180° - 72°$
$= 108°$ ✓
$\angle DAF = 90°$ (angle of a square)
$\angle BAF + \angle DAB + \angle DAF$
$= 360°$ (angles at a point)
$\angle BAF = 360° - 108° - 90°$
$= 162°$ ✓
$AD = AF$ (sides of a square)
$AD = AB$ (sides of a rhombus)
$\therefore\ AB = AF$
$\therefore \angle AFB = \angle ABF$
(angles opposite equal sides, isosceles triangle)

$\angle AFB + \angle ABF + \angle BAF = 180°$
(angle sum ΔABF)
$2\angle AFB + 162° = 180°$
$2\angle AFB = 18°$
$\angle AFB = 9°$
$\angle EFA = 90°$ (angle of a square)
$\angle EFB = 90° + 9°$
$= 99°$ ✓ (3 marks)

10

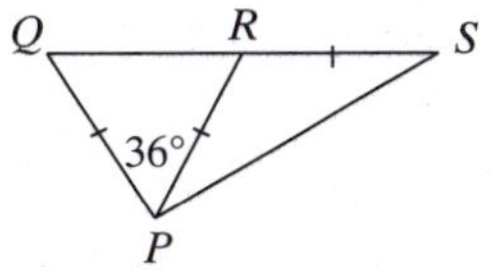

a $\angle PRQ + \angle RQP + \angle QPR$
$= 180°$
(angle sum of ΔPQR)
but $\angle PRQ = \angle RQP$
(angles opposite equal sides)
$2\angle PRQ + 36° = 180°$
$2\angle PRQ = 144°$
$\angle PRQ = 72°$ ✓
$\angle PSR + \angle RPS = \angle QRP$
(exterior angle)
but $\angle PSR = \angle RPS$
(angles opposite equal sides)
$2\angle PSR = 72°$
$\angle PSR = 36°$
$\therefore \angle PSR = \angle QPR$
(both 36°) ✓ (2 marks)

b $\angle RPS = 36°$
(= $\angle PSR$ from part **a**)
$\angle QPS = \angle QPR + \angle RPS$
$= 36° + 36°$
$= 72°$ ✓
$\angle RQP = 72°$
(= $\angle PRQ$ from part **a**)
$\therefore \angle QPS = \angle RQP$
$\therefore$ ΔPSQ is isosceles ✓
(2 marks)
(Total: 25 marks)

ANGLE PROPERTIES
ADVANCED TEST PAGE 60

1 **a**

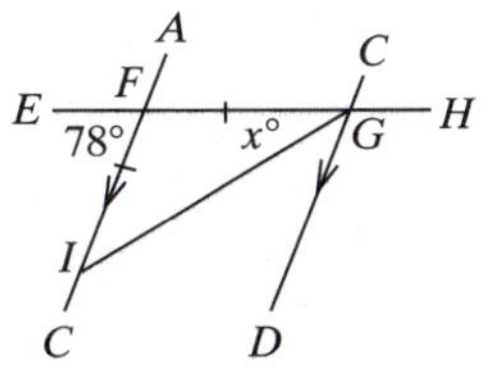

$\angle FIG = \angle FGI$
(base angles isos. Δ)

$\angle FIG + \angle FGI = \angle EFI$
(ext. ∠ of Δ) ✓
$\therefore 2x = 78$
$x = 39$ ✓
(2 marks)

b

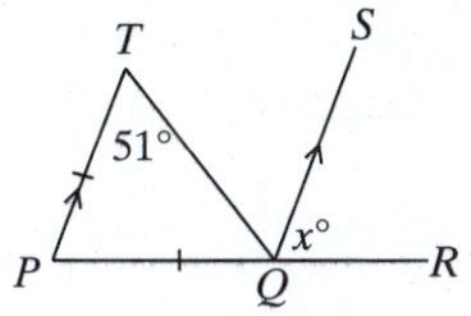

$\angle PQT = \angle PTQ$
(base angles isos. Δ)
$\therefore \angle PQT = 51°$ ✓
$\angle TPQ + \angle PQT + \angle PTQ$
$= 180°$
(∠ sum of Δ)
$\therefore \angle TPQ = 180° - 2 \times 51°$
$= 78°$
$\angle SQR = \angle TPQ$
(corresponding, $PT \parallel QS$)
$\therefore x = 78$ ✓ (2 marks)

c

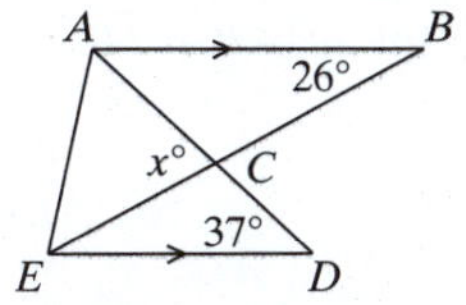

$\angle BAC = \angle CDE$
(alternate, $AB \parallel ED$)
$\therefore \angle BAC = 37°$ ✓
$\angle ACE = \angle BAC + \angle ABC$
(exterior ∠ of Δ)
$\therefore x = 37 + 26$
$= 63$ ✓ (2 marks)

d

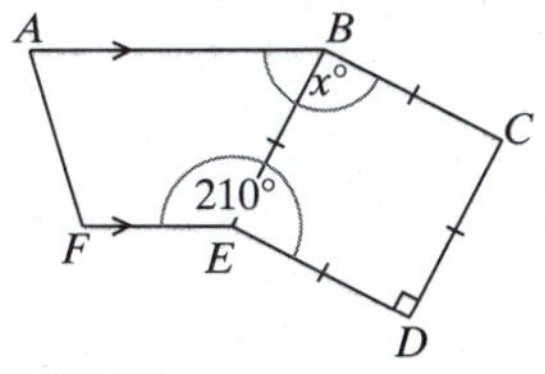

$BCDE$ is a square
(all sides equal, an angle 90°)
$\therefore \angle EBC = \angle BED = 90°$
$\angle FEB = 210° - 90°$
$= 120°$ ✓
$\angle ABE + \angle FEB = 180°$
(co-interior, $AB \parallel FE$)
$\therefore \angle ABE = 180° - 120°$
$= 60°$
$\angle ABC = 60° + 90°$
$\therefore x = 150$ ✓ (2 marks)

WORKED SOLUTIONS

CHECK YOUR SOLUTIONS

2

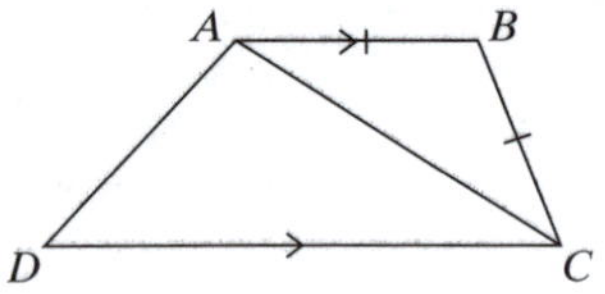

$\angle BAC = \angle BCA$
(base $\angle$s of isos. Δ)
but $\angle BAC = \angle DCA$
(alternate, $AB \parallel DC$)
$\therefore \angle BCA = \angle DCA$
$\therefore AC$ bisects $\angle BCD$ ✓ (1 mark)

3

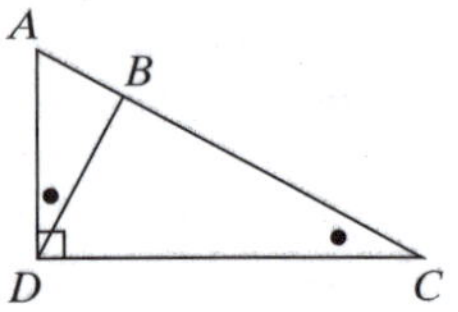

$\angle DAC + \angle ACD + \angle CDA = 180°$
($\angle$ sum Δ)
$\angle DAC + \angle ACD + 90° = 180°$
$\angle DAC + \angle ACD = 90°$
$\therefore \angle DAC + \angle BDA = 90°$ ✓
But $\angle DAB + \angle BDA + \angle ABD = 180°$
($\angle$ sum Δ)
$\therefore \angle ABD = 90°$
$\therefore BD \perp AC$ ✓ (2 marks)

4

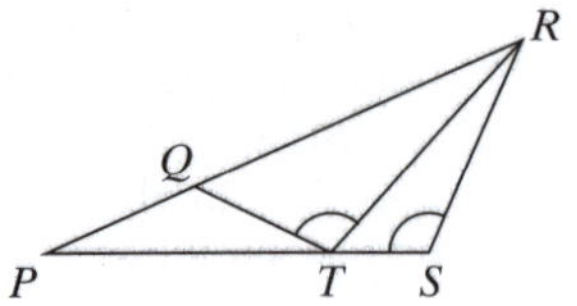

$\angle STR + \angle RTQ + \angle QTP = 180°$
($\angle$s in st. line)
$\angle STR + \angle RST + \angle SRT = 180°$
($\angle$ sum of Δ) ✓
$\therefore \angle RTQ + \angle QTP = \angle RST + \angle SRT$
But $\angle RTQ = \angle RST$ (given)
$\therefore \angle QTP = \angle SRT$ ✓ (2 marks)

5

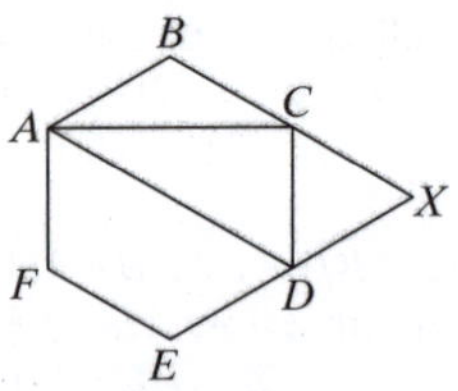

a Angle sum of hexagon
$= 4 \times 180°$
$= 720°$
Each angle $= 720° \div 6$
$= 120°$
$\therefore \angle ABC = 120°$ ✓ (1 mark)

b $AB = BC$
(sides of regular hexagon)
$\therefore \angle ACB = \angle CAB$
(base $\angle$s of isos. Δ)
$\angle ABC + \angle ACB + \angle CAB = 180°$
($\angle$ sum Δ)
$120° + 2\angle ACB = 180°$
$2\angle ACB = 60°$
$\angle ACB = 30°$ ✓
$\angle BCD = 120°$
(angle reg. hexagon)
$\therefore \angle ACD = 120° - 30°$
$= 90°$ ✓ (2 marks)

c $\angle BCD + \angle DCX = 180°$
($\angle$s st. line)
$120° + DCX = 180°$
$\angle DCX = 60°$ ✓
Similarly $\angle CDE = 120°$
and $\angle CDX = 60°$
$\therefore \Delta CXD$ is equilateral
$\therefore CX = CD$
But $BC = CD$
(sides reg. hexagon)
$\therefore BC = CX$
C is the midpoint of BX. ✓
(2 marks)

6

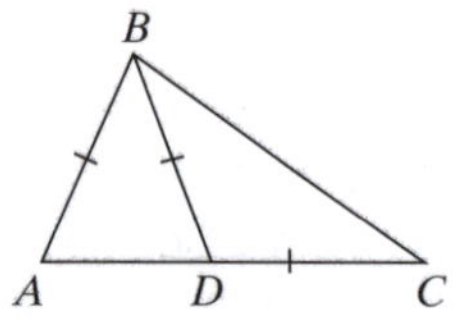

$\angle DBC = \angle BCD$
(base $\angle$s isos. ΔBDC)
$\angle ADB = \angle DBC + \angle BCD$
(exterior $\angle$ of ΔBDC)
$= 2\angle BCD$ ✓
but $\angle BAD = \angle ADB$
(base $\angle$s of isos. ΔABD)
$\therefore \angle BAD = 2\angle BCA$ ✓
(2 marks)

7

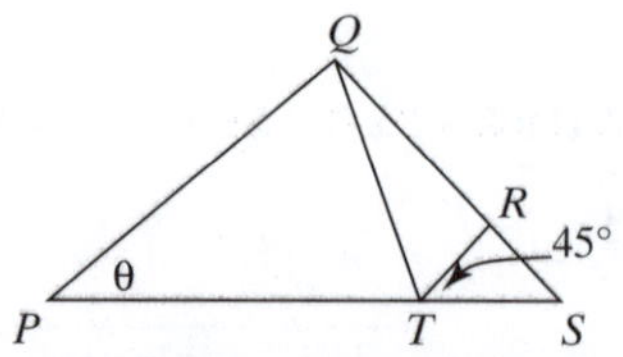

a $\angle QSP = \angle QPS$
(base $\angle$s isos. ΔQPS)
$\therefore \angle QSP = \theta$
$\angle QRT = \angle RTS + \angle RST$
(exterior $\angle$ of ΔRTS)
$= 45° + \theta$ ✓ (1 mark)

b $\angle QTR = \angle QRT$
(base $\angle$s of isos. ΔQRT)
$\angle QTR + \angle QRT$
$= 2(45° + \theta)$
$= 90° + 2\theta$ ✓
$\angle TQR + \angle QTR + \angle QRT$
$= 180°$ ($\angle$ sum ΔQRT)
$\angle TQR = 180° - (90° + 2\theta)$
$= 90° - 2\theta$ ✓
(2 marks)

c $\angle PQS + \angle QPS + \angle QSP$
$= 180°$ ($\angle$ sum ΔQPS)
$\therefore \angle PQS = 180° - 2\theta$
$\angle PQT = 180° - 2\theta - (90° - 2\theta)$
$= 180° - 2\theta - 90° + 2\theta$
$= 90°$ ✓ (1 mark)

8

a All of the sides are equal but not all the angles so the pentagon is not regular. ✓
(1 mark)

b $AE = EB$ (given)
but $AE = ED = DC = BC$ (given)
$\therefore EBCD$ is a square (all sides equal, one angle 90°)
$\therefore \angle BCD = 90°$ ✓ (1 mark)

c ΔEAB is equilateral
(all sides equal)
$\therefore \angle EAB = 60°$ ✓ (1 mark)

d $\angle AED = \angle AEB + \angle BED$
$= 60° + 90°$
$= 150°$
$\therefore \angle AEX = 75°$
(EX bisects $\angle AED$) ✓
(1 mark)

e

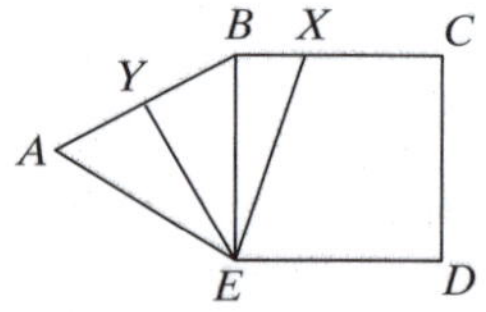

$\angle XEB = 75° - 60°$
$= 15°$
$\therefore \angle YEB = 15°$
($\angle YEB = \angle XEB$, given) ✓
$\angle ABE = 60°$
($\angle$ equilateral ΔYEB)
$\angle EYB + \angle YBE + \angle YEB = 180°$
($\angle$ sum Δ)
$\angle EYB + 60° + 15° = 180°$
$\angle EYB = 105°$ ✓
(2 marks)
(Total: 30 marks)

PROOFS INVOLVING CONGRUENT TRIANGLES
SKILLS CHECK PAGE 62

1

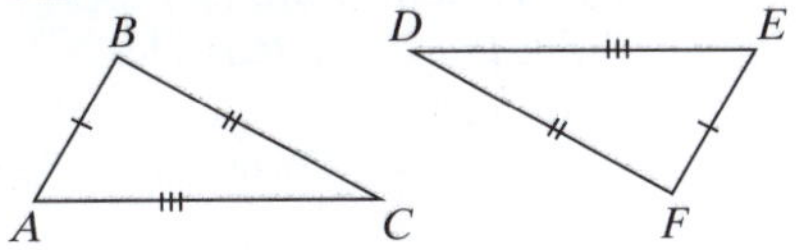

a SSS
[$AB = EF$, $BC = FD$ and $AC = ED$ (all given)]

b $\Delta ABC \equiv \Delta EFD$
[The vertices must be given in that order.]

c **i** $\angle ABC$ corresponds to $\angle EFD$

ii $\angle CAB$ corresponds to $\angle DEF$

iii $\angle BCA$ corresponds to $\angle FDE$

2

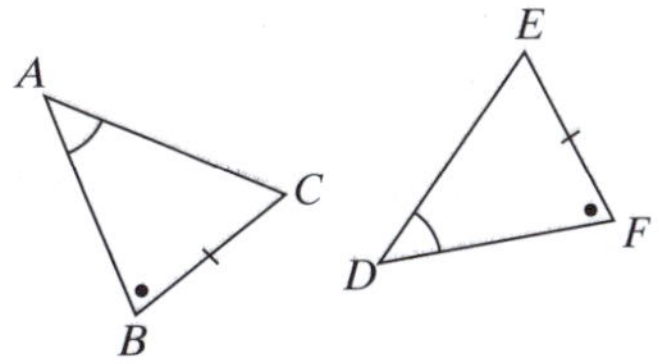

a AAS

b ***DE*** corresponds to AC

c ***DF*** corresponds to AB

d $\angle$***FED*** corresponds to $\angle BCA$

3 **a**

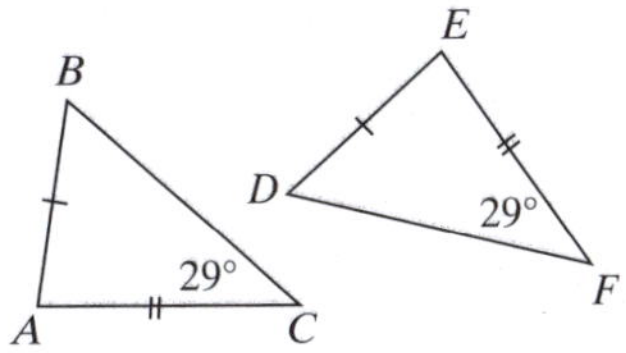

not necessarily congruent
[Two sides and an angle are equal but the angle is not <u>included</u> between the two sides. The triangles will be congruent if $\angle A = \angle E$.]

b

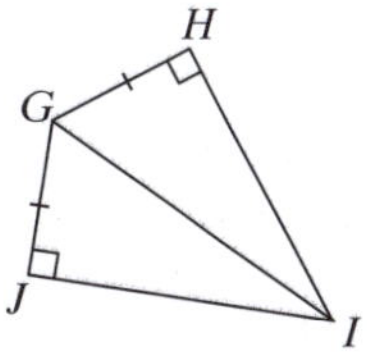

RHS
[In triangles GHI, GJI
$\angle GHI = \angle GJI$
(both 90°, given)
$GI = GI$ (common side)
$GH = GJ$ (given)
$\therefore\ \Delta GHI \equiv \Delta GJI$ RHS]

c

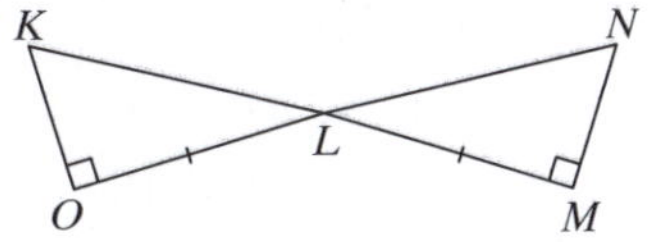

AAS
[In triangles KOL, NML
$\angle KOL = \angle NML$
(both 90°, given)
$\angle KLO = \angle NLM$
(vertically opposite)
$OL = ML$ (given)
$\therefore\ \Delta KOL \equiv \Delta NML$ AAS]

4 **a** An **equilateral** triangle has all sides equal.

b A **scalene** triangle has no sides equal.

c An **isosceles** triangle has two sides equal.

5 **a** The sum of the interior angles of a triangle is **180**°.

b Each angle of an equilateral triangle is equal to **60**°.

c The exterior angle of a triangle is equal to **the sum of the interior opposite angles.**

d If two sides of a triangle are equal the angles opposite those sides are **equal**.

e If two angles of a triangle are equal the sides opposite those angles are **equal**.

6 **a**

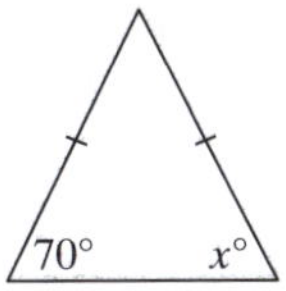

$x = 70$ (angles opposite equal sides, isosceles triangle)

b

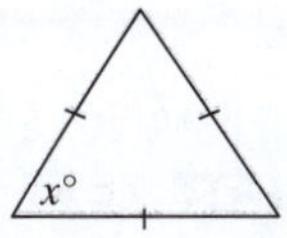

$x = 60$ (equilateral triangle)

c

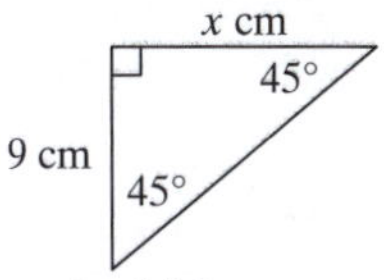

$x = 9$ (sides opposite equal angles, isosceles triangle)

d

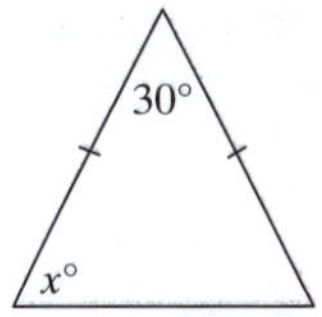

base angles both measure x°
(angles opposite equal sides, isosceles triangle)
$2x + 30 = 180$
(angle sum of a Δ)
$2x = 150$
$x = 75$

7 **a** A **rhombus** is a parallelogram with two adjacent sides equal.

b A **square** is a rectangle with two adjacent sides equal.

c A **trapezium** is a quadrilateral with at least one pair of opposite sides parallel.

d A **rectangle** is a parallelogram with one angle a right angle. [In fact all angles are right angles, but we only need to know that one angle is a right angle in order to know that the parallelogram must be a rectangle.]

e A **parallelogram** is a quadrilateral with both pairs of opposite sides parallel.

WORKED SOLUTIONS

PROOFS INVOLVING CONGRUENT TRIANGLES
INTERMEDIATE TEST PAGE 63

1

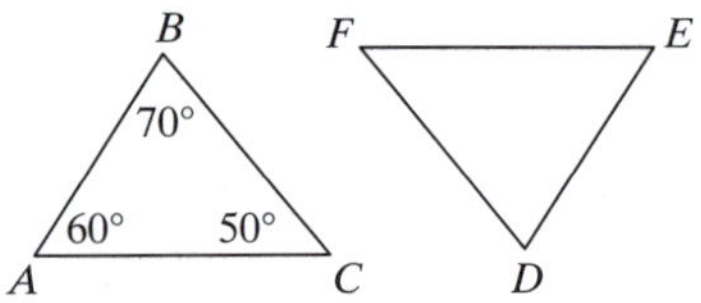

$\Delta ABC \equiv \Delta FDE$
$\angle FED = \angle ACB$
(corresponding angles of congruent triangles)
$\therefore \angle FED = 50°$
$\therefore$ **[A]** ✓ (1 mark)

2

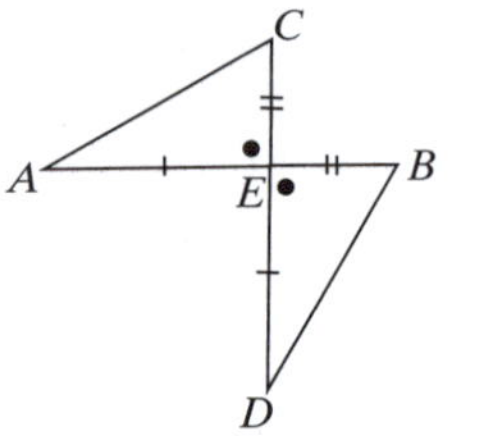

$AB = CD$ (given)
but $AE = DE$ (given)
$\therefore BE = CE$
$\angle AEC = \angle DEB$
(vertically opposite)
$\therefore \Delta AEC \equiv \Delta DEB$ SAS
$\therefore$ **[B]** ✓ (1 mark)

3 All congruent triangles must be similar but similar triangles need not be congruent.
$\therefore$ **[B]** ✓ (1 mark)

4

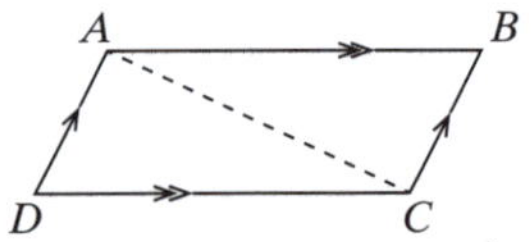

a $ABCD$ is a parallelogram. (both pairs of opposite sides are parallel) ✓ (1 mark)

b Alternate angles ($AB \parallel DC$) ✓ (1 mark)

c Alternate angles ($BC \parallel AD$) ✓ (1 mark)

d Common side ✓ (1 mark)

e AAS [from parts **b**, **c** and **d**] ✓ (1 mark)

f $\angle ABC = \angle CDA$ ✓
(corresponding angles of congruent triangles) ✓
(2 marks)

g Opposite angles of a parallelogram are equal. ✓ (1 mark)

5

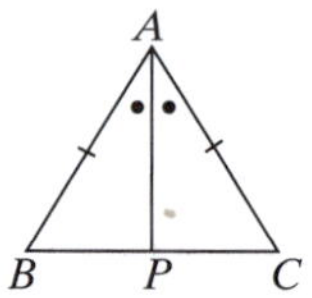

a In ΔABP, ΔACP
$AB = AC$ (given)
$\angle BAP = \angle CAP$
(AP bisects $\angle BAC$)
$AP = AP$
(common side) ✓
$\therefore \Delta ABP \equiv \Delta ACP$ SAS ✓
$\therefore \angle ABP = \angle ACP$
(corresponding angles of congruent triangles) ✓
(3 marks)

b The angles opposite the equal sides of a triangle are equal. ✓ (1 mark)

6 **a**

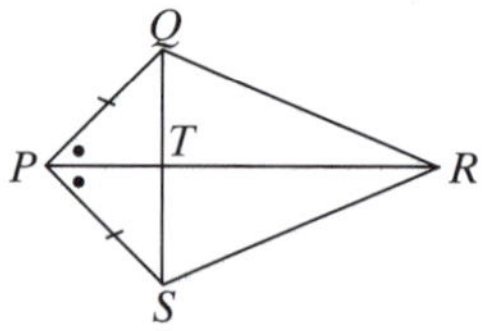

In ΔPQT, ΔPST
$PQ = PS$ (given)
$\angle QPT = \angle SPT$ (given) ✓
$PT = PT$
(common side) ✓
$\therefore \Delta PQT \equiv \Delta PST$ SAS ✓
(3 marks)

b $\angle PTQ = \angle PTS$
(corresponding angles of congruent triangles) ✓
but $\angle PTQ + \angle PTS = 180°$
(angles in a straight line)
$\therefore \angle PTQ = \angle PTS = 90°$ ✓
The diagonals meet at right angles. (2 marks)

7

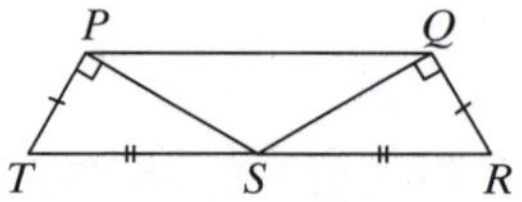

a In ΔTPS, ΔRQS
$\angle TPS = \angle RQS$
(both 90°, given)
$PT = \text{QR}$ (given) ✓
$TS = RS$
(S is the midpoint of TR) ✓
$\therefore \Delta TPS \equiv \Delta RQS$ RHS ✓
(3 marks)

b $PS = QS$
(corresponding sides of congruent triangles) ✓
$\therefore \Delta PQS$ is an isosceles triangle. ✓ (2 marks)

(Total: 25 marks)

PROOFS INVOLVING CONGRUENT TRIANGLES
ADVANCED TEST PAGE 64

1

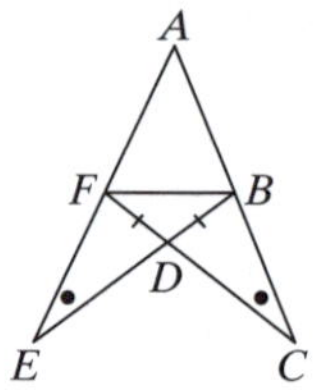

In ΔFBE, ΔBFC
$\angle FEB = \angle BCF$ (given)
$\angle FBE = \angle BFC$
(base $\angle$s isos. ΔFBD)
$FB = BF$ (common side) ✓
$\therefore \Delta FBE \equiv \Delta BFC$ AAS ✓
(2 marks)

2

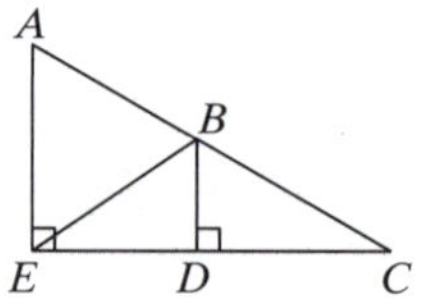

a In ΔEBD, ΔCBD
$BD = BD$ (common side)
$\angle BDE = \angle BDC$
(both 90°, $BD \perp EC$)
$ED = CD$
(D is midpoint of EC) ✓
$\therefore \Delta EBD \equiv CBD$ SAS ✓
(2 marks)

b Let $\angle BCD = \theta$
$\therefore \angle BED = \theta$
(corresp. $\angle$s cong. Δs)
$\therefore \angle AEB = 90° - \theta$ ✓
$\angle EAC + \angle AEC + \angle ECA = 180°$ ($\angle$ sum Δ)
$\angle EAC + 90° + \theta = 180°$
$\angle EAC = 90° - \theta$
So $\angle EAB = \angle AEB$
$\therefore \Delta ABE$ is isosceles ✓
[Or use similar triangles or the proportions of the sides.] (2 marks)

3

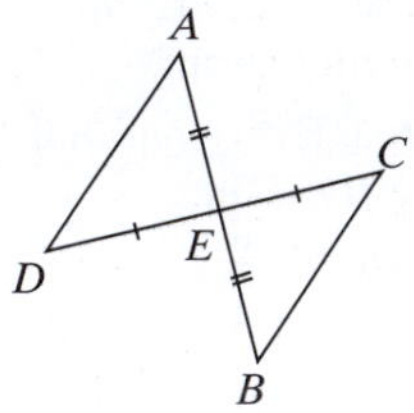

a In ΔAED, ΔBEC
$DE = CE$ (given)
$\angle AED = \angle BEC$
(vertically opp.)
$AE = BE$ (given) ✓
$\therefore \Delta AED \equiv \Delta BEC$ SAS ✓
(2 marks)

b $\angle DAE = \angle CBE$
(corresp. $\angle$s cong. Δs)
$\therefore AD \parallel CB$
(alt. $\angle$s are equal) ✓
(1 mark)

4

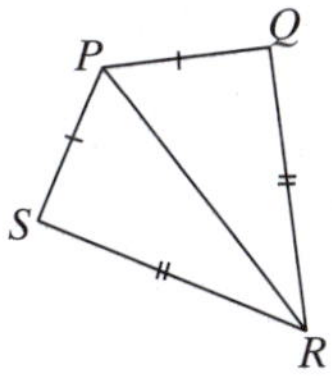

a In ΔPQR, ΔPSR
$PQ = PS$ (given)
$QR = SR$ (given)
$PR = PR$ (common side) ✓
$\therefore \Delta PQR \equiv \Delta PSR$ SSS ✓
(2 marks)

b $\angle QPR = \angle SPR$
(corresp. $\angle$s cong. Δs)
$\therefore$ PR bisects $\angle SPQ$ ✓
$\angle QRP = \angle SRP$
(corresp. $\angle$s cong. Δs)
$\therefore$ PR bisects $\angle SRQ$
$\therefore$ diagonal PR bisects the angles through which it passes. ✓ (2 marks)

5

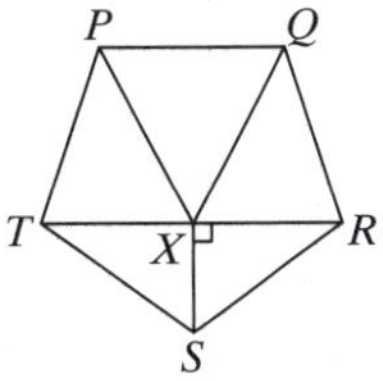

a In ΔTXS, ΔRXS
$\angle TXS = \angle RXS$
(both 90°, $SX \perp TR$)
$TS = RS$
(sides of regular pentagon)
$XS = XS$ (common side) ✓
$\therefore \Delta TXS \equiv \Delta RXS$ RHS ✓
(2 marks)

b Angle sum pentagon
$= 3 \times 180°$
$= 540°$
$\therefore \angle RST = 540° \div 5$
$= 108°$ ✓
$\angle TSX = \angle RSX$
(corresp. $\angle$s cong. Δs)
$\angle TSX = 108° \div 2$
$= 54°$ ✓ (2 marks)

c $\angle PTS = \angle QRS$
(angles of regular pentagon)
but $\angle STX = \angle SRX$
(corresp. $\angle$s cong. Δs)
$\therefore \angle PTX = \angle QRX$ ✓
(1 mark)

d In ΔPTX, ΔQRX
$PT = QR$
(sides of regular pentagon)
$\angle PTX = \angle QRX$
(from part **c**)
$TX = RX$
(corresp. sides cong. Δs) ✓
$\therefore \Delta PTX \equiv \Delta QRX$ SAS ✓
(2 marks)

e $PX = QX$
(corresp. sides cong. Δs)
$\therefore \Delta PQX$ is isosceles. ✓
(1 mark)

f $\angle QPT = 108°$
(angle of reg. pentagon)
$\angle PTS = 108°$
(angle of reg. pentagon)
$\angle STX + \angle TXS + \angle TSX$
$= 180°$
($\angle$ sum Δ)
$\angle STX + 90° + 54° = 180°$
$\therefore \angle STX = 36°$
$\angle PTX = 108° - 36°$
$= 72°$ ✓
$\angle PTX + \angle QPT = 72° + 108°$
$= 180°$
$\therefore PQ \parallel TR$ (co-interior angles supplementary) ✓
(2 marks)

6

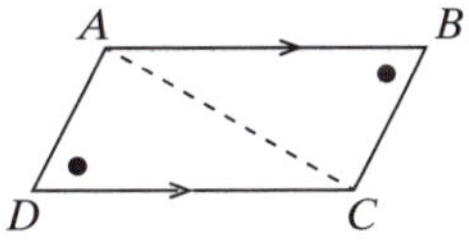

a In ΔABC, ΔCDA
$\angle ABC = \angle CDA$ (given)
$\angle BAC = \angle DCA$
(alt., $AB \parallel DC$)
$AC = CA$
(common side) ✓
$\therefore \Delta ABC \equiv \Delta CDA$ AAS ✓
(2 marks)

b $\angle BCA = \angle DAC$
(corresponding $\angle$s of congruent triangles)
$\therefore AD \parallel BC$
(alternate $\angle$s are equal) ✓
(1 mark)

7

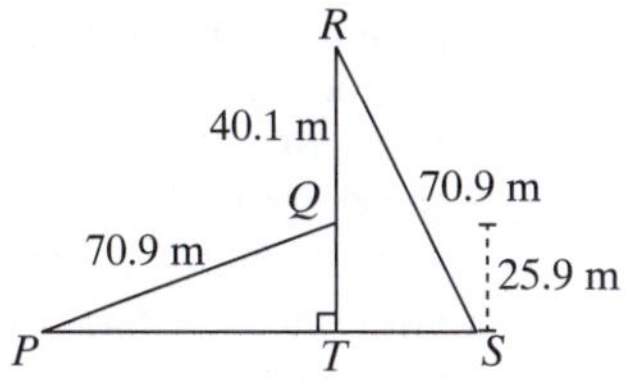

In ΔPQT,
$PT^2 = 70.9^2 - 25.9^2$
$= 4356$
$PT = \sqrt{4356}$ $(PT > 0)$
$= 66$
So the length of PT is 66 m. ✓
$RT = 40.1\text{ m} + 25.9\text{ m}$
$= 66\text{ m}$
In ΔPQT, ΔRST
$\angle PTQ = \angle RTS$
(both 90°, $RT \perp PS$)
$PQ = RS$ (given, both 70.9 m)
$PT = RT$ (both 66 m) ✓
$\therefore \Delta PQT \equiv \Delta RST$ RHS ✓
$\therefore \angle TPQ = \angle TRS$
(corresp. $\angle$s cong. Δs) ✓
(4 marks)
(Total: 30 marks)

LOGICAL REASONING SKILLS CHECK PAGE 66

1 **a**

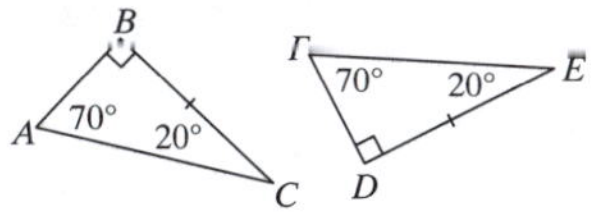

$\Delta ABC \equiv \Delta FDE$
[The vertices must be named in that order because $\angle A$ corresponds to $\angle F$, $\angle B$ to $\angle D$ and $\angle C$ to $\angle E$.]

b

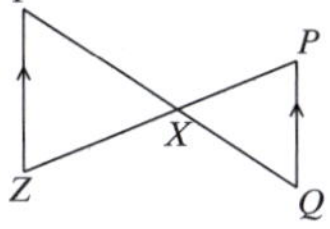

$\Delta XYZ \parallel\!\mid \Delta XQP$
[$\angle XYZ = \angle XQP$ alternate angles, parallel lines]

2 **a**

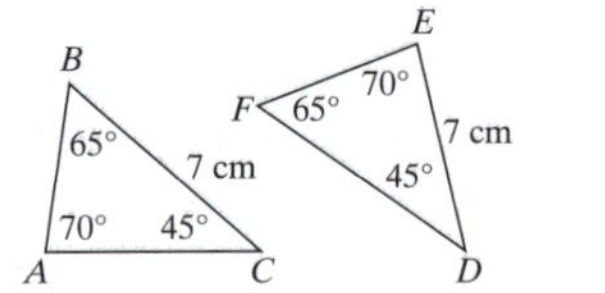

The angle sum of a triangle is 180°. So th missing angles can be found.
$\therefore \Delta ABC \ ||| \ \Delta EFD$
(equiangular)
[The triangles are not congruent because the equal sides are not corresponding sides (they are opposite different angles).]

b

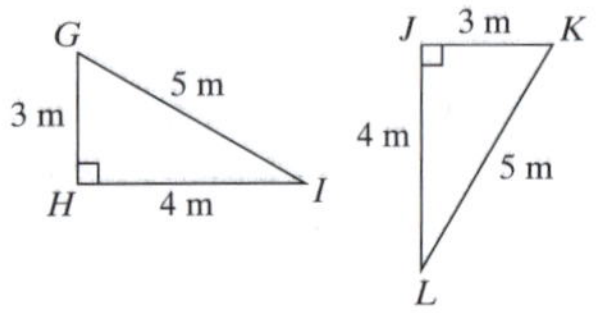

Using Pythagoras' Theorem the missing sides in each right-angled triangle can be found.
$\therefore \Delta GHI \equiv \Delta KJL$
(SSS, RHS or SAS)

c

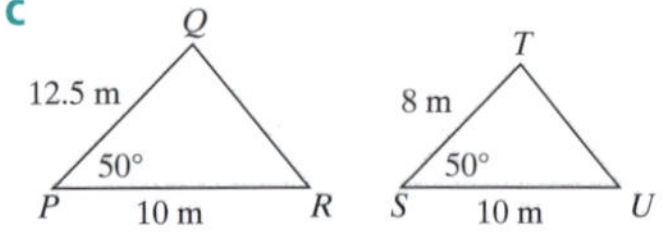

$\angle QPR = \angle UST$ (both 50°)

$\dfrac{ST}{PR} = \dfrac{8}{10} = 0.8$

$\dfrac{SU}{PQ} = \dfrac{10}{12.5} = 0.8$

$\therefore \dfrac{ST}{PR} = \dfrac{SU}{PQ}$

$\therefore \Delta STU \ ||| \ \Delta PRQ$
(two sides in proportion, included angles equal)

3 **a**

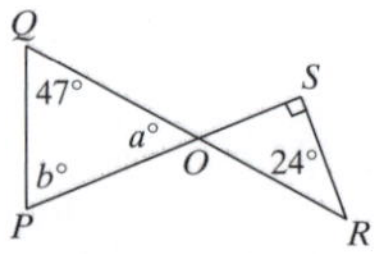

$\angle SOR + 90° + 24° = 180°$
(angle sum Δ)
$\angle SOR = 66°$
$\therefore \ a = 66$ (vertically opp.)
$b + 66 + 47 = 180$
(angle sum Δ)
$b = 67$

b

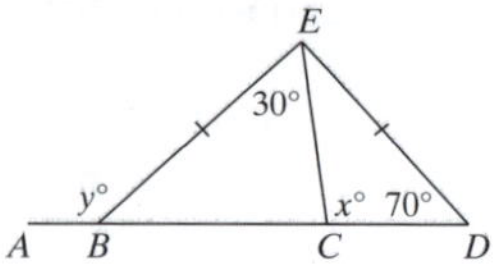

$\angle EBD = \angle EDB$
(base angles isosceles Δ)
$\therefore \angle EBD = 70°$
$\angle ECD = \angle EBC + \angle BEC$
(exterior angle of Δ)
$x = 100$
$y + 70 = 180$
($\angle$s in straight line)
$y = 110$

c

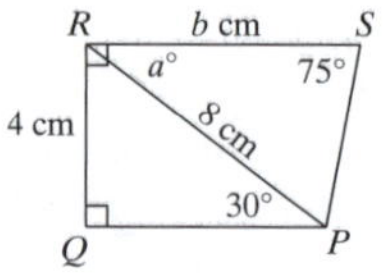

RS is parallel to QP
(both perpendicular to RQ)
$\therefore \angle SRP = \angle QPR$
(alternate angles, || lines)
$a = 30$
$\angle SPR + 30° + 75° = 180°$
(angle sum Δ)
$\angle SPR = 75°$
$\therefore RS = RP$
(sides opposite equal angles, isosceles triangle)
$b = 8$

d

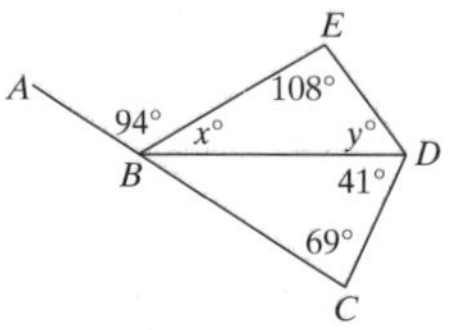

$\angle DBC + \angle BCD + \angle CDB$
$= 180°$ (angle sum Δ)
$\angle DBC + 69° + 41° = 180°$
$\angle DBC = 70°$
$94 + x + 70 = 180$
(angles in a straight line)
$x = 16$
$y + 108 + 16 = 180$
(angle sum Δ)
$y = 56$

4 **a** True. Every square is a rhombus.

b False. Every rhombus is not a square.

c True. Every rectangle is a parallelogram.

d True. Every rhombus is a parallelogram.

e True. Every parallelogram is a trapezium.

f False. Every trapezium is not a parallelogram.

g True. Every square is a parallelogram.

h True. Every square is a rectangle.

5 **a** $x + 27 + 105 + 127 = 360$
(angles in a quad.)
$x + 259 = 360$
$x = 101$

b $x = 6$ (opposite sides of a parallelogram)

c

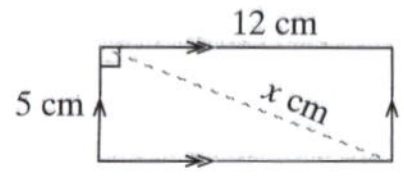

Opposite sides of a rectangle are equal.
By Pythagoras' Theorem:
$x^2 = 12^2 + 5^2$
$= 169$
$x = \sqrt{169}$
$= 13$

d $x = 15$ (diagonals of a rectangle are equal)

6 **a** $a + 69 = 180$
(co-interior angles, parallel lines)
$a = 111$
$b + 56 = 180$
(co-interior angles, parallel lines)
$b = 124$

b $x = 9$
(opposite sides of a parallelogram are equal)
$y = 72$
(opposite angles of a parallelogram are equal)
$(z + 28) + 72 = 180$
(co-interior angles, parallel lines)
$z + 100 = 180$
$z = 80$

LOGICAL REASONING
INTERMEDIATE TEST PAGE 67

1 $b^\circ + c^\circ = d^\circ$
So $c^\circ = d^\circ - b^\circ$
Now $a^\circ + b^\circ = c^\circ$
$a^\circ = c^\circ - b^\circ$
$= (d^\circ - b^\circ) - b^\circ$
$= d^\circ - 2b^\circ$
$\therefore$ **[A]** ✓ (1 mark)

2 The correct statement is "Every rhombus is a trapezium."
[Because every rhombus does have a pair of opposite sides that are parallel.]
$\therefore$ **[B]** ✓ (1 mark)

3 $\Delta AEB \parallel\!| \Delta ADC$ (equiangular)
$\frac{EB}{DC} = \frac{AB}{AC}$
$\frac{EB}{21} = \frac{18}{30}$
$EB = 12.6$
Let $AE = x$ cm
$\frac{AE}{AD} = \frac{AB}{AC}$
$\frac{x}{x + 15} = \frac{18}{30}$
$= \frac{3}{5}$
$5x = 3x + 45$
$2x = 45$
$x = 22.5$
So $EB = 12.6$ cm
and $AE = 22.5$ cm
$\therefore$ **[C]** ✓ (1 mark)

4

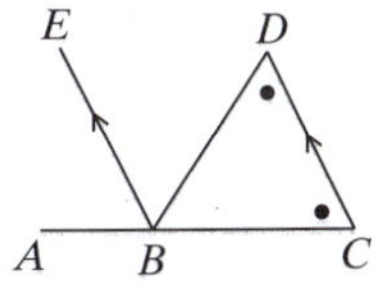

$\angle EBD = \angle BDC$
(alternate, $EB \parallel DC$)
$\angle ABE = \angle BCD$
(corresponding, $EB \parallel DC$) ✓
but $\angle BDC = \angle BCD$ (given)
$\therefore \angle EBD = \angle ABE$
$\therefore BE$ bisects $\angle ABD$ ✓
(2 marks)

5

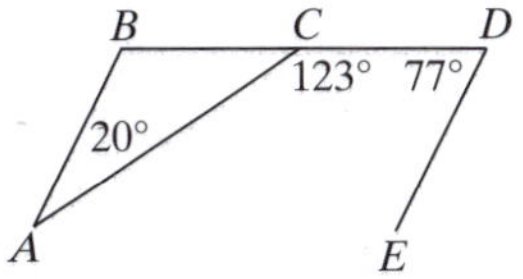

$\angle ACD = \angle ABC + \angle BAC$
(exterior angle ΔABC)
$\therefore \angle ABC = 123^\circ - 20^\circ$
$= 103^\circ$ ✓
$\angle ABC + \angle BDE = 103^\circ + 77^\circ$
$= 180^\circ$
$\therefore AB \parallel ED$
(co-interior angles add to 180°)
✓ (2 marks)

6

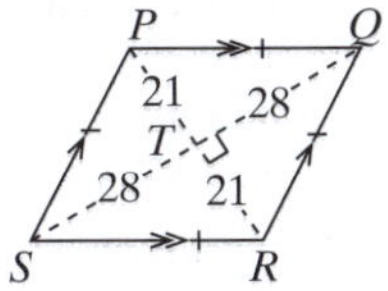

Let T be the point of intersection of the diagonals.
$PQRS$ is a rhombus.
$\therefore \angle QTR = 90^\circ$
(diagonals meet at 90°)
$QT = 28$ cm and $TR = 21$ cm ✓
(diagonals bisect each other)
Let x cm be the length of side QR.
$x^2 = 28^2 + 21^2$
$= 784 + 441$
$= 1225$
$x = \sqrt{1225}$
$= 35$
Each side of the rhombus is 35 cm long. ✓
Perimeter $= 4 \times 35$ cm
$= 140$ cm ✓
(3 marks)

7 **a**

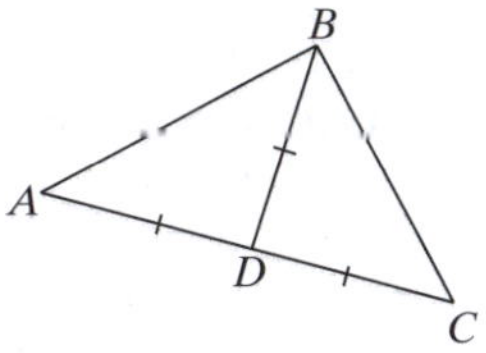

$\angle ADB + \angle CDB = 180^\circ$
(angles in a straight line) ✓
but $\angle ADB = \angle CDB$ (given)
$\therefore \angle ADB = 90^\circ$ ✓
(2 marks)

b $\angle BAD = 45^\circ$ ✓
$\angle BAD = \angle ABD$
($\angle$s opposite equal sides)
but $\angle BAD + \angle ABD = 90^\circ$
(sum of other angles of a right-angled triangle) ✓
(2 marks)

c $\angle ABD = \angle BAD = 45^\circ$
Similarly $\angle DCB = \angle CBD$
$= 45^\circ$
$\therefore \Delta ABC$ is isosceles ✓
$\angle ABC = 45^\circ + 45^\circ$
$= 90^\circ$
$\therefore \Delta ABC$ is right-angled ✓
So ΔABC is a right-angled, isosceles triangle. (2 marks)

8

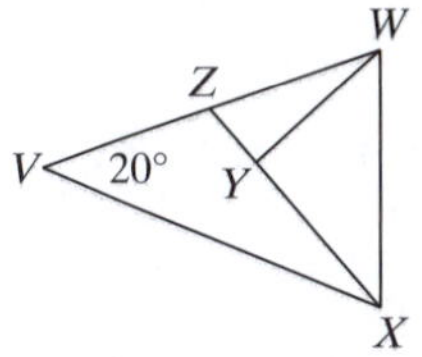

a $\angle VXZ = \angle XVZ$
(angles opposite equal sides, $VZ = ZX$)
$\therefore \angle VXZ = 20^\circ$ ✓
$\angle VXW = \angle VWX$
(angles opposite equal sides, $VX = VW$)
$\angle VXW + \angle VWX$
$+ \angle XVW = 180^\circ$
(angle sum ΔVXW)
$2\angle VXW + 20^\circ = 180^\circ$
$\angle VXW = 80^\circ$ ✓
$\angle YXW = 80^\circ - 20^\circ$
$= 60^\circ$ ✓
(3 marks)

b $\angle YWX = \angle YXW$
(angles opposite equal sides, $YX = YW$)
$\therefore \angle YWX = 60^\circ$ ✓
$\therefore \Delta YWX$ is equilateral
(all angles are 60°)
$\therefore YW = XW$ ✓ (2 marks)

9

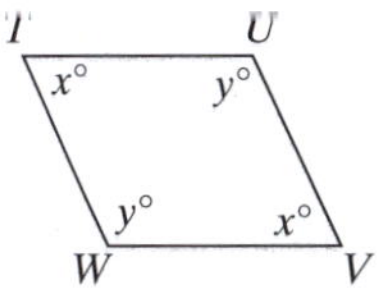

a $x + y + x + y = 360$
(angle sum of a quadrilateral) ✓
$2(x + y) = 360$
$x + y = 180$ ✓
(2 marks)

b $TU \parallel WV$
(co-interior angles are supplementary) ✓
$TW \parallel UV$
(co-interior angles are supplementary) ✓
(2 marks)
(Total: 25 marks)

LOGICAL REASONING
ADVANCED TEST PAGE 68

1

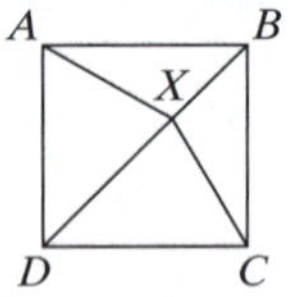

a In ΔABX, ΔCBX
$AB = CB$ (sides of square)
$\angle ABX = \angle CBX$
(diagonals bisect angles)
$BX = BX$ (common side) ✓
$\therefore \Delta ABX \equiv \Delta CBX$ SAS ✓
(2 marks)

b $\angle BCX = \angle BAX$
(corresp. $\angle$s cong. Δs)
$\therefore \angle BCX = 22°30'$
$\angle ABC = 90°$ ($\angle$ of square)
$\angle XBC = 45°$
(diagonals bisect angles) ✓
$\angle DXC = \angle XBC + \angle BCX$
(ext. $\angle$ of Δ)
$\angle DXC = 45° + 22°30'$
$= 67°30'$ ✓ (2 marks)

c $\angle BCD = 90°$ ($\angle$ of square)
$\angle XCD = 90° - 22°30'$
$= 67°30'$
$\therefore \angle DXC = \angle XCD$
$\therefore \Delta DXC$ is isosceles ✓
(1 mark)

2

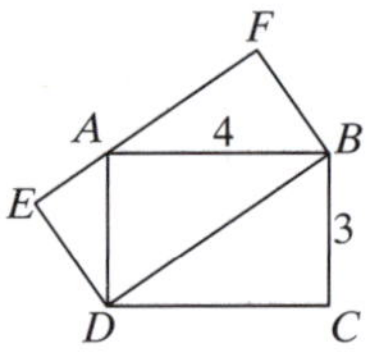

a $\angle FAB = \angle ABD$
(alternate, $AF \parallel DB$)
but $\angle ABD = \angle CDB$
(alternate, $AB \parallel DC$)
$\therefore \angle FAB = \angle CDB$ ✓
(1 mark)

b In ΔAFB, ΔDCB
$\angle FAB = \angle CDB$ (proven)
$\angle AFB = \angle DCB$
(both 90°, $\angle$s of rectangles) ✓
$\therefore \Delta AFB \parallel\!| \Delta DCB$
(equiangular) ✓ (2 marks)

c $DC = AB$ (opp. sides of rect.)
$DC = 4$ cm
$BD^2 = BC^2 + DC^2$
$= 3^2 + 4^2$
$= 25$
$\therefore BD = 5$ cm ✓

$\frac{FB}{CB} = \frac{AB}{DB}$ (sides of sim. Δs in prop.)
$\frac{FB}{3} = \frac{4}{5}$
$FB = 2.4$ cm ✓
Rectangle $ABCD$:
Area $= (4 \times 3)$ cm^2
$= 12$ cm^2
Rectangle $EFBD$:
Area $= (5 \times 2.4)$ cm^2
$= 12$ cm^2
$\therefore$ both rectangles have the same area. ✓ (3 marks)

3

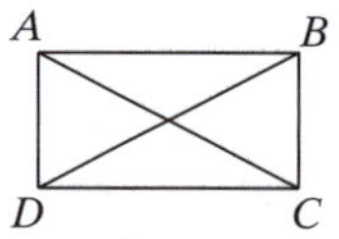

In ΔABD, ΔBAC
$AD = BC$
(opp. sides of rectangle)
$\angle BAD = \angle ABC$
(both 90°, $\angle$s of rectangle)
$AB = BA$ (common side) ✓
$\therefore \Delta ABD \equiv \Delta BAC$ SAS ✓
$\therefore BD = AC$
(corresp. sides cong. Δs)
$\therefore$ the diagonals are equal. ✓
(3 marks)

4

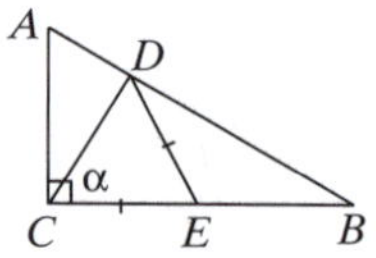

Let $\angle DCE = \alpha$
$\angle CDE = \angle DCE$
($\angle$s opp. equal sides isos. Δ)
$\therefore \angle CDE = \alpha$
$\angle ACD + \angle DCE = 90°$
$\therefore \angle ACD = 90° - \alpha$
$\therefore \angle EDB = 90° - \alpha$
($\angle ACD = \angle EDB$, given)
$\angle CDB = \angle CDE + \angle EDB$
$= \alpha + 90° - \alpha$
$= 90°$ ✓
$\angle DBC + \angle BCD + \angle CDB = 180°$ ($\angle$ sum Δ)
$\angle DBC + \alpha + 90° = 180°$
$\angle DBC = 90° - \alpha$
$\therefore \angle DBC = \angle EDB$ ✓
$\therefore BE = DE$
(sides opp. equal $\angle$s isos. Δ)
But $DE = CE$
$\therefore BE = CE$
$\therefore E$ is the midpoint of BC. ✓
(3 marks)

5

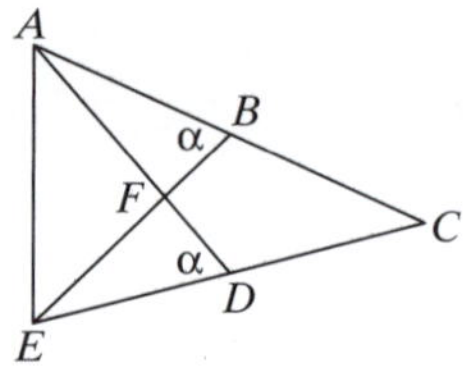

$\angle ABF + \angle FBC = 180°$
($\angle$s straight line)
$\angle FBC = 180° - \alpha$
$\angle FDE + \angle FDC = 180°$
($\angle$s straight line)
$\angle FDC = 180° - \alpha$ ✓
$\angle BFD + \angle FDC + \angle BCD + \angle FBC = 360°$
($\angle$ sum quadrilateral)
$\angle BFD + 180° - \alpha + \angle BCD + 180° - \alpha = 360°$
$\angle BFD + \angle BCD = 2\alpha$
But $\angle AFE = \angle BFD$
(vert. opp.)
$\therefore \angle AFE + \angle BCD = 2\alpha$ ✓
(2 marks)

6

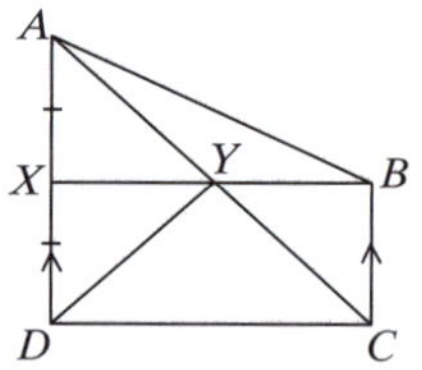

In ΔAXY, ΔDXY
$AX = DX$
(X is midpoint of AD)
$\angle AXY = \angle DXY$
(both 90°, $AD \parallel BC$, $XB \perp BC$) ✓
$XY = XY$ (common side)
$\therefore \Delta AXY \equiv \Delta DXY$ SAS ✓
$\therefore \angle XYA = \angle XYD$
(corresp $\angle$s cong. Δs)
but $\angle XYA = \angle BYC$
(vert. opp.)
$\therefore \angle XYD = \angle BYC$ ✓
(3 marks)

7

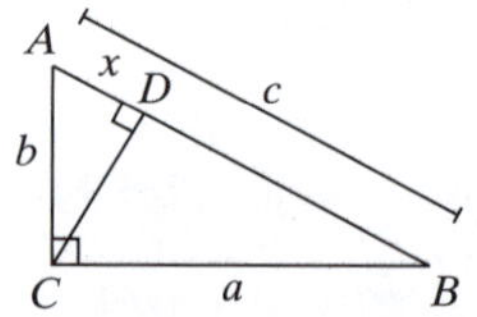

a In ΔADC, ΔACB
$\angle DAC = \angle CAB$
(common angle)
$\angle ADC = \angle ACB$
(both 90°, given) ✓
$\therefore \Delta ADC \parallel\!| \Delta ACB$
(equiangular) ✓ (2 marks)

b $\dfrac{AC}{AB} = \dfrac{AD}{AC}$
(corresponding sides, similar triangles) ✓
$\dfrac{b}{c} = \dfrac{x}{b}$
$b^2 = cx$ ✓ (2 marks)

c In ΔBDC, ΔBCA
$\angle DBC = \angle CBA$
(common angle)
$\angle CDB = \angle ACB$
(both 90°, given) ✓
$\therefore \Delta BDC \;|||\; \Delta BCA$
(equiangular) ✓ (2 marks)

d $AB = c$, $AD = x$,
$\therefore BD = c - x$
$\dfrac{BC}{BA} = \dfrac{BD}{BC}$
(corresponding sides, similar Δs)
$\dfrac{a}{c} = \dfrac{c-x}{a}$ ✓ (1 mark)

e $a^2 = c^2 - cx$ (from part **d**)
but $cx = b^2$ (from part **b**)
$\therefore a^2 = c^2 - b^2$
$\therefore c^2 = a^2 + b^2$ ✓ (1 mark)
(Total: 30 marks)

ANGLES OF ELEVATION AND DEPRESSION, AND BEARINGS
SKILLS CHECK PAGE 70

1 **a** $\sin 29°37'$
$= 0.494194771\ldots$
$= 0.4942$ (4 d.p.)

b $\tan 56°21'$
$= 1.502275145\ldots$
$= 1.5023$ (4 d.p.)

c $\cos 72°40'$
$= 0.297930281\ldots$
$= 0.2979$ (4 d.p.)

2 **a** $\cos\theta = 0.35$
$\theta = 69.5126848\ldots°$
$= 69°31'$
(nearest minute)

b $\tan\theta = 1.7$
$\theta = 59.53445508\ldots°$
$= 59°32'$
(nearest minute)

c $\sin\theta = 0.2916$
$\theta = 16.9537699\ldots°$
$= 16°57'$
(nearest minute)

3 **a**

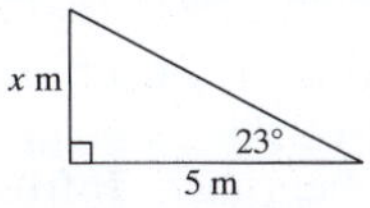

$\tan 23° = \dfrac{x}{5}$
$x = 5 \times \tan 23°$
$= 2.122374\ldots$
$= 2.1$ (1 d.p.)

b

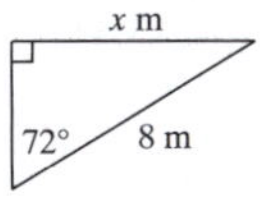

$\sin 72° = \dfrac{x}{8}$
$x = 8 \times \sin 72°$
$= 7.60845\ldots$
$= 7.6$ (1 d.p.)

c

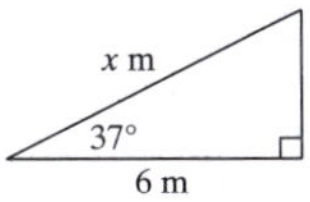

$\cos 37° = \dfrac{6}{x}$
$x = \dfrac{6}{\cos 37°}$
$= 7.5128139\ldots$
$= 7.5$ (1 d.p.)

4 **a**

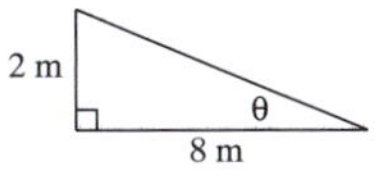

$\tan\theta = \dfrac{2}{8}$
$\theta = 14.03624\ldots°$
$= 14°$
(nearest degree)
[Using $\tan^{-1}$ on a calculator]

b

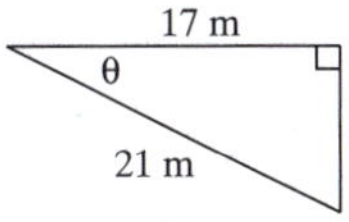

$\cos\theta = \dfrac{17}{21}$
$\theta = 35.950567\ldots°$
$= 36°$
(nearest degree)

c

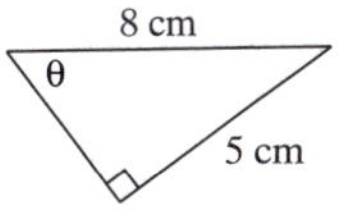

$\sin\theta = \dfrac{5}{8}$
$\theta = 38.682187\ldots°$
$= 39°$
(nearest degree)

5 Angles of elevation and depression are always measured from the **horizontal**.

6

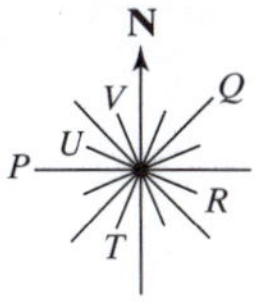

a P West
b Q NE [halfway between N and E]
c R ESE [halfway between E and SE]
d T SSW
e U WNW
f V NNW

7 **a** 058°

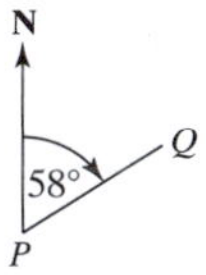

b 135°

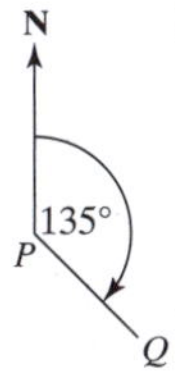

c

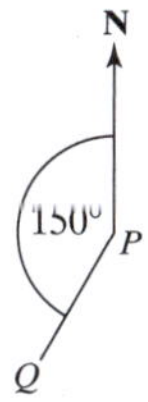

$360° - 150° = 210°$
$\therefore$ the bearing of Q from P is 210°

d

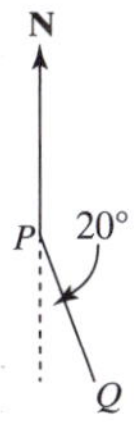

$180° - 20° = 160°$
$\therefore$ the bearing of Q from P is 160°

e

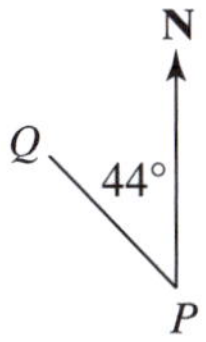

$360° - 44° = 316°$
$\therefore$ the bearing of Q from P is 316°

f

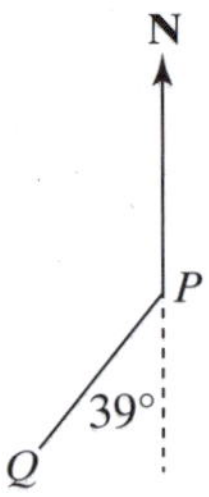

$180° + 39° = 219°$

$\therefore$ the bearing of Q from P is 219°

g

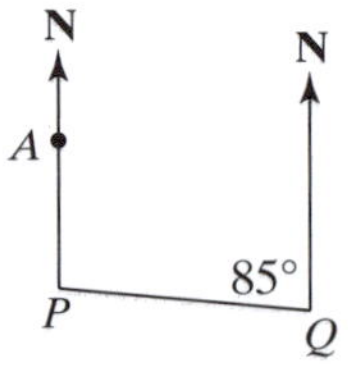

$\angle APQ + 85° = 180°$

(co-interior angles, || lines)

$\angle APQ = 95°$

$\therefore$ the bearing of Q from P is 095°

h

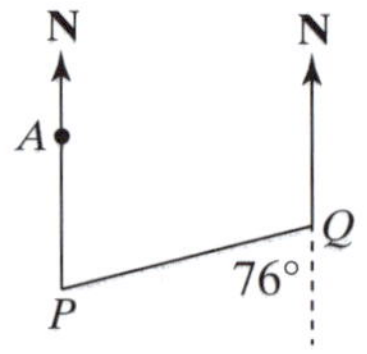

$\angle APQ = 76°$

(alternate angles, || lines)

$\therefore$ the bearing of Q from P is 076°

ANGLES OF ELEVATION AND DEPRESSION, AND BEARINGS
INTERMEDIATE TEST PAGE 71

1

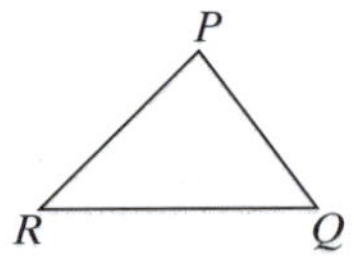

[Straight back does not mean due north or at right angles. The distance PR is greater than the distance PQ and this only happens in triangle D.]

$\therefore$ **[D]** ✓ (1 mark)

2

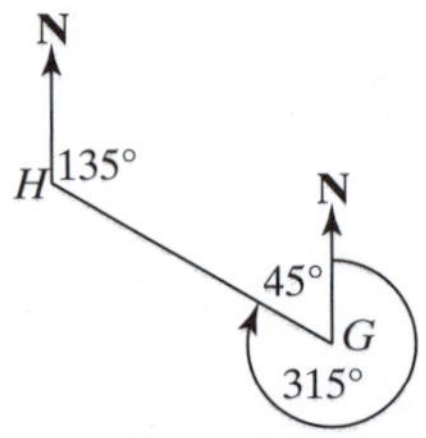

G is south-east of H so the bearing of G from H is 135°. [90° + 45°]

The bearing of H from G is 315°. [360° – 45°]

$\therefore$ **[D]** ✓ (1 mark)

3

$\angle BAC + \angle ABC + \angle BCA = 180°$ (angle sum of Δ)

$\angle BAC + 110° + 30° = 180°$

$\angle BAC + 140° = 180°$

$\angle BAC = 40°$

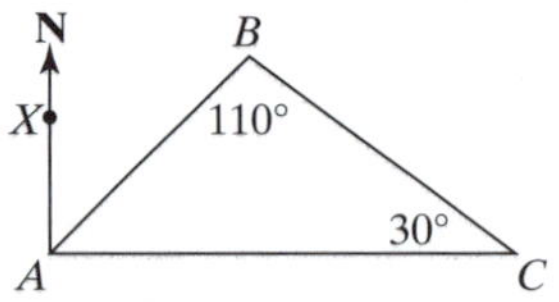

Let X be a point due north of A.

$\angle XAC = 90°$ (C is due east of A)

$\angle XAB = 90° - 40°$

$= 50°$

The bearing of B from A is 050°.

$\therefore$ **[C]** ✓ (1 mark)

4

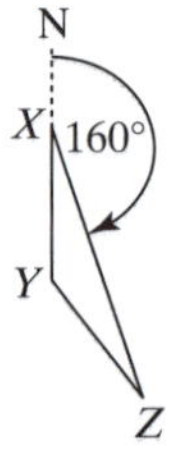

Y is north-west of Z so Z is south-east of Y.

$\angle XYZ = 135°$

[SE is halfway between south and east so its bearing is (90 + 45)° = 135°]

$\angle XYZ + \angle XZY = 160°$ (exterior angle of Δ)

$\angle XZY = 160° - 135°$

$= 25°$

$\therefore$ **[B]** ✓ (1 mark)

5

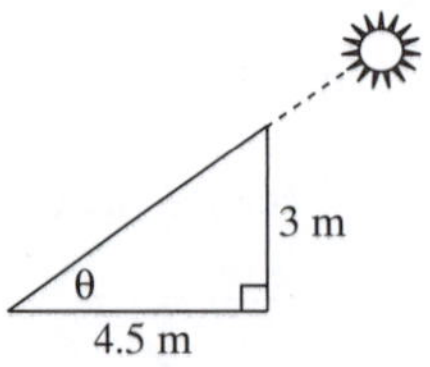

Let θ be the angle of elevation.

$\tan\theta = \dfrac{3}{4.5}$ ✓

$\theta = 33.690\,067\,53\ldots°$ ✓

$= 34°$ (nearest degree)

The angle of elevation is 34°. ✓ (3 marks)

6

$\angle ACB = 90° - 29°$

$= 61°$

Let the distance of the boat from the base of the cliff be x m.

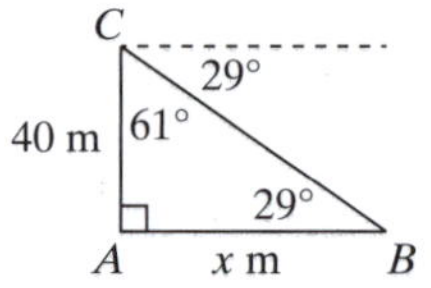

$\tan 61° = \dfrac{x}{40}$ ✓

$x = 40 \times \tan 61°$

$= 72.161\,910\,21\ldots$ ✓

$= 72$ (nearest whole number)

The boat is 72 m from the cliff, to the nearest metre. ✓

[Or: $\angle CBA = 29°$

(alternate angles, parallel lines)

$\tan 29° = \dfrac{40}{x}$

$x = \dfrac{40}{\tan 29°}$

$= 72.161\,910\,21\ldots$] (3 marks)

7

$\angle JLK = 360° - 304°$

(angles in a revolution)

$= 56°$ ✓

Let the distance from J to L be x km.

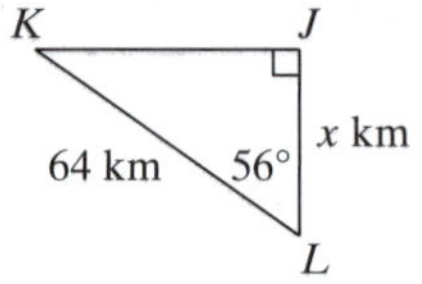

$\cos 56° = \dfrac{x}{64}$ ✓

$x = 64 \times \cos 56°$

$= 35.788\,345\,82\ldots$

$= 35.8$ (1 d.p.)

L is approximately 35.8 km south of J. ✓ (3 marks)

8

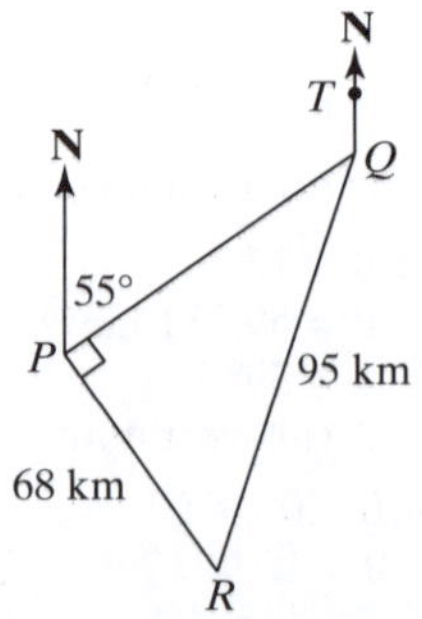

a $\angle QPR = 145° - 55°$

$= 90°$ ✓ (1 mark)

b $\sin Q = \dfrac{68}{95}$ ✓

$Q = 45.707\,9368\ldots°$ ✓

$= 46°$

(nearest degree)

$\angle PQR = 46°$ to the nearest degree. ✓ (3 marks)

c Let T be a point due north of Q.

$\angle PQT = 180° - 55°$

(co-interior angles, || lines)

$= 125°$ ✓

$\angle RQT = 46° + 125°$

$= 171°$ ✓

The bearing of R from Q

$= 360° - 171°$

$= 189°$

(angles in a revolution) ✓

[Or: Let T be a point south of Q.

$\angle PQT = 55°$ (alternate angles)

$\angle TQR = 55° - 46° = 9°$.

So the bearing is $180° + 9°$]

(3 marks)

(Total: 20 marks)

ANGLES OF ELEVATION AND DEPRESSION, AND BEARINGS
ADVANCED TEST PAGE 72

1 **a** Q is north west of P.

Bearing of Q from P

$= (360 - 45)°$

$= 315°$ ✓ (1 mark)

b

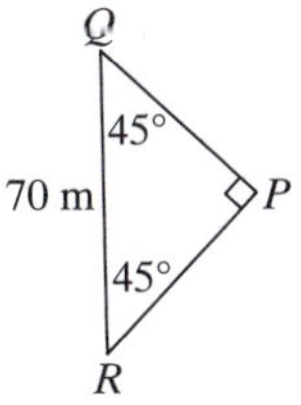

✓ (1 mark)

c $\sin 45° = \dfrac{PR}{70}$

$PR = 70 \sin 45°$ ✓

$= 49.497\,47\ldots$

$= 49$ m (nearest m) ✓

(2 marks)

2 **a** Let the distance between the buildings be x m.

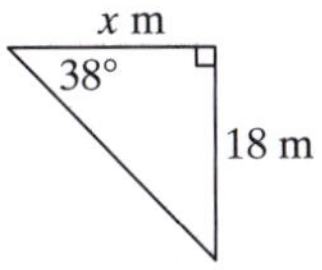

$\tan 38° = \dfrac{18}{x}$

$x = \dfrac{18}{\tan 38°}$ ✓

$= 23.0389\ldots.$

The distance between the buildings is 23 m to the nearest metre. ✓ (2 marks)

b Let y m be the difference in height between the two buildings.

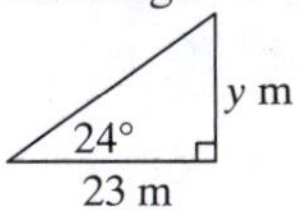

$\tan 24° = \dfrac{y}{23.0389\ldots}$

$y = 23.0389\ldots \times \tan 24°$ ✓

$= 10.2576\ldots$

To the nearest metre, the second building is 10 m taller than the first.

The height of the second building is 28 m to the nearest metre. ✓ (2 marks)

3 **a** Let X be a point due north of A and let Y be a point due north of B.

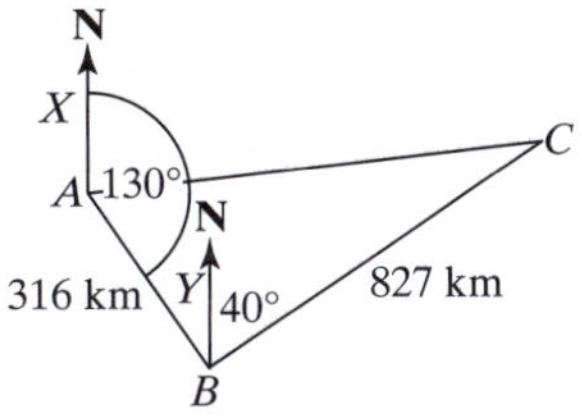

$\angle XAB + \angle ABY = 180°$

(co-int. ∠s, parallel lines)

$\angle ABY = 180° - 130°$

$= 50°$

$\angle ABC = 50° + 40°$

$= 90°$ ✓ (1 mark)

b $\tan A = \dfrac{827}{316}$

$A = \tan^{-1}\left(\dfrac{827}{316}\right)$ ✓

$= 69.0879\ldots°$

$= 69°$ (nearest degree)

✓ (2 marks)

c $AC^2 = 316^2 + 827^2$

$= 783\,785$

$AC = \sqrt{783\,785}$ $(AC > 0)$

$= 885.316\ldots$

$= 885$ km (nearest km)

✓ (1 mark)

d $\angle XAC = 130° - 69°$

$= 61°$

Let Z be a point due south of C.

$\angle ZCA = 61°$ (alternate, parallel lines)

Bearing of A from C

$= 180° + 61°$

$= 241°$ ✓ (1 mark)

4 **a**

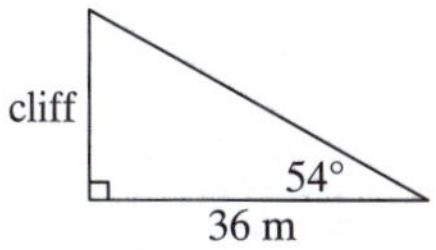

Let x m be the height of the cliff.

$\tan 54° = \dfrac{x}{36}$

$x = 36 \tan 54°$ ✓

$= 49.5497\ldots$

$= 50$ (nearest unit)

The cliff is 50 m high to the nearest metre. ✓ (2 marks)

b 2×36 m $= 72$ m

Let θ be the angle of elevation when the boat is 72 m from the cliff.

$\tan\theta = \dfrac{49.5497\ldots}{72}$

$\theta = \tan^{-1}\left(\dfrac{49.5497\ldots}{72}\right)$ ✓

$= 34.53539\ldots°$

$= 35°$ (nearest degree)

The angle of elevation would be 35° to the nearest degree. ✓ (2 marks)

5

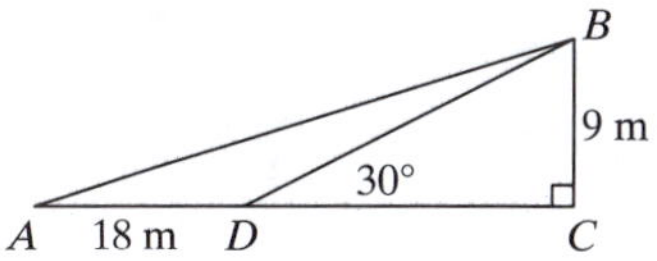

a In ΔBCD,

$\sin 30° = \dfrac{9}{BD}$

$BD = \dfrac{9}{\sin 30°}$

$= 18$ m ✓ (1 mark)

b AD and BD are both 18 m so ΔABD is isosceles and $\angle BAD = \angle ABD$.

$\angle BDC$ is an exterior angle of $\angle BAD$.

So $\angle BAD + \angle ABD = 30°$

$\therefore\ \angle BAD = 30° \div 2$

$= 15°$ ✓ (1 mark)

c In ΔABC,

$\sin 15° = \frac{9}{AB}$

$AB = \frac{9}{\sin 15°}$ ✓

$= 34.773\,329\ldots$

$= 34.8$ (1 d.p.)

So the length of AB is 34.8 m to the nearest 10 cm.

✓ (2 marks)

6 **a**

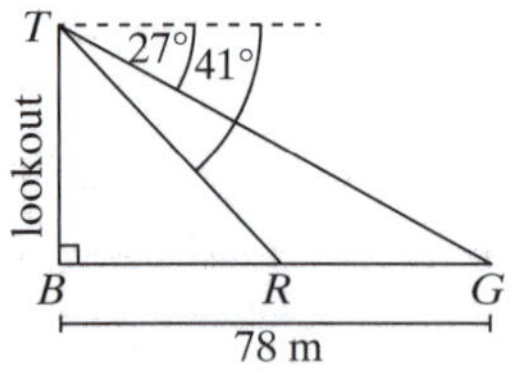

$\angle TGB = 27°$ (alt. $\angle$s, || lines)

$\tan 27° = \frac{TB}{78}$

$TB = 78 \tan 27°$ ✓

$= 39.7429\ldots$

$= 40$ m (nearest m)

The lookout is 40 m high, to the nearest metre. ✓

(2 marks)

b $\angle BTR = 90° - 41°$

$= 49°$

$\tan 49° = \frac{BR}{39.7429\ldots}$

$BR = 39.7429\ldots \times \tan 49°$ ✓

$= 45.719\ldots$

$= 46$ m (nearest m)

Now $78 - 46 = 32$

So, the distance between the red and green markers is 32 m to the nearest metre.

✓ (2 marks)

7 **a**

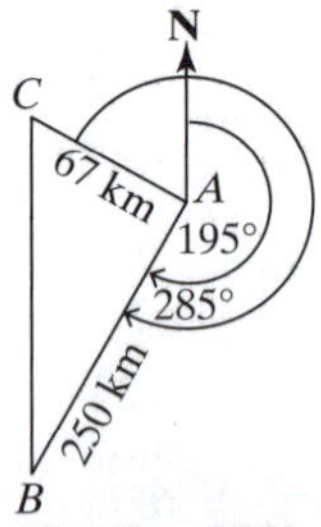

✓ (1 mark)

b $\angle BAC = 285° - 195°$

$= 90°$ ✓ (1 mark)

c $\tan B = \frac{67}{250}$

$B = \tan^{-1}\left(\frac{67}{250}\right)$ ✓

$= 15.002\,711\,54\ldots°$

$= 15°$ (nearest degree) ✓

Let X be a point due south of A.

$\angle XAB = 195° - 180°$

$= 15°$

So $\angle CBA = \angle XAB$

$\therefore CB \parallel AX$ (alt. $\angle$s equal)

$\therefore B$ is due south of C. ✓

(3 marks)

(Total: 30 marks)

COMPOUND EVENTS
SKILLS CHECK PAGE 74

1 **a** Number of counters

$= 7 + 5 + 4$

$= 16$

$P(\text{blue}) = \frac{7}{16}$

b $P(\text{red}) = \frac{5}{16}$

c $P(\text{white}) = \frac{4}{16}$

$= \frac{1}{4}$

d $P(\text{not blue}) = 1 - P(\text{blue})$

$= 1 - \frac{7}{16}$

$= \frac{9}{16}$

2 **a** $P(\text{red}) = \frac{26}{52}$

$= \frac{1}{2}$

b $P(\text{club}) = \frac{13}{52}$

$= \frac{1}{4}$

c $P(\text{ace}) = \frac{4}{52}$

$= \frac{1}{13}$

d There is only one queen of hearts.

$P(\text{queen of hearts}) = \frac{1}{52}$

e There are two red sevens.

$P(\text{red } 7) = \frac{2}{52}$

$= \frac{1}{26}$

3 **a** There are 4 cards.

$P(\text{A}) = \frac{1}{4}$

b $P(\text{D}) = \frac{1}{4}$

c $P(\text{E}) = 0$ [There is no E. It is impossible.]

d $P(\text{B or C}) = \frac{2}{4}$

$= \frac{1}{2}$

e $P(\text{A, B, C or D}) = 1$

[It is certain.]

4 $P(\text{not germinate})$

$= 1 - P(\text{will germinate})$

$= 1 - 92\%$

$= 8\%$

[1 whole = 100%]

5 **a** $P(\text{red}) = \frac{1}{2}$

$P(\text{not red}) = 1 - \frac{1}{2}$

$= \frac{1}{2}$

b $P(\text{club}) = \frac{1}{4}$

$P(\text{not a club}) = 1 - \frac{1}{4}$

$= \frac{3}{4}$

c $P(\text{ace of spades}) = \frac{1}{52}$

$P(\text{not ace of spades})$

$= 1 - \frac{1}{52}$

$= \frac{51}{52}$

6 **a** The events are independent. [It doesn't matter what number appears on the first die, it will not affect the second die.]

b The events are dependent. [The chance of getting a red jelly bean second, say, will depend on the colour of the first jelly bean.]

7

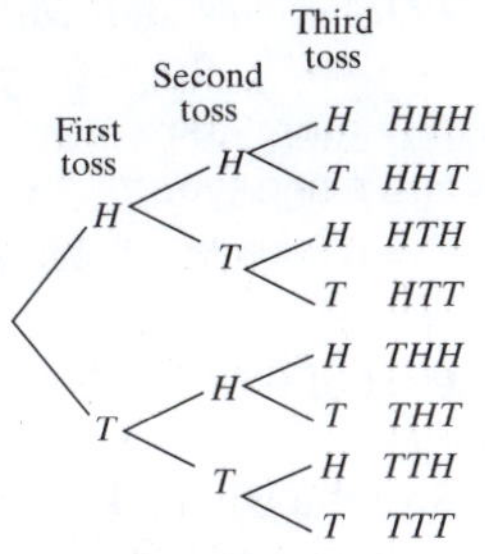

a $P(HHH) = \frac{1}{8}$

b $P(TTT) = \frac{1}{8}$

c $P(\text{exactly 2 heads}) = \frac{3}{8}$

d $P(\text{exactly one head}) = \frac{3}{8}$

e $P(\text{at least one head}) = \frac{7}{8}$

8 $P(H) = \frac{1}{2},\ P(5) = \frac{1}{6}$

$P(H \text{ and } 5) = \frac{1}{2} \times \frac{1}{6}$

$= \frac{1}{12}$

9 **a** $P(\text{white}) = \frac{4}{9}$

b 8 tees remain, 3 of which are white.

$P(\text{white}) = \frac{3}{8}$

c $P(\text{white}) = \frac{4}{9}$

d $P(WW) = \frac{4}{9} \times \frac{3}{8}$

$= \frac{1}{6}$

10 **a** $P(4) = \frac{1}{6}$

b $P(4) = \frac{1}{4}$

[The sample space is 3, 4, 5, 6.]

c $P(4) = \frac{1}{6}$

[The events are independent.]

COMPOUND EVENTS
INTERMEDIATE TEST PAGE 75

1 There are two red fives in a standard pack.

$P(\text{red } 5) = \frac{2}{52}$

$= \frac{1}{26}$

$\therefore$ [B] ✓ (1 mark)

2 $P(\text{not picking a } 7) = 1 - P(7)$

$= 1 - \frac{2}{5}$

$= \frac{3}{5}$

$\therefore$ [C] ✓ (1 mark)

3 $P(R \text{ and } B) = P(R) \times P(B)$

$= \frac{3}{5} \times \frac{1}{3}$

$= \frac{1}{5}$

$\therefore$ [A] ✓ (1 mark)

4 $P(HTH) = P(H) \times P(T) \times P(H)$

$= \frac{1}{2} \times \frac{1}{2} \times \frac{1}{2}$

$= \frac{1}{8}$

$\therefore$ [A] ✓ (1 mark)

5 The statement is false. ✓ The statement would only be true if winning a prize or not winning a prize were equally likely but if you only buy one ticket you only have one chance, out of the total number of tickets, of winning a particular prize. ✓

(2 marks)

6 Number of jellybeans

$= 2 + 3 + 1 + 4$

$= 10$

[If there are no black jellybeans left in the bowl each person must have taken a black one.]

$P(BBB) = \frac{3}{10} \times \frac{2}{9} \times \frac{1}{8}$ ✓

$= \frac{1}{120}$ ✓ (2 marks)

7 **a** $P(\text{first prize}) = \frac{5}{100}$

$= \frac{1}{20}$ ✓ (1 mark)

b $P(\text{all prizes})$

$= \frac{5}{100} \times \frac{4}{99} \times \frac{3}{98}$ ✓✓

$= \frac{1}{16\,170}$ ✓ (3 marks)

8 The statement is false. ✓ The probability of a family of three children being three girls is much less than that of a family having, in any order, two girls and a boy. But the birth of each child is an independent event and once the first two are girls, the third child has an (almost) equal chance of being a girl or a boy. ✓ (2 marks)

9 **a**

×	1	2	3	4	5	6
1	1	2	3	4	5	6
2	2	4	6	8	10	12
3	3	6	9	12	15	18
4	4	8	12	16	20	24
5	5	10	15	20	25	30
6	6	12	18	24	30	36

✓

$P(\text{odd score}) = \frac{9}{36}$

$= \frac{1}{4}$ ✓

$P(\text{even score} > 16) = \frac{9}{36}$

$= \frac{1}{4}$

Both Kelly and Chris have an equal chance of winning. ✓

(3 marks)

b $P(\text{neither wins}) = 1 - (\frac{1}{4} + \frac{1}{4})$

$= \frac{1}{2}$

$\therefore P(\text{second roll}) = \frac{1}{2}$ ✓

$P(\text{Chris wins on second roll})$

$= \frac{1}{2} \times \frac{1}{4}$

$= \frac{1}{8}$ ✓ (2 marks)

c $P(\text{no-one has won after 3 rolls}) = \frac{1}{2} \times \frac{1}{2} \times \frac{1}{2}$

$= \frac{1}{8}$ ✓ (1 mark)

(Total: 20 marks)

COMPOUND EVENTS
ADVANCED TEST PAGE 76

1 **a** Number of pink

$= 18 - 12$

$= 6$

$P(\text{first is pink}) = \frac{6}{18}$

$= \frac{1}{3}$ ✓ (1 mark)

b $P(\text{third is pink}) = \frac{6}{18}$

$= \frac{1}{3}$

[Every marshmallow has an equal chance of being chosen third.] ✓ (1 mark)

c $P(\text{all are pink})$

$= \frac{6}{18} \times \frac{5}{17} \times \frac{4}{16}$

$= \frac{5}{204}$ ✓ (1 mark)

d $P(\text{all are white})$

$= \frac{12}{18} \times \frac{11}{17} \times \frac{10}{16}$

$= \frac{55}{204}$ ✓

$P(\text{at least 1 is pink})$

$= 1 - P(\text{all are white})$

$= 1 - \frac{55}{204}$

$= \frac{149}{204}$ ✓ (2 marks)

2 **a**

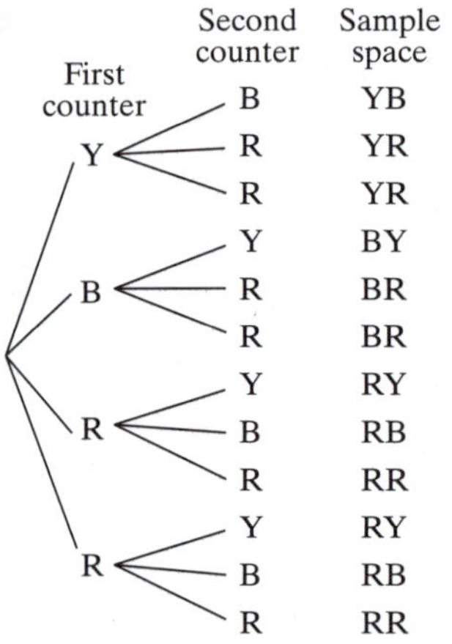

✓✓ (2 marks)

b $P(RR) = \frac{2}{12} = \frac{1}{6}$ ✓ (1 mark)

c Because the counters are not replaced it is not possible to get two blue counters or two yellow counters. The only possibility of having two counters of the same colour is if they are both red. ✓ (1 mark)

d If the first counter was red there are now only six different possibilities for the two counters. There are now two chances in six, or one chance in three of getting two red counters. ✓ (1 mark)

e If the first counter is replaced there are now another four possibilities; YY, BB, RR and RR. There is now a better chance that both counters are red.

$P(RR) = \frac{4}{16} = \frac{1}{4}$ ✓ (1 mark)

3 **a** $P(\text{no six}) = \frac{5}{6} \times \frac{5}{6} \times \frac{5}{6}$

$= \frac{125}{216}$ ✓

$P(\text{at least 1 six}) = 1 - \frac{125}{216}$

$= \frac{91}{216}$ ✓ (2 marks)

b $P(\text{same number each time})$

$= \frac{1}{6} \times \frac{1}{6}$

$= \frac{1}{36}$

[It doesn't matter which number is rolled first; there is one chance in six that the same number will occur in the second roll and, for each of those, one chance in six that it will occur again in the third roll.] ✓ (1 mark)

c $P(\text{five each time})$

$= \frac{1}{6} \times \frac{1}{6} \times \frac{1}{6}$

$= \frac{1}{216}$ ✓ (1 mark)

d Neither is correct. The events are independent so the result of one roll will not affect the result of another. Because the die is fair, it is equally likely to show 5 as any other number. ✓ (1 mark)

4 **a** $P(\text{John wins second prize})$

$= \frac{10}{500} \times 100\%$

$= 2\%$

[Every ticket has an equal chance of winning second prize.] ✓ (1 mark)

b $P(\text{John wins all prizes})$

$= \frac{10}{500} \times \frac{9}{499} \times \frac{8}{498} \times 100\%$

$= 0.000\,579\,47\ldots\%$

$= 0.0006\%$ (1 sig.fig.) ✓ (1 mark)

c Cost of tickets $= 10 \times \$1$

$= \$10$

So John will make a profit if he wins any prize.

$P(\text{not winning a prize})$

$= \frac{490}{500} \times \frac{489}{499} \times \frac{488}{498}$

$= 0.941\,07\ldots$ ✓

$P(\text{winning at least 1 prize})$

$= 1 - 0.941\,27\ldots$

$= 0.058\,923\ldots$

$= 5.9\%$ (2 sig figs) ✓ (2 marks)

5

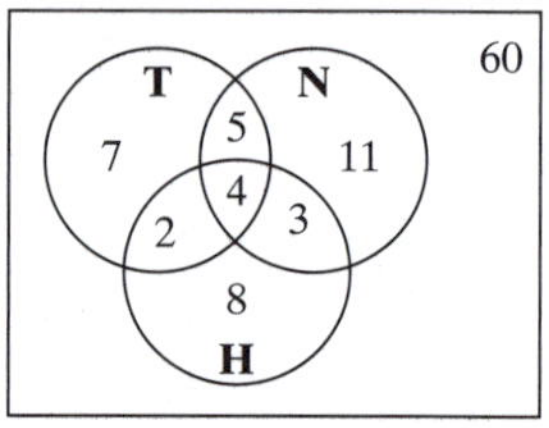

a Number representing in netball

$= 11 + 5 + 4 + 3$

$= 23$

Total students in group

$= 23 + 7 + 2 + 8 + 60$

$= 100$

$P(\text{students represented in netball}) = \frac{23}{100}$ ✓ (1 mark)

b $5 + 4 = 9$ students who represented in netball also represented in tennis.

$P(\text{netball representative also represented in tennis})$

$= \frac{9}{23}$ ✓ (1 mark)

c Only two students represented the school in both tennis and hockey but not netball. So it would be impossible to select three such students. ✓ (1 mark)

d Number representing in hockey

$= 4 + 3 + 8 + 2$

$= 17$

P(three represent in hockey)

$= \frac{17}{100} \times \frac{16}{99} \times \frac{15}{98}$

$= \frac{34}{8085}$ ✓ (1 mark)

6 a

×	1	2	3	4	5	6
1	1	2	3	4	5	6
2	2	4	6	8	10	12
3	3	6	9	12	15	18
4	4	8	12	16	20	24
5	5	10	15	20	25	30
6	6	12	18	24	30	36

✓ (1 mark)

b $P(4) = \frac{3}{36} = \frac{1}{12}$ ✓ (1 mark)

c P(4 three times)

$= \frac{1}{12} \times \frac{1}{12} \times \frac{1}{12}$

$= \frac{1}{1728}$ ✓ (1 mark)

d $P(\text{score} < 10) = \frac{17}{36}$

$P(\text{score} > 10) = \frac{17}{36}$

Sally is equally likely to get a score less than 10 as a score greater than 10. ✓ (1 mark)

e To get a score of more than 100 Sally could get 3 scores of 36 or 2 scores of 36 and 1 score of 30.

P(3 scores of 36)

$= \frac{1}{36} \times \frac{1}{36} \times \frac{1}{36}$

$= \frac{1}{46\,656}$

P(2 scores of 30, 1 of 36)

$= 3 \times \frac{2}{36} \times \frac{2}{36} \times \frac{1}{36}$

$= \frac{1}{3888}$ ✓

[The score of 36 could occur first, second or third so there are three different favourable outcomes.]

$P(\text{score} > 100)$

$= \frac{1}{46\,656} + \frac{1}{3888}$

$= \frac{13}{46\,656}$ ✓ (2 marks)

(Total: 30 marks)

QUARTILES AND INTERQUARTILE RANGE
SKILLS CHECK PAGE 78

1 a $1 + 2 + 2 + 5 + 6 + 7 + 7 + 7 + 8 = 45$

Mean $= \frac{45}{9}$

$= 5$

b Mode = 7

[there are more 7's than any other number]

c Median = 6

[There are 9 scores. The middle score is the 5th score, which is 6]

d Range = 8 − 1

= 7

2 a In order the scores are: 2, 4, 5, 6, 7, 7, 8, 9

Median $= \frac{6 + 7}{2}$

$= 6.5$

[the average of the two middle scores]

b In order: 37, 58, 58, 62, 63, 67

Median $= \frac{58 + 62}{2}$

$= 60$

3 Interquartile range = 72 − 57

= 15

[upper quartile lower quartile]

4 73, 75, 79, 82, 85, 86, 87, 87, 88, 91, 94

a Median = 86

[The middle score is the sixth score: 86]

b Lower quartile = 79

[The middle score of those below the median, 73, 75, 79, 82, 85.]

c Upper quartile = 88

5 13, 13, 14, 15, 15, 15, 16, 16, 16, 16, 17, 17, 18, 18, 18, 18, 18, 19, 19, 20

a Lower quartile = 15

b Upper quartile = 18

c Interquartile range

= 18 − 15

= 3

6

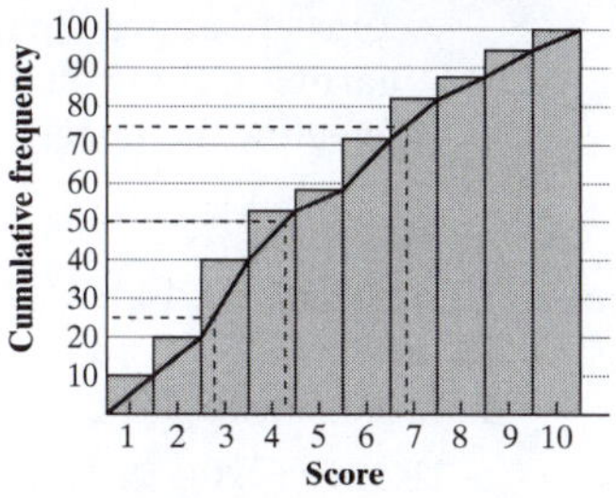

a Median = 4

b Lower quartile = 3

c Upper quartile = 7

[The median is 50%, the lower quartile is 25% and the upper quartile 75%.]

7 78, 56, 32, 25, 48, 52, 63, 50, 61

[First place the scores in ascending order.]

25, 32, 48, 50, 52, 56, 61, 63, 78

a Upper extreme = 78

b Lower extreme = 25

c Median = 52

d Upper quartile $= \frac{61 + 63}{2}$

$= 62$

e Lower quartile $= \frac{32 + 48}{2}$

$= 40$

f Range = 78 − 25

= 53

g Interquartile range

= 62 − 40

= 22

8 a 100% [All scores lie between the lower extreme and upper extreme.]

b 50% [Half of the scores are between the lower extreme and the median.]

c 25% [One quarter of scores are above the upper quartile.]

d 25% [One quarter of scores are between the lower quartile and the median.]

e 50% [Half of the scores are between the lower and upper quartiles.]

f 75% [One quarter of scores are above the upper quartile so three quarters are below the upper quartile.]

9 20, 13, 9, 12, 16, 14, 17, 17
In order:
9, 12, 13, 14, 16, 17, 17, 20

a Range = 20 – 9
= 11

b Lower quartile = 12.5
Upper quartile = 17
Interquartile range
= 17 – 12.5
= 4.5

QUARTILES AND INTERQUARTILE RANGE
INTERMEDIATE TEST PAGE 79

1 In order:
19, 20, 24, 25, 27, 27, 32
Median = 25
∴ [B] ✓ (1 mark)

2 In order: 5, 6, 8, 9, 10, 10, 12
Median = 9
Range = 12 – 5 = 7
Lower extreme = 5
Lower quartile = 6
Upper quartile = 10
Interquartile range = 10 – 6 = 4
The incorrect statement is 'The lower extreme is 6.'
∴ [C] ✓ (1 mark)

3 4, 5, 5, 5, 6, 7, 7, 8, 9, 9, 10, 10
Mean = 85 ÷ 12
= 7.083333...
Median = 7
Mode = 5
Lower quartile = 5
Upper quartile = 9
Interquartile range = 9 – 5 = 4
The mean is the largest.
∴ [A] ✓ (1 mark)

4 3, 5, 9, 10, 12, 13, 14, 16, 19
Median = 12
Lower quartile $= \frac{5+9}{2} = 7$
Upper quartile $= \frac{14+16}{2} = 15$
Interquartile range = 15 – 7
= 8
∴ [C] ✓ (1 mark)

5 In order:
55, 56, 57, 59, 59, 61, 64, 65, 67, 69, 70, 71, 71, 73, 76, 77, 78, 79
Lower extreme = 55
Upper extreme = 79
Median $= \frac{67+69}{2} = 68$
Lower quartile = 59
Upper quartile = 73
Now 59 – 55 = 4
68 – 59 = 9
73 – 68 = 5
79 – 73 = 6
So the biggest difference is between the lower quartile and the median.
∴ [B] ✓ (1 mark)

6 **a** 78% is the upper quartile so about one-quarter of students scored more than 78%. ✓
Number of students
= 80 000 ÷ 4
= 20 000 ✓ (2 marks)

b 50% is the lower quartile so about three-quarters of the students scored more than 50%. ✓
Number of students
$= \frac{3}{4} \times 80\,000$
= 60 000 ✓ (2 marks)

7

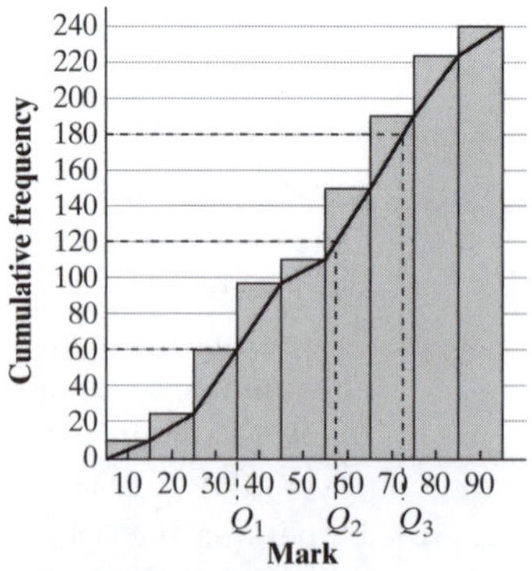

a Median = 57 ✓ (1 mark)

b Lower quartile = 35 ✓
Upper quartile = 72
So the middle 50% of marks are between 35 and 72. ✓ (2 marks)

8 32, 35, 36, 38, 38, 39, 42, 44, 45, 45, 46, 48, 56, 75, 88

a Median = 44 ✓ (1 mark)

b Range = 88 – 32
= 56 ✓ (1 mark)

c Lower quartile = 38 ✓
Upper quartile = 48 ✓
Interquartile range
= 48 – 38
= 10 ✓ (3 marks)

d The range gives the total spread of all the times. ✓ The interquartile range gives the spread of the middle 50% of scores. ✓ Any outliers (75, 88) are excluded in the interquartile range so it gives a better picture of the usual spread of times. The range gives the actual complete picture. ✓ (3 marks)

(Total: 20 marks)

QUARTILES AND INTERQUARTILE RANGE
ADVANCED TEST PAGE 80

1 47, 49, 52, 54, 58, 60, 60, 61, 63, 65, 69, 70, 74, 75, 77, 79, 81, 85, 92.

a median = 65
[There are 19 scores, the middle score is the 10th score and the 10th score is 65.] ✓ (1 mark)

b lower quartile = 58
[The median of the scores below 65.] ✓ (1 mark)

c Upper quartile = 77
[The median of the scores above 65.] ✓ (1 mark)

d Interquartile range
= 77 – 58
= 19 ✓ (1 mark)

e Range = 92 – 47
= 45
19 is not half of 45, so Stacey is not correct. 50% of the scores lie within the interquartile range but this is not necessarily 50% of the range. ✓ (1 mark)

2 **a** interquartile range
= upper quartile
– lower quartile
26 = 154 – lower quartile
lower quartile = 154 – 26
= 128
The lower quartile was 128 cm. ✓ (1 mark)

b 154 cm is the upper quartile. So, about 25% of students are taller than 154 cm. ✓ (1 mark)

c No, Megan is not correct. The median would only be 141 cm if the data is symmetrical and that is not necessarily the case. ✓ (1 mark)

d Knowing the interquartile range we have information about the spread of scores. Knowing the upper quartile as well gives us information about the location of the middle 50% of scores. ✓ (1 mark)

3 In order the scores are:
7, 19, 26, 28, 28, 31, 32, 34, 35, 37, 39, 40, 41, 43, 44, 46, 48, 51, 59, 69

a mean = 37.85 (by calculator) ✓ (1 mark)

b mode = 28 ✓ (1 mark)

c median = $\frac{37 + 39}{2}$
= 38 ✓ (1 mark)

d range = 69 – 7
= 62 ✓ (1 mark)

e upper quartile = $\frac{44 + 46}{2}$
= 45
lower quartile = $\frac{28 + 31}{2}$
= 29.5
interquartile range
= 45 – 29.5
= 15.5 ✓ (1 mark)

f 1½ × interquartile range
= 1.5 × 15.5
= 23.25
Now 29.5 – 23.25 = 6.25
No scores are less than 6.25 so no scores are less than a position 1½ times the interquartile range below the lower quartile. ✓ (1 mark)

g 45 + 23.25 = 68.25
The score 69 is more than a position 1½ times the interquartile range above the upper quartile. ✓ (1 mark)

h If the score of 69 is ignored:
mean decreases to 36.2 (1 d.p.); ✓
mode is unchanged;
median decreases to 37;
range decreases by 10 to 52; ✓
lower quartile is now 28;
upper quartile is now 44;
interquartile range
= 44 – 28
= 16
So the interquartile range increases. ✓ (3 marks)

4 [37, 48, 62, 79, 96]

a range = 96 – 37
= 59 ✓ (1 mark)

b interquartile range
= 79 – 48
= 31 ✓ (1 mark)

c The middle 50% of marks lie between 48 and 79. ✓ (1 mark)

d **i** 100% are between the upper and lower extremes. ✓ (1 mark)

ii 50% are between the median and lower extreme. ✓ (1 mark)

iii 75% are between the lower quartile and upper extreme. ✓ (1 mark)

e **i** About 25% are less than lower quartile
Number who scored less than 48
= 0.25 × 32 000
= 8000 ✓ (1 mark)

ii About 50% scored more than the median.
Number who scored more than 62
= 32 000 ÷ 2
= 16 000 ✓ (1 mark)

iii About 75% scored less than the upper quartile
Number scoring less than 79
= 0.75 × 32 000
= 24 000 ✓ (1 mark)

5 **a** The class centres go up by 8 each time.
Missing class centre
= 35 + 8
= 43 ✓ (1 mark)

b

Class centre	Frequency	Cumulative frequency
11	9	9
19	12	21
27	13	34
35	15	49
43	20	69
51	21	90
59	17	107
67	13	120

✓ (1 mark)

c

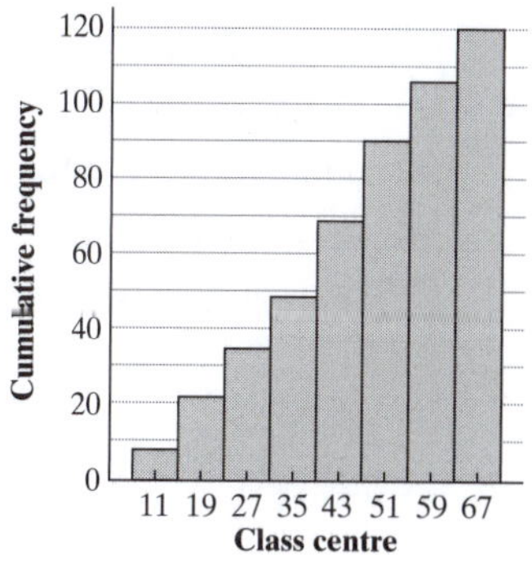

✓✓ (2 marks)

d

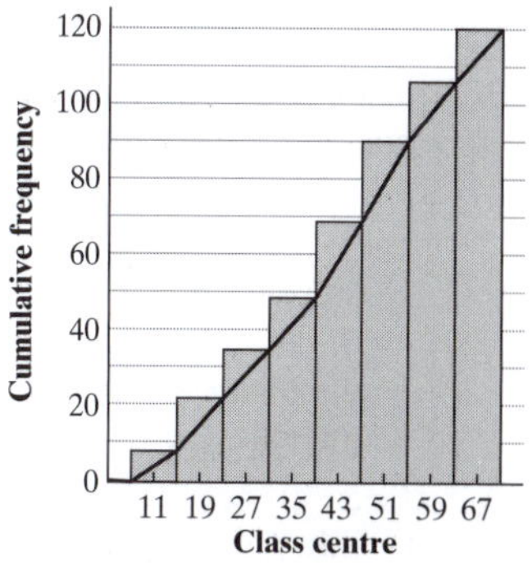

✓ (1 mark)

e median = 44 ✓ (1 mark)

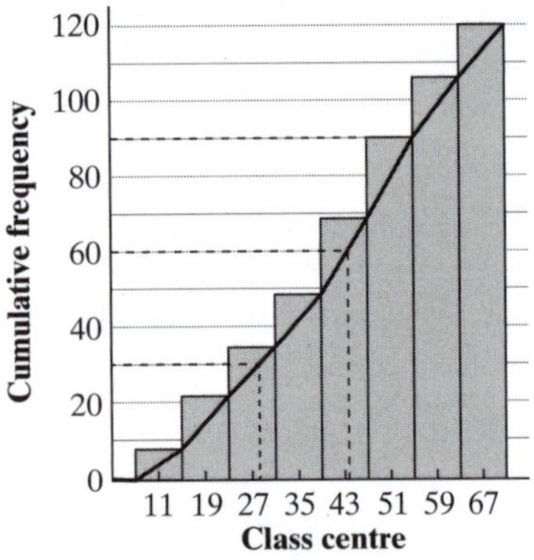

f upper quartile = 55
lower quartile = 29
interquartile range
= 55 – 29
= 26 ✓ (1 mark)
(Total: 35 marks)

BOX PLOTS AND SCATTER PLOTS
SKILLS CHECK PAGE 82

1

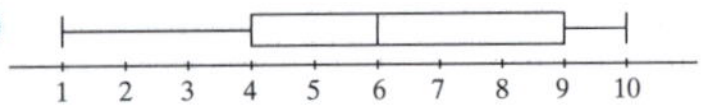

a median = 6
b lower extreme = 1
c upper quartile = 9
d upper extreme = 10
e lower quartile = 4
f interquartile range = 9 – 4
= 5

2

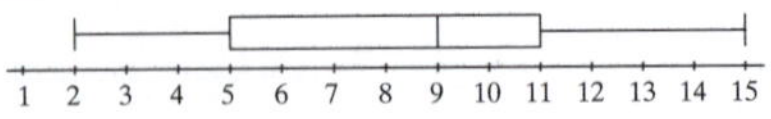

3

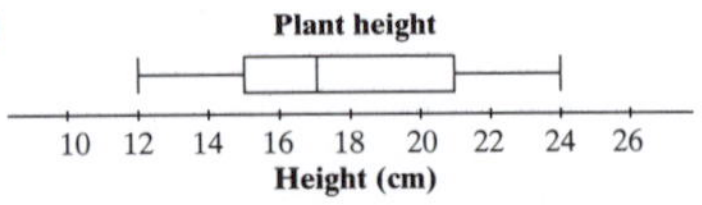

[The lower extreme is 12, lower quartile 15, median 17, upper quartile 21 and upper extreme 24.]

a 50% are between 15 cm and 21 cm. [Half are between the lower and upper quartiles.]

b 50% are more than 17 cm. [Half are above the median.]

c 75% are less than 21 cm. [Three quarters are below the upper quartile.]

d 25% are between 17 cm and 21 cm. [One quarter are between the median and upper quartile.]

e 75% are more than 15 cm in height. [Three quarters are above the lower quartile.]

f 25% are between 21 cm and 24 cm. [One quarter are above the upper quartile.]

4 a Symmetrical

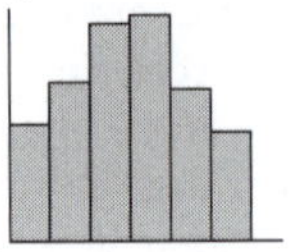

b Positively skewed

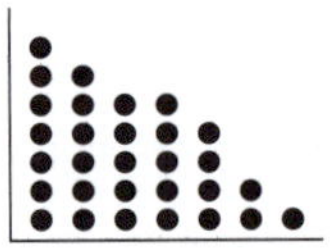

[There are more lower scores than higher ones.]

c Negatively skewed

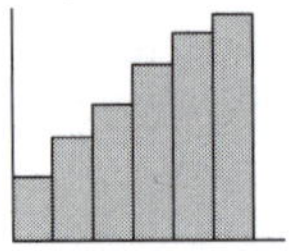

[There are more higher scores than lower ones.]

d Symmetrical

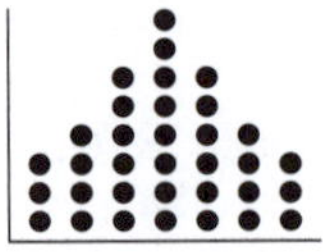

5

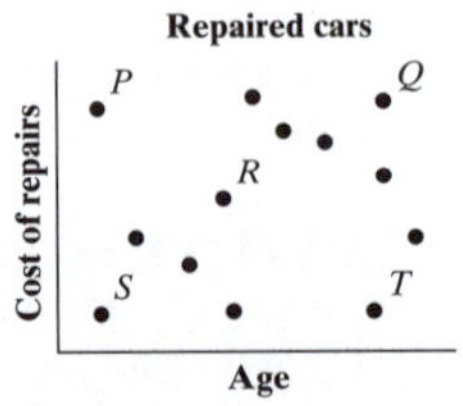

a Point Q would represent an old car which had had a lot of repairs. [The age increases as you move to the right.]

b Point S would represent a new car which had had very few problems. [Both the age and the cost would be low.]

c Point T would represent an older car that had had few repairs.

d Point R would represent an average age car with an average cost of repairs. [In the centre of both axes.]

e Point P would represent a car that is not old but which has cost a lot to repair.

6 a Strong positive association

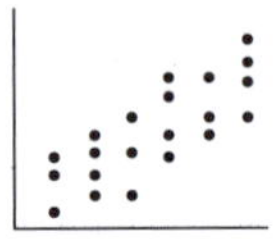

[The points lean to the right and are not widely spread from a straight line.]

b Weak negative

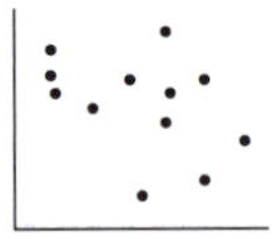

[They lean to the left but are quite wide spread.]

c No association

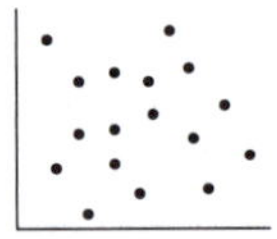

[No mathematical pattern.]

d Strong negative

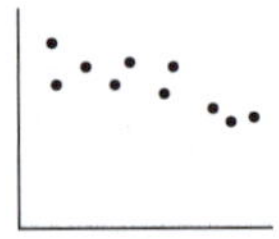

BOX PLOTS AND SCATTER PLOTS
INTERMEDIATE TEST PAGE 83

1

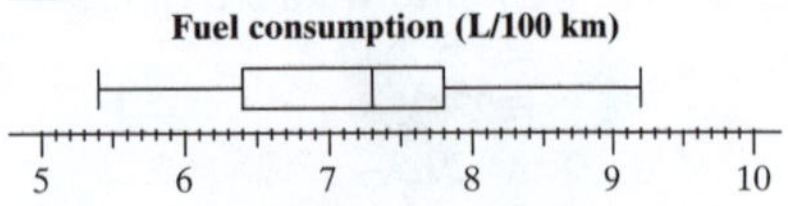

Lower quartile = 6.4
Upper quartile = 7.8

Interquartile range
= 7.8 – 6.4
= 1.4
∴ **[A]** ✓ (1 mark)

2 7.8 L/100 km is the upper quartile. So 75% of cars had a lower fuel consumption.
∴ **[C]** ✓ (1 mark)

3 6.4 L/100 km is the lower quartile. So one quarter had a lower fuel consumption.
Number of cars = 84 ÷ 4
= 21
∴ **[B]** ✓ (1 mark)

4 Mick would be represented by the point D.

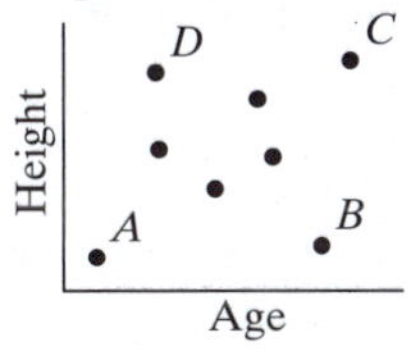

[He is one of the youngest so cannot be B or C. He is quite tall, so cannot be A.]
∴ **[D]** ✓ (1 mark)

5 **a**

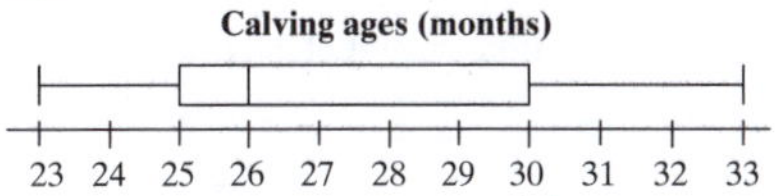

✓✓ (2 marks)

b The data is positively skewed. The median is much closer to the lower quartile than the upper quartile and the lower quartile is closer to the lower extreme than the upper quartile is to the upper extreme. ✓
We could conclude that most of the cows are more likely to calve in the lower range of ages. ✓ (2 marks)

6

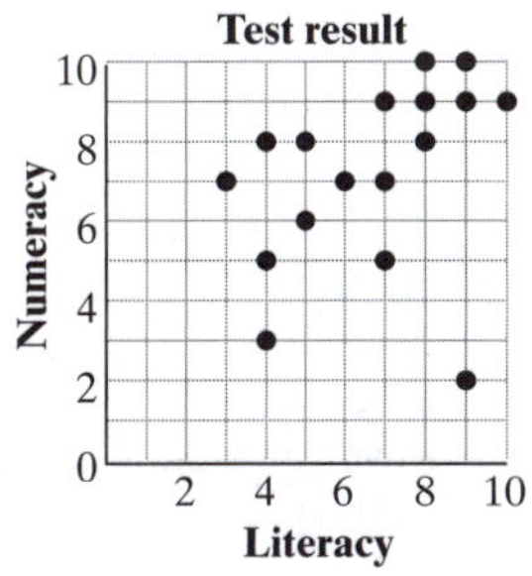

a Sally scored 6 in numeracy. She scored 5 in literacy. ✓
[Only one student scored 6 in numeracy so that must be Sally.] (1 mark)

b Sally is partly correct.
The association is positive because the dots form a pattern that leans to the right. ✓
It is moderately strong because the dots do tend to form a straight line. ✓
(2 marks)

c One person scored 9 in literacy but only 2 in numeracy. ✓ (1 mark)

d Using the results from the other tests, Jim should be given a mark of 7½. ✓

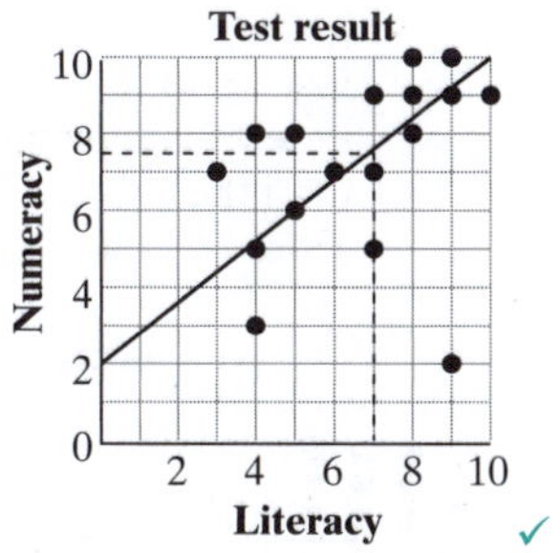

✓

[Other answers of around 7 or 8 are correct, as long as you justify your answer.]
(2 marks)

7

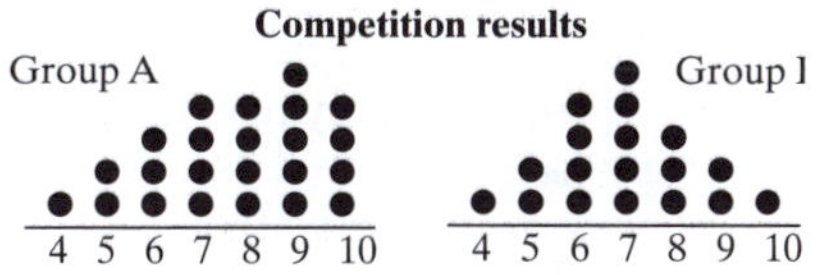

a 5 number summary:
Group A: [4, 6, 8, 9, 10] ✓
Group B: [4, 6, 7, 8, 10] ✓

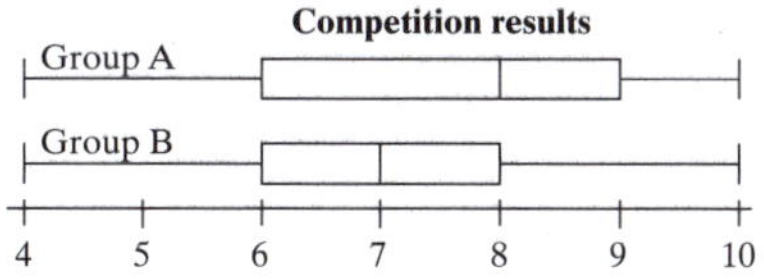

✓✓ (4 marks)

b The dot plots allow you to see every score. The box plots allow you to see the five number summary. ✓
From both plots we can see that the data in group B is symmetrical while the data in group A is negatively skewed. The box plot probably allows us to see this more clearly. The box plot also allows us to see that 50% of group A did better than 75% of group B. ✓ (2 marks)

(Total: 20 marks)

BOX PLOTS AND SCATTER PLOTS
ADVANCED TEST PAGE 84

1 In order the ages are:
7, 8, 9, 10, 11, 12, 12, 13, 13, 14, 15, 15, 15, 16, 16, 18

a median = 13
upper quartile = 15
lower quartile = 10.5 ✓
The five number summary is [7, 10.5, 13, 15, 18] ✓
(2 marks)

b

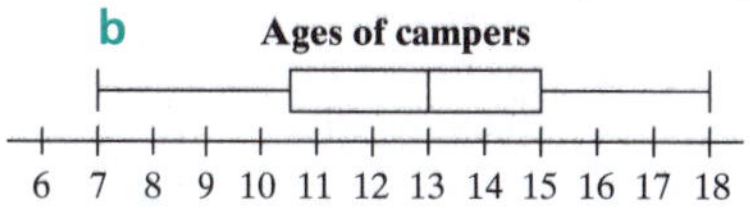

✓✓ (2 marks)

c 12 of the 16 students are older than 10.
Percentage older than 10
$= \frac{12}{16} \times 100\%$
= 75% ✓ (1 mark)

2

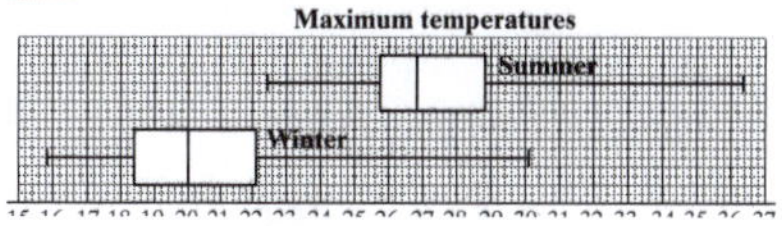

a median summer temperature = 26.8°C ✓
(1 mark)

b Range in winter
= 30.1° – 15.8°
= 14.3° ✓ (1 mark)

c The middle 50% of summer maximums are between 25.7°C and 28.8°C. ✓
(1 mark)

d Difference
= 36.4° – 30.1°
= 6.3° ✓ (1 mark)

e 20° is about the median winter temperature, so about half the days were hotter than 20 °C.
Number of days in winter
= 30 + 31 + 31
= 92
So about 46 days were hotter than 20 °C. ✓(1 mark)

f 25.7 °C is the lower quartile in summer. So about 25% of days would have a maximum temperature less than 25.7°.
Number of days in summer
= 31 + 31 + 28
= 90
[Assuming it is not a leap year.]
Now 0.25 × 90 = 22.5
So around 23 days would have a maximum temperature less than 25.7 °C. ✓ (1 mark)

g The two box plots are fairly similar in shape. The box plot for summer is a little more positively skewed. ✓ As expected the winter temperatures are consistently colder than summer with more than 75% of winter maximums less than that of the coldest summer day. ✓ (2 marks)

h i Spring temperatures ranged between 18.9 °C and 36.7 °C. The temperatures are fairly consistent with the middle 50% between 23.1 °C and 27.2 °C, only varying by about 4°. The hottest spring day was hotter than the hottest summer day. ✓ (1 mark)

ii By drawing a box plot the shape of the distribution could allow us to see any symmetry or skewness. We could also better see similarities and differences with summer and winter. ✓ (1 mark)

3 a

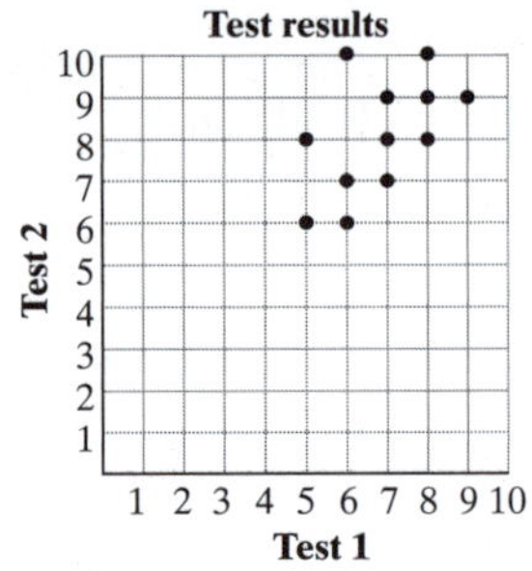

✓✓ (2 marks)

b It is easy to see any patterns or trends. ✓ (1 mark)

c All the scores are readily available and the individuals with those scores can be seen. ✓ (1 mark)

d There is a reasonably strong positive association between the scores from test 1 and test 2. The scores in test 2 are generally better than the corresponding scores in test 1. ✓ (1 mark)

e

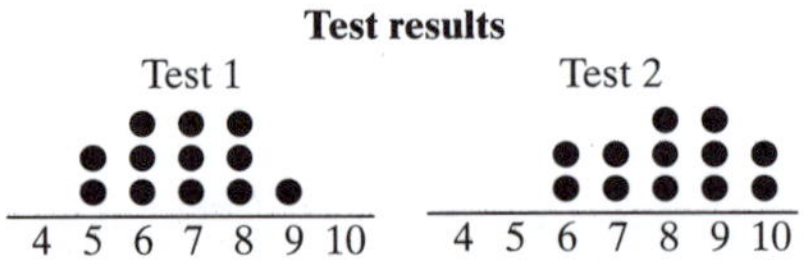

✓✓ (2 marks)

f Both dot plots are reasonably symmetrical with that for test 1 being very slightly positively skewed and that for test 2 being slightly negatively skewed. The scores in test 2 are consistently higher than test 1. The spread of scores is the same for both tests. ✓ (1 mark)

4 a

Score	Frequency
3	1
4	2
5	1
6	3
7	6
8	4
9	2
10	1
Total	20

✓✓ (2 marks)

b

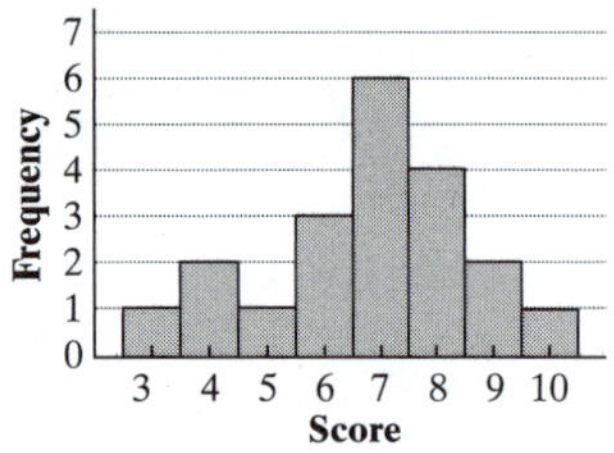

✓✓ (2 marks)

c median = 7
upper quartile = 8
lower quartile = 6 ✓
Five number summary:
[3, 6, 7, 8, 10] ✓ (2 marks)

d

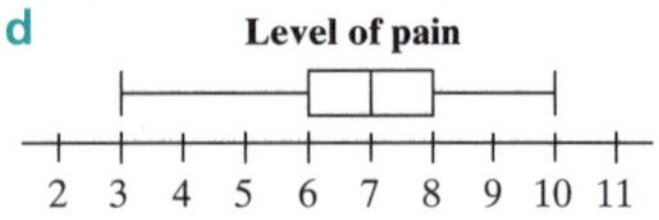

✓✓ (2 marks)

e The box plot appears reasonably symmetrical whereas the histogram is negatively skewed. ✓ From the box plot we know that the middle 50% of scores are between 6 and 8. The mode can be seen from the histogram. ✓ (2 marks)

5 a

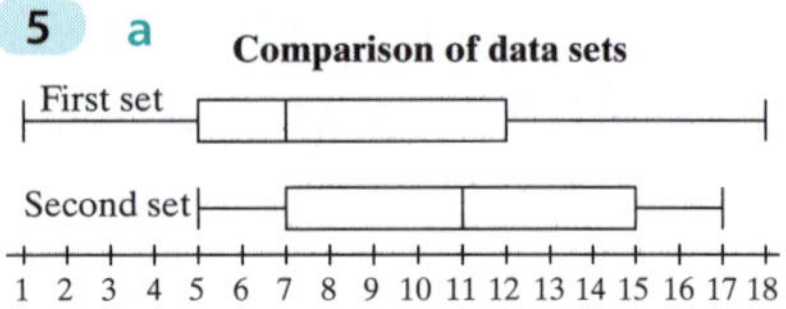

✓✓ (2 marks)

b The box plot for the first set is positively skewed. ✓ The box plot for the second set is symmetrical. ✓ (2 marks)

c The first set has a much greater spread of scores but the middle 50% of scores are more consistent than for the second set. ✓ 25% of scores in the first set are less than the lowest score in the second set. ✓ (2 marks)

d No, Alan is not necessarily correct. The same percentage of scores in the first set are between the lower quartile and median as the percentage of scores in the second set between the lower extreme and lower quartile, but unless both sets have exactly the same number of scores the numbers will not be the same. ✓ (1 mark)

(Total: 40 marks)

DATA INVOLVING TIME; EVALUATING REPORTS

SKILLS CHECK PAGE 86

1

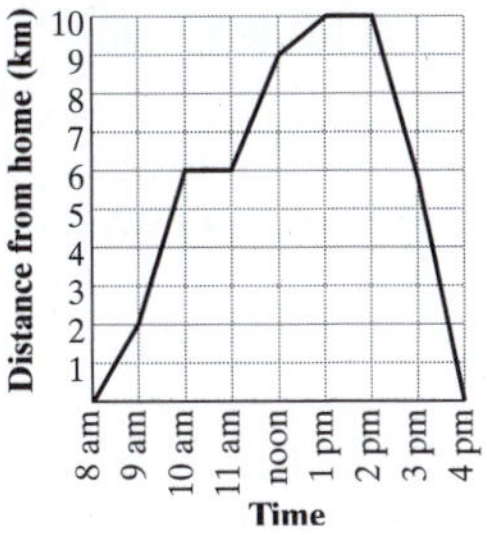

a At 9 am Yasmin was 2 km from home.

b At 1 pm, Yasmin was 10 km from home.

c At 11 am the distance was 6 km. At noon the distance was 9 km. Yasmin travelled 3 km during that hour.

d Yasmin left home at 8 am.

e She got back home at 4 pm.

f Yasmin was 6 km from home between 10 am and 11 am and again at 3 pm.

g Yasmin was stopped between 10 am and 11 am and between 1 pm and 2 pm.

h This is shown by a horizontal line on the graph.

i Yasmin began her trip home at 2 pm.

j This is shown on the graph by a negative gradient. Before 2 pm she was continually getting further away from home (or stopped) but after 2 pm she gets closer to home as time passes.

k Yasmin was travelling faster between 9 o'clock and 10 o'clock.

l It is possible to tell when she is travelling faster because the graph has a greater gradient then. The steeper the line (whether the gradient is positive or negative), the faster she is travelling.

m Yasmin travelled 20 km. [10 km away and 10 km back home.]

2 a B [The population is constant, it doesn't change, so the line goes straight across.]

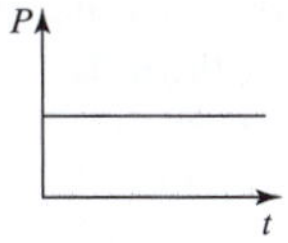

b E [The line goes up when it is increasing, it is straight because the increase is constant.]

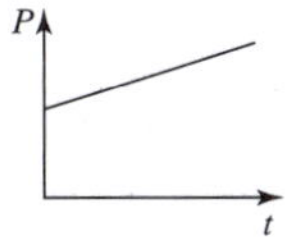

c C [The line goes down because the population is decreasing. It is straight because the decrease is constant.]

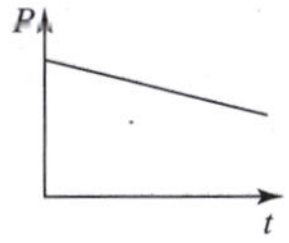

d D [The line goes up. The gradient increases when the rate increases.]

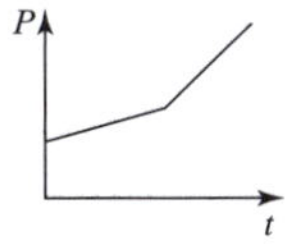

e A [The line goes up. The gradient decreases when the rate decreases.]

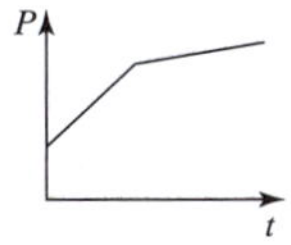

3

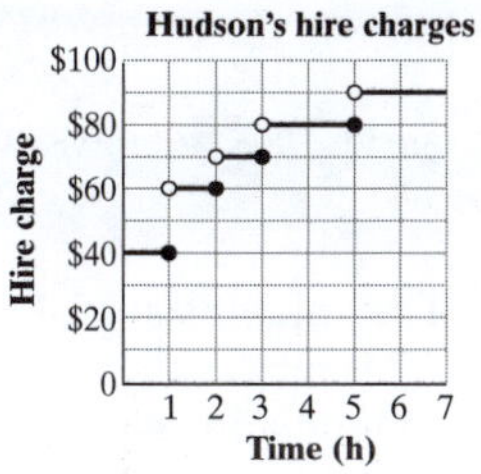

a The cost for 4 hours is $80.

b 80 minutes = 1 hour and 20 minutes
The cost = $60
[The cost for any time longer than 1 hour up to and including 2 hours is $60.]

c The cost for 5 hours = $80

d Penny had the skates for more than 2 hours but less than, or equal to, 3 hours.

e The cost for 5 hours = $80. The cost for more than 5 hours is $90. Mervyn had to pay an extra $10.

4 a

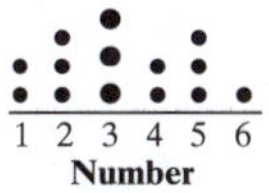

The dots for 3 are larger than all the others. [This makes it seem as if there are many more of 3 and there are actually the same as for 5.]

b

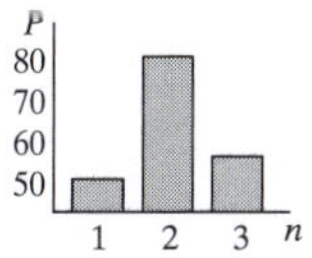

The vertical axis doesn't begin at 0. [When $n = 2$ this makes P seem much bigger than the value of P for other values of n.]

c

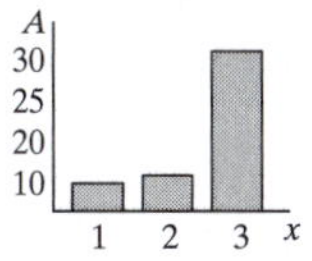

The scale on the vertical axis is not even. [It goes up by 10's then by 5. This makes A appear much larger for $x = 3$ then it would actually be.]

d 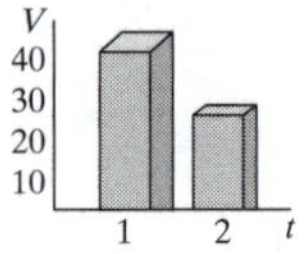

The columns are 3-D. [This makes the taller column appear larger than it should actually be.]

DATA INVOLVING TIME; EVALUATING REPORTS
INTERMEDIATE TEST PAGE 87

1 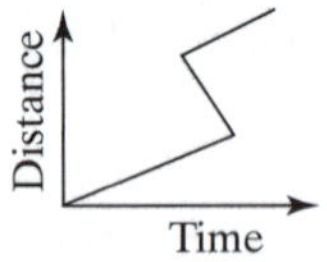

B could not represent an actual journey. [It is impossible to be at two (or more) different distances at the same time.]
∴ [B] ✓ (1 mark)

2 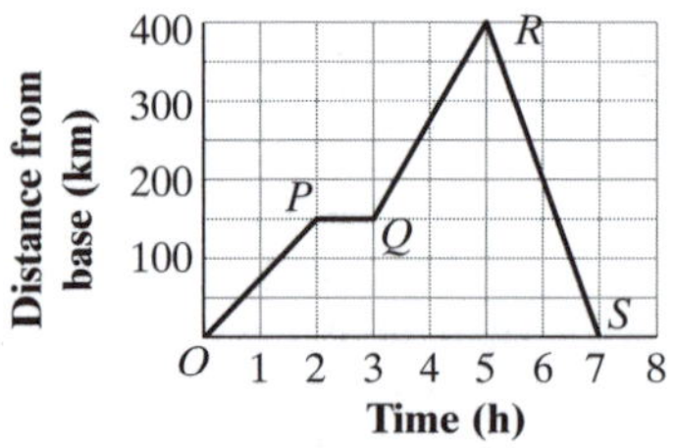

The train is travelling fastest between R and S.
[The line is steepest between R and S.]
∴ [D] ✓ (1 mark)

3 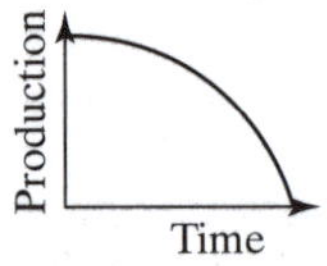

[The graphs in C and D are both decreasing, but D is decreasing at a decreasing rate.]
∴ [C] ✓ (1 mark)

4 The water level will be shown by the graph in A.

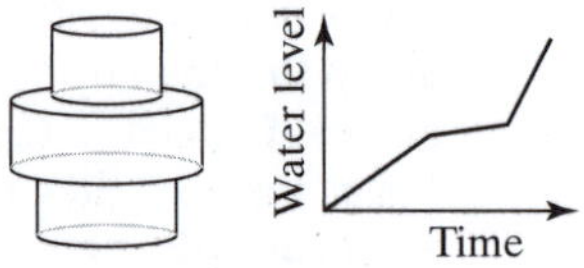

[The water level will rise at a constant rate at first, then at a slower rate for the larger middle section, then at the fastest rate for the top (smaller) section.]
∴ [A] ✓ (1 mark)

5 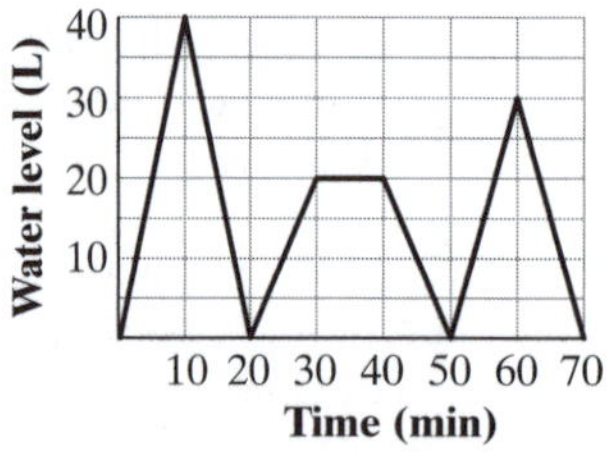

The water level increases constantly until it reaches 40 litres after 10 minutes. It then immediately empties, the water level decreasing to zero after a further 10 minutes. ✓ It increases again, slower than before, reaching 20 litres in 10 minutes and remains at this level for a further 10 minutes. It then decreases until empty again after a total of 50 minutes. ✓ It fills again to 30 litres in 10 minutes and then immediately empties, the whole process taking a total of 70 minutes. ✓ (3 marks)

6 Some possible responses:
The sample size might have been very small. [5 'yes' responses and 1 'no' could have produced these results.]
The report shown might have been biased and produced a biased response. The sample was not a random sample, only those who felt strongly would have replied.
The sample might not have been representative of the whole population. The viewers of the program might, for example, be mainly older people. ✓✓✓ (3 marks)

7 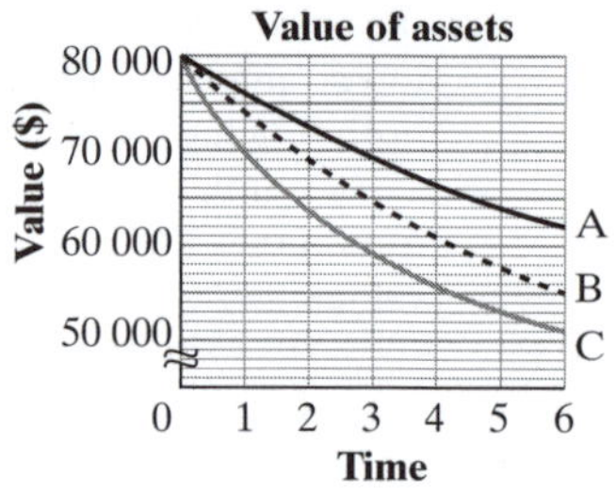

a The initial value was \$80000. ✓ (1 mark)

b After 5 years the value of asset C is \$53000. ✓ (1 mark)

c Asset B was valued at \$61000 after 4 years. ✓ (1 mark)

d After 1 year the value of asset A was \$76000. It had lost \$4000 in value. ✓ (1 mark)

e The graph could be misleading because although the scale on the vertical axis begins at 0 there is a gap in it. This makes the differences between the three assets appear larger than it really is. (But it is not misleading in that the gap is clearly marked.) ✓ (1 mark)

(Total: 15 marks)

DATA INVOLVING TIME; EVALUATING REPORTS
ADVANCED TEST PAGE 88

1 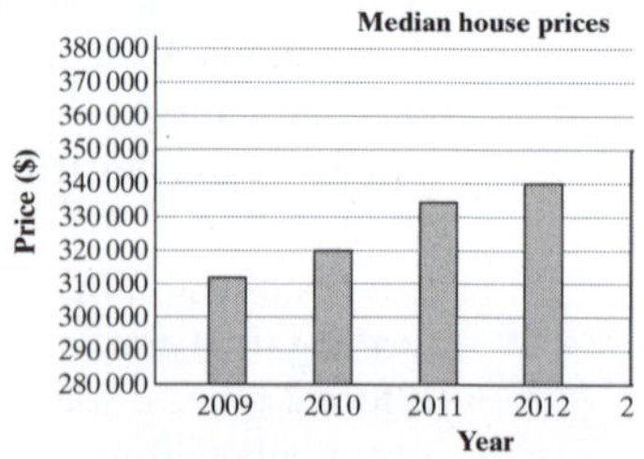

a The mean can be easily influenced by one or two prices that are significantly different to most others. The median gives a better idea of the location of prices. ✓ (1 mark)

b The vertical axis does not begin at 0. ✓ (1 mark)

c

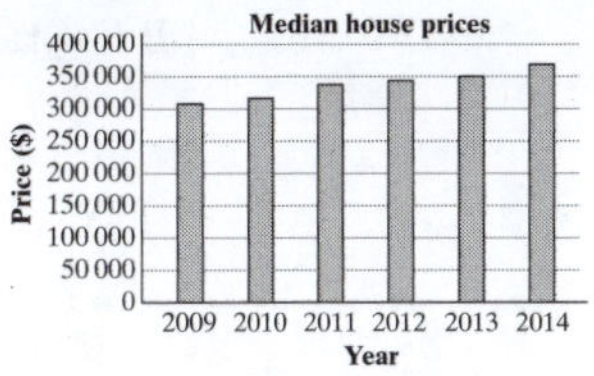

✓✓✓ (3 marks)

d The original graph better illustrates their point of view. ✓ (1 mark)

e No, there has been an upward trend over the last few years and an increase over the last year, but it could not be described as 'huge'. ✓✓ (2 marks)

2 **a** The independent variable is time. ✓ (1 mark)

b Around 1981. ✓ (1 mark)

c Between 1930 and 1945. ✓ This was the time of the great depression followed by the second world war. Economic hardship and the ravages of war would have meant that the birth rate was lower and the death rate higher than otherwise might have been expected. ✓ (2 marks)

d In 1960, Australia's population was about 10 400 000.
Percentage living in Queensland

$$= \frac{1\,500\,000}{10\,400\,000} \times 100\%$$
$$= 14\% \text{ (nearest per cent)}$$

✓ (1 mark)

e Queensland's population tripled in those 50 years. ✓ In 2010, Australia's population was about 22 million. So that was a little more than double the population in 1960. So the population in Queensland was increasing at a much faster rate than the rest of Australia. ✓ (2 marks)

f Yes, it is a reasonable prediction based on the continuing trend. However it is only a prediction and different things could have a bearing on what does happen. ✓ (1 mark)

3

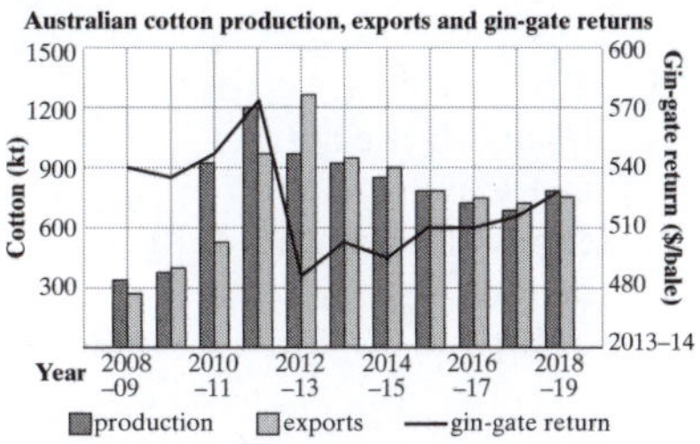

a Exports were highest in 2012–13. ✓ (1 mark)

b About 1 200 000 tonnes of cotton were produced in 2011–12. ✓ (1 mark)

c In 2010–11, production was about 950 000 tonnes and exports about 500 000 t.
Percentage

$$= \frac{500\,000}{950\,000} \times 100\%$$
$$= 52.631\,57\ldots\%$$

So exports were around 53% of production. ✓ (1 mark)

d The cotton that was exported must have been cotton that was produced in earlier years. ✓ (1 mark)

e The expectation is that there will not be much difference between production and exports and that there will not be much change in either of these over the next few years, remaining around 750 000 tonnes. ✓ (1 mark)

f In 2008-09 the gin-gate return was about \$540/bale. ✓ (1 mark)

g The gin-gate return decreased by about \$90/bale. ✓ (1 mark)

h The scale on the right side does not begin at zero. This means that fluctuations in the graph appear much larger than they might otherwise have been. ✓ A choice was also made to have increments of \$30 on this axis. A different choice would produce a different looking graph, with larger or smaller fluctuations. ✓ (2 marks)

(Total: 25 marks)

SAMPLE EXAM PAPER 1

Part A: Multiple Choice PAGE 92

1 $7x^2 + 5x - x^2 = 6x^2 + 5x$
$\therefore$ [B] ✓ (1 mark)

2 $r = 4, h = 7$

$$V = \pi r^2 h$$
$$= \pi \times 4^2 \times 7$$
$$= 351.858\,3772\ldots$$

Volume = 352 cm³ to the nearest cm³.
$\therefore$ [A] ✓ (1 mark)

3

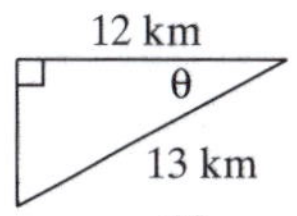

$$\cos\theta = \frac{12}{13}$$
$$\theta = 22.619\,864\,95\ldots°$$
$$= 23° \text{ (nearest degree)}$$

$\therefore$ [A] ✓ (1 mark)

4 If $t = -2$, $4t^2 = 4 \times (-2)^2$
$= 4 \times 4$
$= 16$
$\therefore$ [C] ✓ (1 mark)

5

$$\frac{x}{2} - 3 < 1$$
$$\frac{x}{2} < 4$$
$$x < 8$$

$\therefore$ [A] ✓ (1 mark)

6

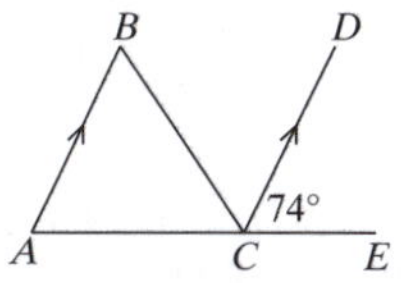

$\angle ACD + \angle DCE = 180°$
(angles in a straight line)
$\angle ACD + 74° = 180°$
$\angle ACD = 106°$
$\angle ACB = \frac{1}{2}\angle ACD$
(CB bisects $\angle ACD$)
$= 106° \div 2$
$= 53°$
$\therefore$ [C] ✓ (1 mark)

7 $\frac{x}{2} + \frac{x}{3} = \frac{3x}{6} + \frac{2x}{6}$

$= \frac{5x}{6}$

∴ [D] ✓ (1 mark)

8 $P = \$800,\ R = 0.05,\ n = 3$

$A = P(1 + R)^n$

$= \$800(1.05)^3$

$= \$926.10$

Compound interest

$= \$926.10 - \$800$

$= \$126.10$

∴ [B] ✓ (1 mark)

9 $a^2b^3 \div ab^2 = ab$

$[a^2 \div a = a^{2-1},\ b^3 \div b^2 = b^{3-2}]$

∴ [A] ✓ (1 mark)

10 Gradient $= \frac{2}{3}$

The gradient of a perpendicular line is the negative reciprocal.

$m = -\frac{3}{2}$

∴ [D] ✓ (1 mark)

11 $(x - 3)(x - 7)$

$= x^2 - 7x - 3x + 21$

$= x^2 - 10x + 21$

∴ [D] ✓ (1 mark)

12 $2(2a + 3) + 4(a - 2)$

$= 4a + 6 + 4a - 8$

$= 8a - 2$

∴ [D] ✓ (1 mark)

13

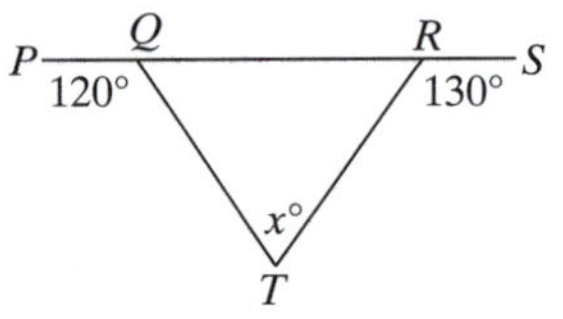

$\angle PQT + \angle TQR = 180°$

(angles in a straight line)

$\angle TQR = 180° - 120°$

$= 60°$

$\angle SRT = \angle TQR + \angle RTQ$

(exterior ∠ of Δ)

$130° = 60° + x°$

$x = 70$

∴ [C] ✓ (1 mark)

14

$x + 3y = 10$ [i]

$x - 3y = 6$ [ii]

[i] + [ii] $2x = 16$

$x = 8$

∴ [D] ✓ (1 mark)

15 $s = 5,\ A = 6s^2$

$= 6 \times 5^2$

$= 6 \times 25$

$= 150$

The surface area is 150 cm^2.

∴ [C] ✓ (1 mark)

16 $8x^2 - 6xy = 2x(4x - 3y)$

[A and B are correct factorisations, but the expressions are not fully factorised.]

∴ [C] ✓ (1 mark)

17

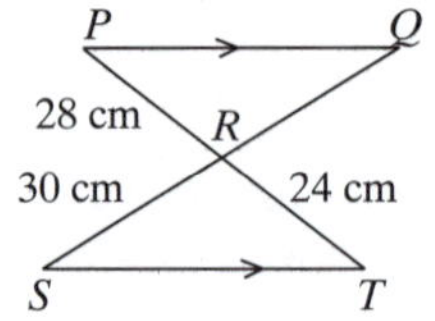

In ΔPQR, ΔTSR

$\angle QPR = \angle STR$

(alternate angles, || lines)

$\angle PQR = \angle TSR$

(alternate angles, || lines)

∴ ΔPQR ||| ΔTSR

(equiangular)

$\frac{QR}{SR} = \frac{PR}{TR}$

(corresponding sides of similar Δs)

$\frac{QR}{30} = \frac{28}{24}$

$QR = \frac{28}{24} \times 30$

$= 35$

The length of QR is 35 cm.

∴ [D] ✓ (1 mark)

18 $3(x + 2) - 5x = 18$

$3x + 6 - 5x = 18$

$-2x + 6 = 18$

$-2x = 12$ (not 24)

The mistake occurs between line 3 and line 4.

∴ [C] ✓ (1 mark)

19 $y = \frac{x}{4}$ is the equation of a straight line.

$[y = \frac{1}{4}x$. Gradient $= \frac{1}{4}$,

y-intercept = 0]

∴ [A] ✓ (1 mark)

20 $x^2 + y^2 = 25$ or $x^2 + y^2 = 5^2$

The radius is 5.

[The circle in B has radius 25.]

∴ [B] ✓ (1 mark)

21

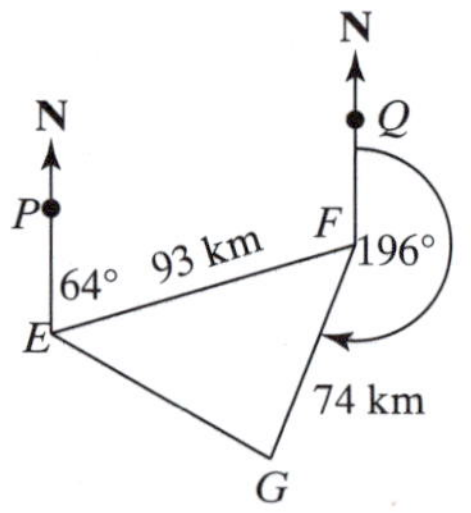

Let P be a point due north of E and Q a point due north of F.

$\angle QFE = 180° - 64°$

(alternate ∠s, || lines)

$= 116°$

$\angle EFG = 360° - 116° - 196°$

(∠s in a revolution)

$= 48°$

B is the best diagram.

∴ [B] ✓ (1 mark)

22

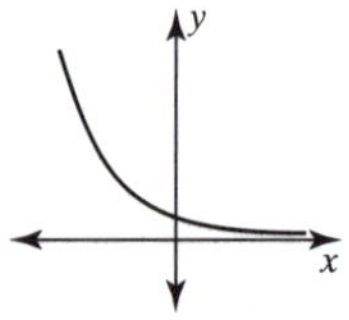

The graph of $y = 2^{-x}$ is B.

[When $x = 0,\ y = 1$;

when $x = 1,\ y = \frac{1}{2}$]

∴ [B] ✓ (1 mark)

23 $\tan\theta = 0.369$

$\theta = 20.254\,060\,82\ldots°$

[Using $\tan^{-1}$ on the calculator]

$= 20° 15'$

(nearest minute)

∴ [B] ✓ (1 mark)

24

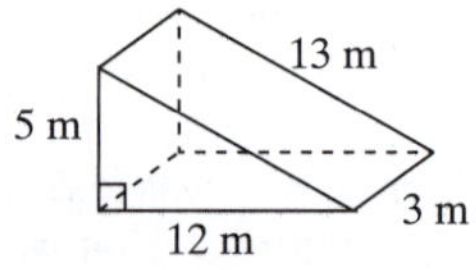

Triangular prism.

Area of triangle:

$A = \frac{1}{2}bh$

$= \frac{1}{2} \times 12 \times 5$

$= 30$

The area of the triangular face is 30 m^2.

$V = AH$

$= 30 \times 3$

$= 90$

The volume is 90 m^3.

$\therefore$ **[A]** ✓ (1 mark)

25 $\dfrac{2x+3y}{6x+9y} = \dfrac{2x+3y}{3(2x+3y)}$

$= \dfrac{1}{3}$

$\therefore$ **[D]** ✓ (1 mark)

(Total Part A: 25 marks)

SAMPLE EXAM PAPER 1

Part B: Short Answer PAGE 94

26 $C = \dfrac{5}{9}(F - 32)$

When $F = 95$, $C = \dfrac{5}{9}(95 - 32)$

$= 35$

The temperature is 35 °C ✓

(1 mark)

27 $\dfrac{8x-1}{5} \geq 3$

$8x - 1 \geq 15$

$8x \geq 16$

$x \geq 2$ ✓

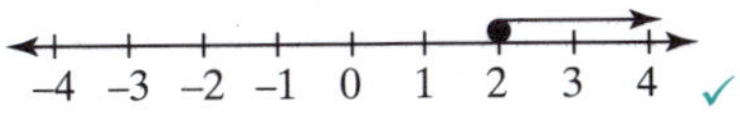

✓ (2 marks)

28 11, 11, 12, 13, 14, 14, 14, 16, 16, 17, 17, 17, 17, 18, 19, 19, 20, 20

a Median $= \dfrac{16+17}{2}$

$= 16.5$ ✓ (1 mark)

b Range $= 20 - 11$

$= 9$ ✓ (1 mark)

c Lower quartile = 14

Upper quartile = 18 ✓

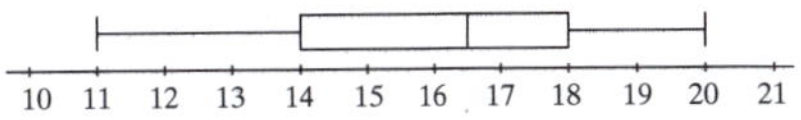

✓✓ (3 marks)

29 **a**

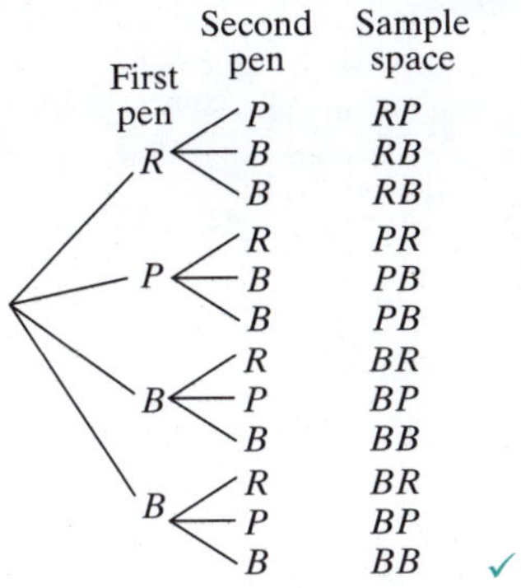

✓ (1 mark)

b $P(BB) = \dfrac{2}{12}$

$= \dfrac{1}{6}$ ✓ (1 mark)

c $P(\text{at least 1 blue}) = \dfrac{10}{12}$

$= \dfrac{5}{6}$ ✓

(1 mark)

30

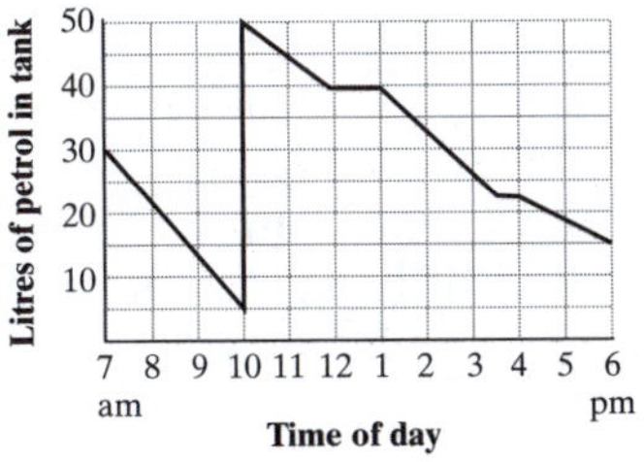

a 45 litres ✓

[At 10 am the amount of fuel in the tank increased from 5 litres to 50 litres.]

(1 mark)

b The car used (25 + 35) litres

= 60 litres of fuel. ✓

Fuel consumption

= 60 L to travel 750 km

= (60 ÷ 7.5) L/100 km

= 8 L/100 km ✓ (2 marks)

31

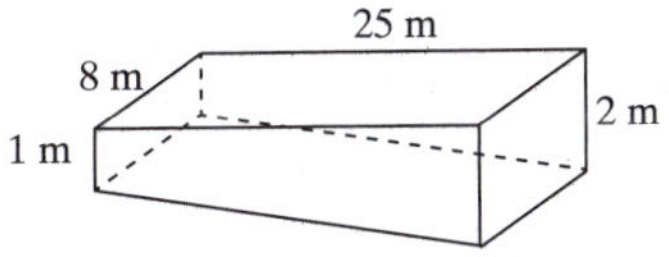

a The pool is a trapezoidal prism.

$A = \dfrac{1}{2}h(a+b)$

$= \dfrac{1}{2} \times 25 \times (1+2)$

$= 37.5$

$V = AH$

$= 37.5 \times 8$

$= 300$

The volume of the pool is 300 m^3. ✓ (1 mark)

b 1 m^3 holds 1000 litres

The pool will hold 300 000 litres. ✓ (1 mark)

c The base is a rectangle. The width is 8 m. Let the length of the rectangle be x m.

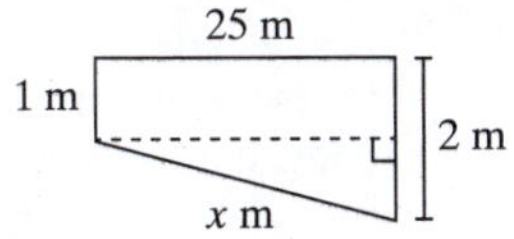

By Pythagoras' Theorem

$x^2 = 25^2 + 1^2$

$= 625 + 1$

$= 626$

$x = \sqrt{626}$

$= 25.019\,992\,01\ldots$

$= 25.0$ (1 d.p.) ✓

Jennifer is correct. The length of the rectangle is very close to 25 m. ✓

(2 marks)

d Two trapezoidal sides each of area 37.5 m^2.

Shallow end:

Area = 8 m × 1 m

$= 8\text{ m}^2$

Deep end:

Area = 8 m × 2 m

$= 16\text{ m}^2$

Bottom of pool:

Area ≈ 25 m × 8 m

$= 200\text{ m}^2$ ✓

Total area

$= (2 \times 37.5 + 8 + 16 + 200)\text{ m}^2$

$= 299\text{ m}^2$ ✓ (2 marks)

32

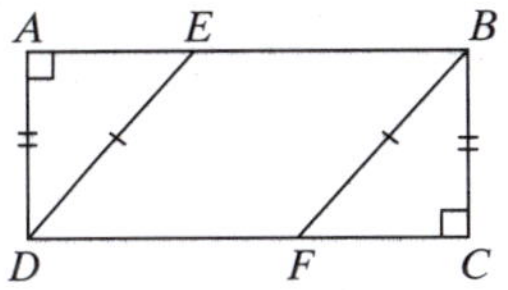

a In ΔAED, ΔCFB

$\angle EAD = \angle FCB$

(both 90°, angles of a rectangle) ✓

$ED = FB$ (given)

$AD = CB$

(opposite sides of a rectangle) ✓

$\therefore \Delta AED \equiv \Delta CFB$

RHS ✓ (3 marks)

b *EBFD* is a parallelogram. ✓

$AE = CF$

(corresponding sides of congruent triangles)

but $AB = DC$

(opposite sides of a rectangle)

$\therefore EB = DF$

$\therefore$ *EBFD* is a parallelogram (both pairs of opposite sides equal) ✓

[Or: $\angle AED = \angle CFB$

(corresponding angles of congruent triangles)

but $\angle AED = \angle EDF$

(alternate, || lines)

$\therefore \angle CFB = \angle EDF$

$\therefore ED \parallel BF$

(corresponding $\angle$s are equal)

Or: show that both pairs of opposite angles are equal]

(2 marks)

(Total Part B: 25 marks)

SAMPLE EXAM PAPER 2

Part A: Multiple Choice PAGE 96

1

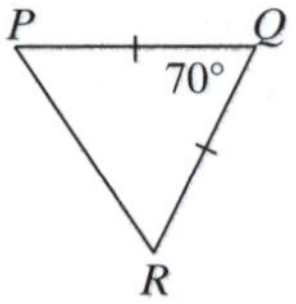

$\angle QPR + \angle PRQ + \angle PQR = 180°$

(angle sum of a triangle)

$\angle QPR + \angle PRQ + 70° = 180°$

$\angle QPR + \angle PRQ = 110°$

But $\angle QPR = \angle PRQ$

(angles opposite equal sides, isosceles triangle)

$2\angle PRQ = 110°$

$\angle PRQ = 55°$

$\therefore$ **[B]** ✓ (1 mark)

2 $y = \frac{x}{2}$ is the equation of a straight line with gradient $\frac{1}{2}$ and y-intercept 0.

$\therefore$ **[C]** ✓ (1 mark)

3

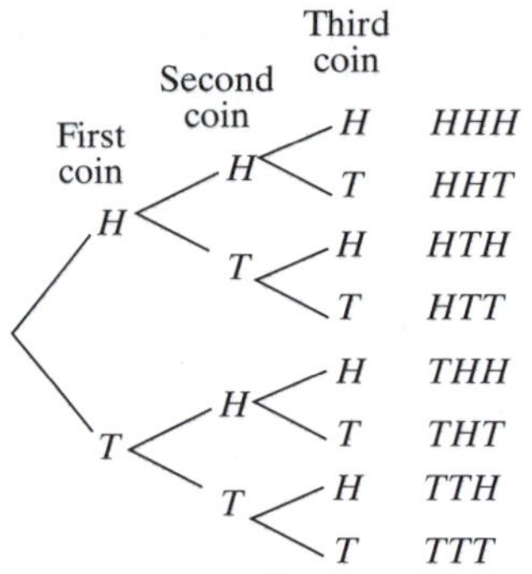

$P(\text{one head}) = \frac{3}{8}$

$\therefore$ **[D]** ✓ (1 mark)

4 $-x > 2$

$x < -2$

[dividing by –1, the inequality sign must be reversed]

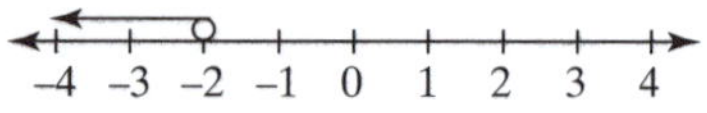

$\therefore$ **[D]** ✓ (1 mark)

5 $\frac{x}{2} + \frac{2x}{5} = \frac{5x}{10} + \frac{4x}{10}$

$= \frac{9x}{10}$

$\therefore$ **[D]** ✓ (1 mark)

6

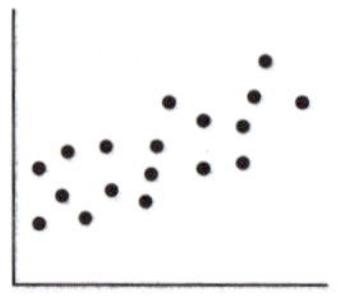

The scatter plot in B shows a strong positive association. [The association is negative in A, there is no association in C and the association in D is not linear.]

$\therefore$ **[B]** ✓ (1 mark)

7

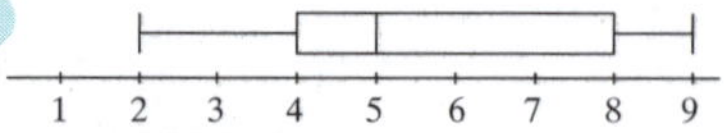

Range = 9 – 2 = 7

Median = 5

Lower quartile = 4

Upper quartile = 8

Interquartile range = 8 – 4 = 4

Incorrect statement is C.

$\therefore$ **[C]** ✓ (1 mark)

8 $r = 9, h = 50$

$A = 2\pi r^2 + 2\pi rh$

$= 2 \times \pi \times 9^2 + 2 \times \pi \times 9 \times 50$

$= 3336.371\,398\ldots$

The surface area is closest to 3336 cm^2.

$\therefore$ **[C]** ✓ (1 mark)

9 $P = \$4000,\ R = 0.02,$

$n = 4 \times 3 = 12$

$A = P(1 + R)^n$

$= \$4000(1.02)^{12}$

$= \$5072.967\,178\ldots$

$= \$5073$ (nearest dollar)

$I = A - P$

$\approx \$5073 - \$4000$

$= \$1073$

$\therefore$ **[D]** ✓ (1 mark)

10 $\frac{2ab^2}{8a^2b^2} = \frac{1}{4a}$

[Dividing both numerator and denominator by $2ab^2$.]

$\therefore$ **[A]** ✓ (1 mark)

11

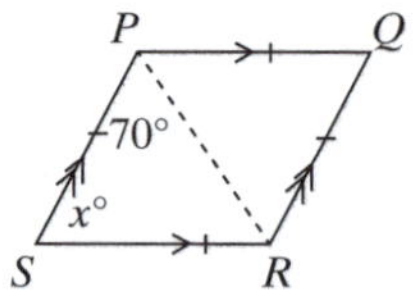

PQRS is a rhombus (a quadrilateral with a pair of adjacent sides equal).

$\angle SPQ = 140°$

(diagonals bisect angles)

$\angle SPQ + \angle PSR = 180°$

(co-interior angles, || lines)

$140 + x = 180$

$x = 40$

$\therefore$ **[A]** ✓ (1 mark)

12 $x^2 + 5x - 6 = (x + 6)(x - 1)$

[Look for two numbers that multiply to –6 and add to 5.]

$\therefore$ **[C]** ✓ (1 mark)

13

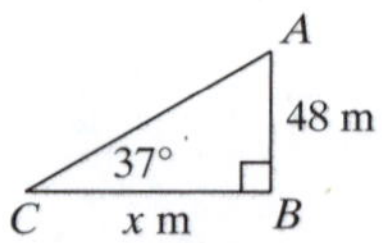

$\tan 37° = \frac{48}{x}$

$x = \frac{48}{\tan 37°}$

$= 63.698\ldots$

The distance is closest to 64 m.

The value of x is 64.

$\therefore$ **[A]** ✓ (1 mark)

14

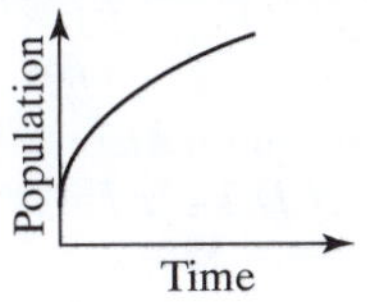

The graph in A shows the population increasing at a decreasing rate.
[In B it is increasing at an increasing rate. In C and D it is decreasing.]
∴ **[A]** ✓ (1 mark)

15 $A = 108, a = 7, h = 12$

$$A = \frac{1}{2}h(a+b)$$
$$108 = \frac{1}{2} \times 12 \times (7+b)$$
$$108 = 6(7+b)$$
$$18 = 7 + b$$
$$b = 11$$

[Or try each option in the formula to see which gives a value for A of 108.]
∴ **[D]** ✓ (1 mark)

16 $16m^{-6} \div 8m^2$
$= 2m^{-6-2}$
$= 2m^{-8}$
[A] ✓ (1 mark)

17 $3 - 2x \geq 5$
$-2x \geq 2$
$x \leq -1$
[Because we divide by a negative number the inequality sign must be reversed.]
∴ **[A]** ✓ (1 mark)

18 $y = \frac{1}{2}x + 5$ Gradient $= \frac{1}{2}$
[The gradient of a perpendicular line is the negative reciprocal.]
Perpendicular lines have gradient –2.
Of the options, the line with gradient –2 is $y = -2x + 1$.
∴ **[C]** ✓ (1 mark)

19 The prism with the greatest volume has the greatest capacity.
Try each option.
A: $V = 10 \times 4 \times 3 = 120$
B: $V = 11 \times 11 \times 1 = 121$
C: $V = 8 \times 2 \times 8 = 128$
D: $V = 5 \times 5 \times 5 = 125$
The greatest volume is 128 cm³.
So the prism in option C has the greatest capacity.
∴ **[C]** ✓ (1 mark)

20 1, 4, 5, 6, 6, 7, 7, 7, 8, 9, 9, 10
Median = 7
Lower quartile = 5.5
Upper quartile = 8.5
Interquartile range = 8.5 – 5.5
= 3
∴ **[B]** ✓ (1 mark)

21 $P(\text{both red}) = \frac{1}{4} \times \frac{1}{3}$
$= \frac{1}{12}$
∴ **[A]** ✓ (1 mark)

22 The triangles in option B are not necessarily congruent.
[It is not SAS because the given angle is not the included angle (the one between the two given sides.)
A: Congruent SSS (Two pairs of sides are given equal, the third side is common.)
C: Congruent AAS (Both triangles have angles of 50°, 60° and 70.)
D: Congruent RHS]
∴ **[B]** ✓ (1 mark)

23 $(2a-3)(2a+3) = (2a)^2 - 3^2$
$= 4a^2 - 9$
∴ **[C]** ✓ (1 mark)

24 $\angle PTQ = \angle PSR$
(both 90°, given)
$\angle QPT = \angle RPS$
(common angle)
∴ $\Delta PQT \parallel\!| \Delta PRS$
(equiangular)
Let the length of PT be x m.

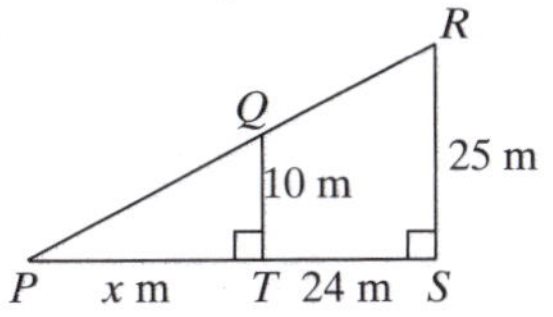

$$\frac{PT}{PS} = \frac{QT}{RS}$$
(corresponding sides of similar triangles)
$$\frac{x}{x+24} = \frac{10}{25}$$
$$25x = 10x + 240$$
$$15x = 240$$
$$x = 16$$
The length of PT is 16 m.
∴ **[D]** ✓ (1 mark)

25 $\frac{2a-4}{a^2-4} = \frac{2(a-2)}{(a+2)(a-2)}$
$= \frac{2}{a+2}$
∴ **[B]** ✓ (1 mark)
(Total Part A: 25 marks)

SAMPLE EXAM PAPER 2
Part B: Short Answer PAGE 99

26 $9m^2 - 6m = 3m(3m-2)$ ✓ (1 mark)

27

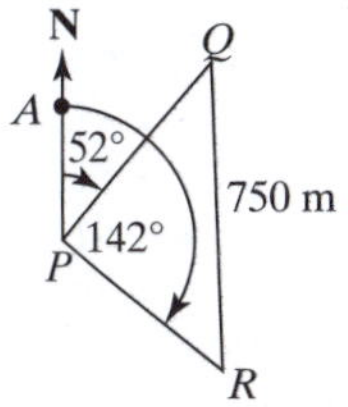

a Let A be a point due north of P. The bearing of Q from P is 052°.
∴ $\angle APQ = 52°$
The bearing of R from P is 142°.
∴ $\angle APR = 142°$
$\angle QPR = 142° - 52°$
$= 90°$ ✓ (1 mark)

b AP is parallel to QR [A is due north of P and Q is due north of R.]
$\angle PQR = \angle APQ$
(alternate ∠s, || lines)
∴ $\angle PQR = 52°$ ✓ (1 mark)

c

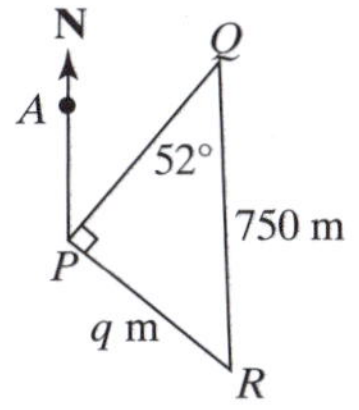

$\sin 52° = \frac{q}{750}$ ✓
$q = 750 \times \sin 52°$
$= 591.0080652\ldots$
$= 591$
(nearest whole number)
The length of PR is 591 m to the nearest metre. ✓ (2 marks)

28 Line $y = 3x - 1$ has gradient 3
Parallel lines have $m = 3$.
The equation of any parallel line is of the form $y = 3x + b$. ✓
This line passes through (2, –4).
So, when $x = 2$, $y = -4$

$y = 3x + b$
$-4 = 3 \times 2 + b$
$-4 = 6 + b$
$b = -4 - 6$
$= -10$
$y = 3x - 10$ ✓ (2 marks)

29 $x^2 - 10x - 24 = 0$
$(x - 12)(x + 2) = 0$
$x - 12 = 0$ or $x + 2 = 0$
$x = 12$ or $x = -2$ ✓✓
(2 marks)

30 $3x + 4y = 9$ [i]
$5x - 2y = 41$ [ii]
[ii] × 2 $10x - 4y = 82$ [iii]
[i] + [iii] $13x = 91$
$x = 7$ ✓
Substitute into [i]:
$3 \times 7 + 4y = 9$
$21 + 4y = 9$
$4y = -12$
$y = -3$ ✓
$\therefore\ x = 7$ and $y = -3$ (2 marks)

31

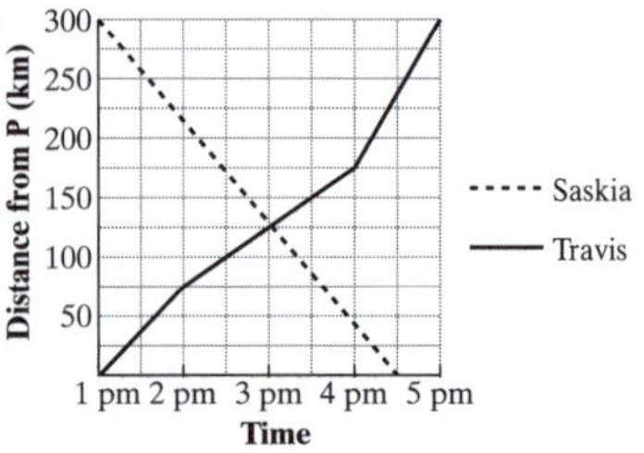

a Same distance at 3 p.m. ✓
That distance is 125 km from *P*. ✓ (2 marks)

b Travis travels 300 km in 4 hours
$S = \dfrac{D}{T}$
$= \dfrac{300}{4}$
$= 75$
Travis's average speed is 75 km/h ✓ (1 mark)

c In the hour between 2.30 pm and 3.30 pm. Travis travels 50 km while Saskia travels about 100 km. ✓
Saskia travels further in the time so she is travelling faster. ✓
[Or find the average speed for both over that time. Or notice that the line representing Saskia is steeper than the line representing Travis.]
(2 marks)

32 $P = \$48\,000$, $R = 0.15$, $n = 5$
$A = P(1 - R)^n$
$= \$48\,000(1 - 0.15)^5$
$= \$48\,000(0.85)^5$
$= \$21\,297.855$ ✓
The value of the tractor is now \$21 300 to the nearest hundred dollars. ✓ (2 marks)

33

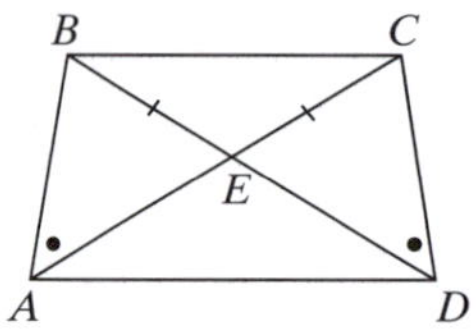

a $BE = CE$ (given)
$\therefore \angle ACB = \angle DBC$
(angles opposite equal sides, isosceles triangle) ✓
(1 mark)

b In ΔBAC, ΔCDB
$\angle BAC = \angle CDB$ (given)
$\angle ACB = \angle DBC$
(proven from part **a**)
$BC = CB$
(common side) ✓
$\therefore \Delta BAC \equiv \Delta CDB$
AAS ✓ (2 marks)

c $AC = DB$
(corresponding sides of congruent triangles) ✓
but $CE = BE$ (given)
$\therefore AE = DE$
$\therefore \angle BDA = \angle CAD$
(angles opposite equal sides, isosceles triangle) ✓
(2 marks)

d Yes. $AD \parallel BC$ ✓
Let $\angle EBC = x°$
$\therefore \angle BCE = x°$
$\angle BEC = 180° - 2x°$
(angle sum of ΔBEC)
$\angle AED = \angle BEC$
(vertically opposite $\angle$s)
$\angle AED = 180 - 2x°$
$\angle EAD + \angle EDA + \angle AED = 180°$
(angle sum ΔAED)
But $\angle EDA = \angle EAD$
$2\angle EDA + 180° - 2x° = 180°$
$2\angle EDA = 2x°$
$\angle EDA = x°$
$\therefore \angle EDA = \angle EBC$
$\therefore AD \parallel BC$
(alternate $\angle$s are equal) ✓
(2 marks)
(Total Part B: 25 marks)

SAMPLE EXAM PAPER 3

Part A: Multiple Choice PAGE 101

1 In order the scores are:
12 12 13 14 14 14 15 15
16 17 18 18 19 21 24 25
median = 15.5
Interquartile range
$= 18.5 - 14$
$= 4.5$
$\therefore$ **[B]** ✓ (1 mark)

2 $3x - y + 11 = 0$ [i]
$y = x + 3$ [ii]
Substitute [ii] into [i]:
$3x - (x + 3) + 11 = 0$
$3x - x - 3 + 11 = 0$
$2x + 8 = 0$
$2x = -8$
$x = -4$
Substitute into [ii]:
$y = -4 + 3$
$= -1$
$x = -4$, $y = -1$
[Or substitute the given values into both equations to see which satisfy both.]
$\therefore$ **[D]** ✓ (1 mark)

3 $-2x^2(3x - 5) - 3x(x - 5)$
$= 6x^3 + 10x^2 - 3x^2 + 15x$
$= 6x^3 + 7x^2 + 15x$
$\therefore$ **[A]** ✓ (1 mark)

4

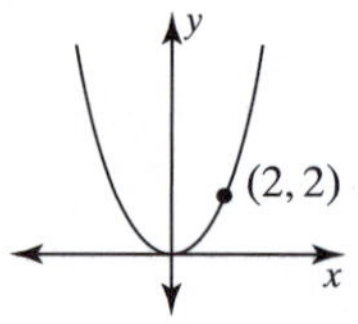

The graph has its vertex at the origin. Equation is of the form

$y = ax^2$. It passes through the point (2, 2).

$2 = a \times 2^2$

$2 = 4a$

$a = \frac{1}{2} \quad \therefore y = \frac{1}{2}x^2$

$\therefore$ **[B]** ✓ (1 mark)

5

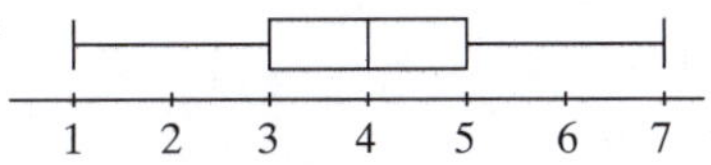

From the box plot the median is 4, the lower quartile is 3 and the upper quartile is 5. Of the options the only dot plot with lower quartile 3 and upper quartile 5 is

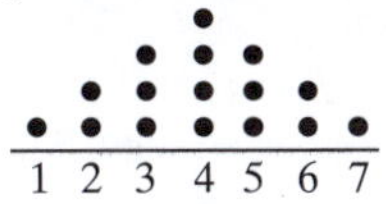

$\therefore$ **[A]** ✓ (1 mark)

6 $\frac{3t}{5} - \frac{2t}{3} = \frac{9t}{15} - \frac{10t}{15}$

$= \frac{-t}{15}$

$\therefore$ **[D]** ✓ (1 mark)

7 $\frac{a^2 - 25}{a^2 - 10a + 25} = \frac{(a+5)(a-5)}{(a-5)^2}$

$= \frac{a+5}{a-5}$

$\therefore$ **[D]** ✓ (1 mark)

8 $2 - 3k \geq -10$

$-3k \geq -12$

$k \leq 4$

[We divide by –3 so the inequality sign must be reversed.]

$\therefore$ **[A]** ✓ (1 mark)

9

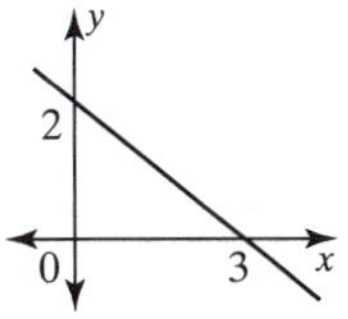

The y-intercept is 2.

[From that point, run across 3 units and down 2 units to another point on the line.]

Gradient $= \frac{\text{rise}}{\text{run}} = \frac{-2}{3}$

Equation is $y = \frac{-2}{3}x + 2$

[Or substitute to see which equation is satisfied by both points (0, 2) and (3, 0).]

$\therefore$ **[A]** ✓ (1 mark)

10 $(3x^{-4})^{-2} \div x^{-5} = 3^{-2}x^{8} \div x^{-5}$

$= \frac{x^{13}}{9}$

$\therefore$ **[B]** ✓ (1 mark)

11

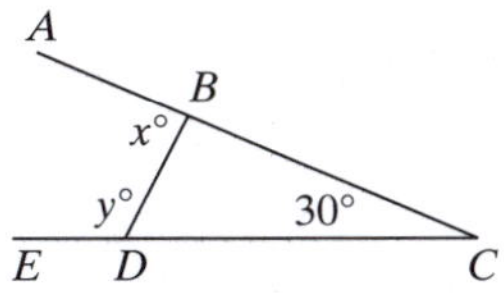

$\angle CBD + \angle BDC + \angle DCB = 180°$

(angle sum of a Δ)

$\angle CBD = 180° - x°$

(angles in a straight line)

$\angle BDC = 180° - y°$

(angles in a straight line)

$180° = 180° - x° + 180° - y° + 30°$

$180° = 390° - (x° + y°)$

$390° - 180° = x° + y°$

$= 210°$

[Or use exterior angles of a triangle.]

$\therefore$ **[D]** ✓ (1 mark)

12 $P(BBB) = \frac{3}{12} \times \frac{2}{11} \times \frac{1}{10}$

$= \frac{1}{220}$

$\therefore$ **[C]** ✓ (1 mark)

13 $4x - 2y + \frac{2}{3}y - \frac{x}{2}$

$= \frac{7x}{2} - \frac{4y}{3}$

$[(4 - \frac{1}{2})x,\ (-2 + \frac{2}{3})y]$

$\therefore$ **[B]** ✓ (1 mark)

14

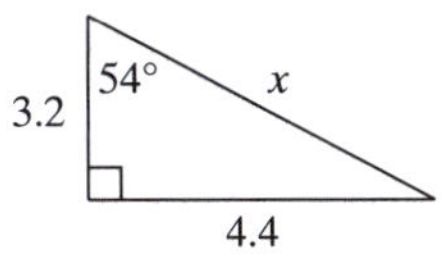

$\sin 54° = \frac{4.4}{x}$

$x = \frac{4.4}{\sin 54°}$

[But that is not one of the options.]

$\cos 54° = \frac{3.2}{x}$

$x = \frac{3.2}{\cos 54°}$

$\therefore$ **[C]** ✓ (1 mark)

15 $R = \frac{1}{a} + \frac{1}{b}$

$R - \frac{1}{a} = \frac{1}{b}$

$\frac{Ra - 1}{a} = \frac{1}{b}$

$b = 1 \div \frac{Ra - 1}{a}$

$= \frac{a}{Ra - 1}$

$\therefore$ **[D]** ✓ (1 mark)

16 Total repaid $= 18 \times \$563.08$

$= \$10\,135.44$

18 months = 6 quarters

$A = P(1 + R)^n$

$\$10\,135.44 = \$9000(1 + R)^6$

$(1 + R)^6 = 1.126\,16$

$1 + R = 1.019\,99\ldots$

$(1 + R > 0)$

$R = 0.02$ (2 d.p.)

The quarterly interest rate is 2%.

Annual rate $= 4 \times 2\%$

$= 8\%$

$\therefore$ **[A]** ✓ (1 mark)

17 $\frac{x^2}{4} - 1 = 0$

$\frac{x^2}{4} = 1$

$x^2 = 4$

$x = \pm 2$

One solution is $x = -2$

[Or substitute into the equation.]

$\therefore$ **[B]** ✓ (1 mark)

18

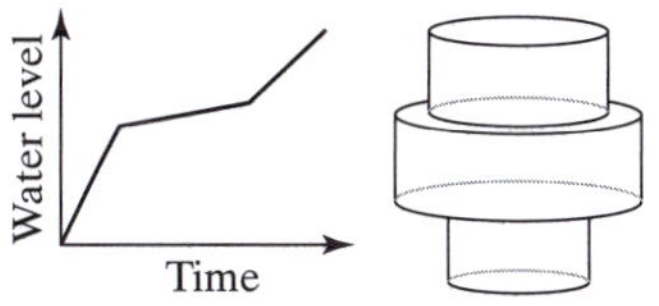

The water level in the container rises quickly, then much slower, then more quickly but not as fast as it first did.

$\therefore$ **[B]** ✓ (1 mark)

19 Volumes in ratio $1:8 = 1:2^3$
Areas in ratio $1:2^2 = 1:4$
The area of the larger figure is four times the area of the smaller figure.
Area of smaller figure
$= (72 \div 4)\ \text{m}^2$
$= 18\ \text{m}^2$
$\therefore$ **[C]** ✓ (1 mark)

20

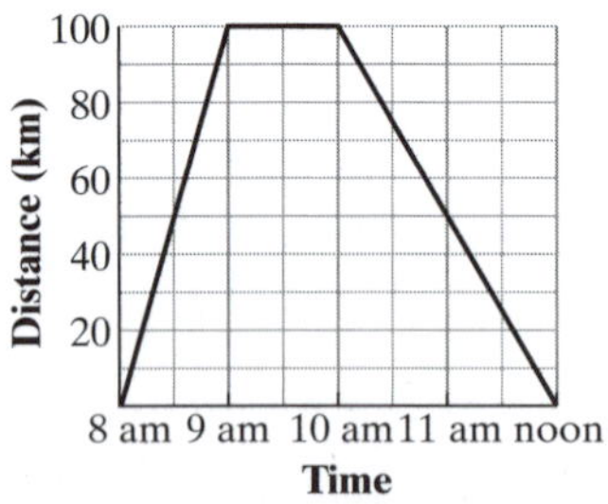

The truck travels furthest between 8 am and 9 am so its speed is greatest then.
$\therefore$ **[A]** ✓ (1 mark)

21

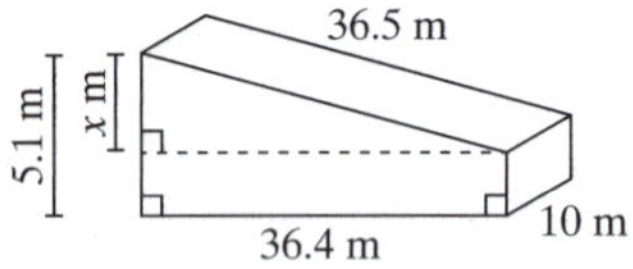

$x^2 = 36.5^2 - 36.4^2$
$= 7.29$
$x = \sqrt{7.29}\quad (x > 0)$
$= 2.7$
Right side $= (5.1 - 2.7)$ m
$= 2.4$ m
$V = Ah$
$= \frac{1}{2} \times 36.4 \times (5.1 + 2.4) \times 10$
$= 1365\ \text{m}^3$
$\therefore$ **[A]** ✓ (1 mark)

22

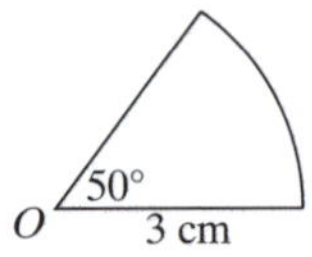

Arc length:
$l = \frac{\theta}{360} \times 2\pi r$
$= \frac{50}{360} \times 2 \times \pi \times 3$
$= 2.617\,993\,878...$
$= 2.6$ (1 d.p.)
Perimeter
≈ 2.6 cm + 3 cm + 3 cm
$= 8.6$ cm
$\therefore$ **[D]** ✓ (1 mark)

23 Angle sum pentagon
$= 3 \times 180°$
$= 540°$
$\angle BCE = 540° \div 5$
$= 108°$
$\angle CBE + \angle BEC + \angle BCE$
$= 180°$ ($\angle$ sum Δ)
$\angle CBE + \angle BEC$
$= 180° - 108°$
$= 72°$
But $\angle CBE = \angle BEC$
(ΔBCE is isosceles)
$\therefore \angle CBE = 72° \div 2$
$= 36°$
$\angle BDE + 90° + 36° = 180°$
($\angle$ sum ΔBDE)
$\angle BDE = 54°$
$\therefore$ **[B]** ✓ (1 mark)

24

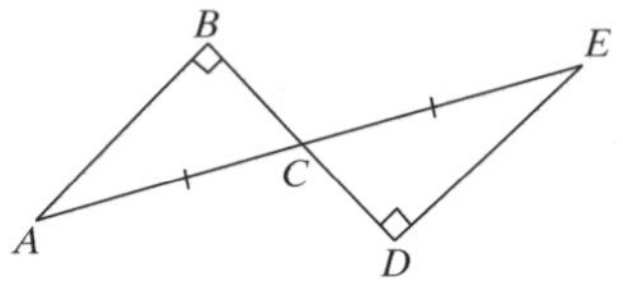

In ΔABC, ΔEDC
$\angle ABC = \angle EDC$
($BD \perp AB$, $BD \perp DE$, given)
$\angle ACB = \angle ECD$
(vertically opposite)
$AC = EC$
(C is the midpoint of AE)
$\therefore \Delta ABC \equiv \Delta EDC$ AAS
$\therefore$ **[D]** ✓ (1 mark)

25 $x^2 + y^2 = 25$
Circle, centre (0, 0), radius 5 units
When $x = 4$ and $y = 3$,
$x^2 + y^2 = 4^2 + 3^2$
$= 16 + 9$
$= 25$
$\therefore$ the point lies on the circle.
$\therefore$ **[B]** ✓ (1 mark)
(Total Part A: 25 marks)

SAMPLE EXAM PAPER 3
Part B: Short Answer PAGE 104

26 **a** $x^2 + xy - x - y$
$= x(x + y) - 1(x + y)$
$= (x + y)(x - 1)$ ✓
(1 mark)

b $x^2 - x^4$
$= x^2(1 - x^2)$
$= x^2(1 + x)(1 - x)$ ✓
(1 mark)

27 $(n - 3)(n + 5) = 105$
$n^2 + 5n - 3n - 15 = 105$
$n^2 + 2n - 120 = 0$ ✓
$(n + 12)(n - 10) = 0$
$n = -12$ or $n = 10$ ✓
(2 marks)

28 **a**

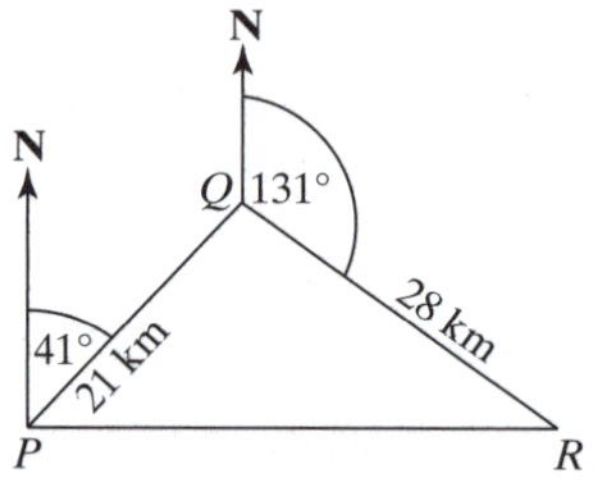

✓ (1 mark)

b Let T be a point due south of Q.
$\angle PQT = 41°$
(alternate angles, || lines)
$\angle RQT = 180° - 131°$
($\angle$s in a straight line)
$= 49°$
$\angle PQR = 41° + 49°$
$= 90°$ ✓ (1 mark)

c By Pythagoras' Theorem:
$PR^2 = 21^2 + 28^2$
$= 441 + 784$
$= 1225$
$PR = \sqrt{1225}$
$= 35$
The distance from P to R is 35 km. ✓ (1 mark)

d $\tan R = \frac{21}{28}$
$R = 36.869\,897\,65...°$
$= 37°$
(nearest degree)
$\therefore \angle PRQ = 37°$ to the nearest degree ✓
[Or use $\sin R = \frac{21}{35}$
or $\cos R = \frac{28}{35}$] (1 mark)

e Let U be a point due north of R.
$\angle QRU = 180° - 131°$
(co-interior angles, || lines)
$= 49°$
$\angle URP = 49° + 37°$
$= 86°$

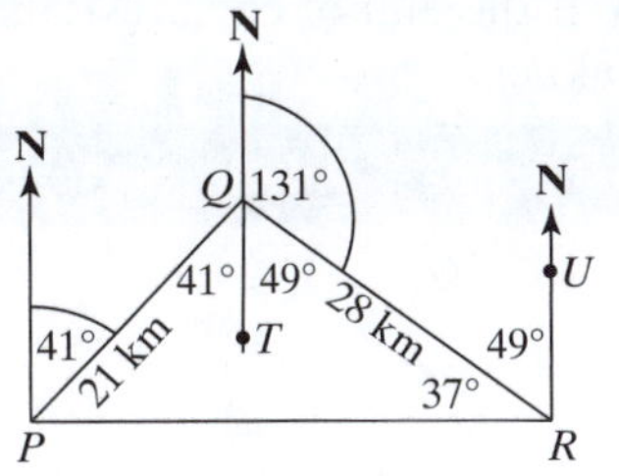

Reflex angle URP
$= 360° - 86°$
$= 274°$
The bearing of P from R is 274°. ✓ (1 mark)

29 **a** Each roll of the die is independent, so it doesn't matter what happened before, if the die is fair then the chance of rolling 1 each time is $\frac{1}{6}$. ✓ (1 mark)

b P(1 four times)
$= \frac{1}{6} \times \frac{1}{6} \times \frac{1}{6} \times \frac{1}{6}$
$= \frac{1}{1296}$ ✓ (1 mark)

30 **a** $a + b = 46$
[The number of girls plus the number of boys equals 46.]
$a = b + 8$ ✓
[The number of girls equals the number of boys plus eight.] (1 mark)

b $a + b = 46$ [i]
$a = b + 8$ [ii]
Substitute [ii] into [i]:
$b + 8 + b = 46$
$2b + 8 = 46$
$2b = 38$
$b = 19$ ✓
Substitute into [ii]:
$a = 19 + 8$
$= 27$ ✓
There are 27 girls and 19 boys. (2 marks)

31

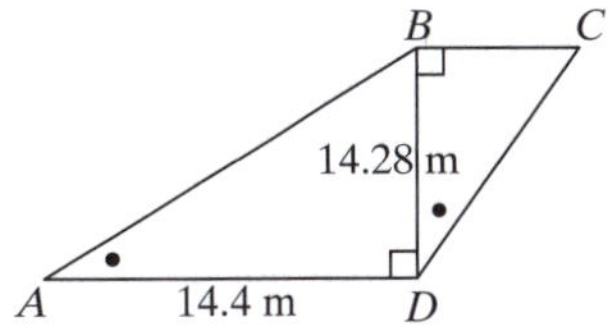

Let the length of AB be d m.
$d^2 = 14.28^2 + 14.4^2$
$= 411.2784$
$d = \sqrt{411.2784}$ $(d > 0)$
$= 20.28$
The length of AB is 20.28 m. ✓
In ΔABD, ΔDCB
$\angle BAD = \angle CDB$ (given)
$\angle ADB = \angle DBC$ (both 90°)
$\Delta ABD \,|||\, \Delta DCB$ (equiangular) ✓
$\frac{CD}{BA} = \frac{BD}{DA}$
(corresponding sides of similar triangles)
$\frac{CD}{20.28} = \frac{14.28}{14.4}$
$CD = \frac{14.28}{14.4} \times 20.28$
$= 20.111$
The length of CD is 20.111 m. ✓ (3 marks)

32 [The area of the shaded face is the area of a square plus the area of a triangle.]

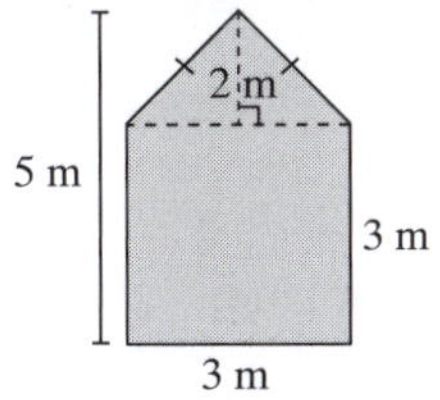

$A = 3 \times 3 + \frac{1}{2} \times 3 \times 2$
$= 12$
The area is 12 m². ✓
Let x m be the length of the equal sides.

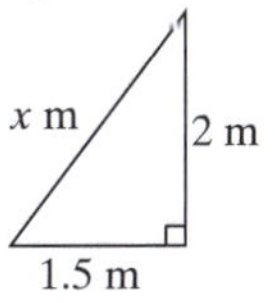

$x^2 = 2^2 + 1.5^2$
$= 4 + 2.25$
$= 6.25$
$x = \sqrt{6.25}$ $(x > 0)$
$= 2.5$
The equal sides are both 2.5 m long. ✓
[The prism has 7 faces, 5 rectangular and 2 pentagonal. Three of the rectangular faces are 12 m long and 3 m wide. The other two rectangular faces are 12 m long and 2.5 m wide. The pentagonal faces both have area of 12 m².]
$A = 3 \times 12 \times 3 + 2 \times 12 \times 2.5 + 2 \times 12$
$= 192$
The surface area is 192 m². ✓ (3 marks)

33 **a** $A(5, -2)$, $B(3, 5)$
$m = \frac{y_2 - y_1}{x_2 - x_1}$
$= \frac{5 - (-2)}{3 - 5}$
$= -\frac{7}{2}$
∴ perpendicular line has
$m = \frac{2}{7}$ ✓
$C(-4, -1)$
$y - y_1 = m(y - x_1)$
$y - (-1) = \frac{2}{7}(x - (-4))$
$7(y + 1) = 2(x + 4)$
$7y + 7 = 2x + 8$
$2x - 7y + 1 = 0$ ✓ (2 marks)

b $2x - 7y + 1 = 0$
When $y = 0$,
$2x + 1 = 0$
$2x = -1$
$x = -\frac{1}{2}$ ✓ (1 mark)

34

a There is no scale on the graph. The use of 3-D images makes the increase in sales appear much larger than it actually is. ✓ (1 mark)

b By doubling each dimension, the capacity is $2 \times 2 \times 2$ or 8 times larger. Each dimension should be increased by x where
$x^3 = 2$
$x = \sqrt[3]{2}$
$= 1.259921\ldots$
$= 1.26$ (2 d.p.) ✓ (1 mark)
(Total Part B: 25 marks)

TEST & EXAM RESULTS

Transfer your **percentage score** that you calculated in the **Your Feedback** box at the end of each **test** and **exam** to the table below. This will help you work out your areas of strength and weakness.

Test Topic	Intermediate Test Score	Advanced Test Score
Number and Algebra: Compound Interest	%	%
Number and Algebra: Products and Quotients	%	%
Number and Algebra: Factorising	%	%
Number and Algebra: Algebraic Fractions	%	%
Number and Algebra: Binomial Products	%	%
Number and Algebra: Factorisation of Quadratic Expressions	%	%
Number and Algebra: Special Binomial Products and Factorisations	%	%
Number and Algebra: Substitution, Equations and Formulae	%	%
Number and Algebra: Further Equations and Inequalities	%	%
Number and Algebra: Quadratic Equations	%	%
Number and Algebra: Simultaneous Equations	%	%
Number and Algebra: Parallel and Perpendicular Lines	%	%
Number and Algebra: Further Graphs	%	%
Measurement and Geometry: Surface Area and Volume	%	%
Measurement and Geometry: Angle Properties	%	%
Measurement and Geometry: Proofs Involving Congruent Triangles	%	%
Measurement and Geometry: Logical Reasoning	%	%
Measurement and Geometry: Angles of Elevation and Depression, and Bearings	%	%
Statistics and Probability: Compound Events	%	%
Statistics and Probability: Quartiles and Interquartile Range	%	%
Statistics and Probability: Box Plots and Scatter Plots	%	%
Statistics and Probability: Data Involving Time; Evaluating Reports	%	%
Sample Exam Papers		
Sample Exam Paper 1		%
Sample Exam Paper 2		%
Sample Exam Paper 3		%

FEEDBACK CHECKLIST TO IMPROVE YOUR TEST & EXAM RESULTS

Do you want to improve your scores in the Sample Exams? Check that:

☑ You are ready.

This is a revision workbook designed specifically for the revision of work already done. It is not a replacement for your textbook or class notes. Cover the topic in class or in your own time by reading your textbook or the ***Excel*** *Mathematics Study Guide*, *Years 9–10*, before using this book for further practice on the topic.

☑ You are revising in the right order.

Maths is a subject that builds on previous knowledge. Often you will need to have a good grasp of an early topic in order to fully understand a later topic.

☑ Your standards are realistic.

You cannot expect to score 100% all the time. Remember that this book has been designed to help you identify your strengths as well as your weaknesses, so it is OK to make mistakes. The key to success is learning from those mistakes.

☑ You are not rushing through the questions.

Many Maths students fail to read questions properly. This accounts for a large number of the mistakes made in tests and exams. Although your time is limited, you must still take enough time to read each question carefully. If necessary, read the question a second (or even third) time. Don't start your answer until you are sure that you understand the question. For the longer questions, take a moment or two to plan your answer.

☑ You are allowing yourself enough time to study.

Preparation for any test or exam requires a sufficient investment of time to be undertaken successfully. This means that you have to plan ahead. Don't wait until a week or so before an exam, for example, to ask yourself the question: How much time will I need to devote to this subject to revise it thoroughly? Don't forget to allow sufficient time for your other subjects—Maths isn't the only subject you will have to sit an exam for!

☑ You are OK.

You may be having some problems unrelated to your schoolwork. It is wise to sort out such problems as quickly as possible. Speak to your teacher, a parent or your school counsellor if you think you need help.

NOTES